Energy, Economics, and the Environment

Energy, Economics, and the Environment

Co-authored by

RUSSELL MILLS AND ARUN N. TOKÉ

With Susan Mills

Vermont Technical College

Prentice-Hall, Inc., Englewood Cliffs, New Jersey 07632

Library of Congress Cataloging in Publication Data

Mills, Russell.
Energy, economics, and the environment.

Includes bibliographies and index.
1. Energy development. 2. Energy consumption—History.
3. Energy industries—Environmental aspects. 4. Power
resources. I. Toké, Arun N. II. Mills, Susan.
III. Title.
HD9502.A2M55 1985 333.79 84-8201
ISBN 0-13-277468-2

Editorial/production supervision: *Mary Carnis*
Cover design: *Jeanette Jacobs*
Manufacturing buyer: *Anthony Caruso*

Printed in the United States of America

10 9 8 7 6 5 4 3 2 1

ISBN 0-13-277468-2

Prentice-Hall International, Inc., *London*
Prentice-Hall of Australia Pty. Limited, *Sydney*
Editora Prentice-Hall do Brasil, Ltda., *Rio de Janeiro*
Prentice-Hall Canada Inc., *Toronto*
Prentice-Hall of India Private Limited, *New Delhi*
Prentice-Hall of Japan, Inc., *Tokyo*
Prentice-Hall of Southeast Asia Pte. Ltd., *Singapore*
Whitehall Books Limited, *Wellington, New Zealand*

This book is dedicated to Nathan
and to the future generation he represents.

Contents

Preface

In the field of energy analysis, there often seems to be more answers than questions. Nearly everyone has an "answer" for the energy crisis. Unfortunately, many of these same people are unfamiliar with the basic questions, or have chosen not to ask them. It is easy to see why this is so. The energy problem is highly complex, with many intertwined threads. It is tempting to solve the problem with a single stroke, like Alexander with the Gordian knot, rather than painstakingly sort out the issues. But as we have discovered so often during the past few years, in the field of energy such quick solutions often do more harm than good.

In a sense, the very multiplicity of energy solutions is promising. For every energy resource, there are hundreds of individuals and groups developing and promoting that resource. But often the technology develops more rapidly than our ability to use it wisely. In our excitement over perfecting a technological solution, we sometimes forget to ask what that solution will accomplish for us, or what the consequences of the solution will be.

This text deals with the issues that need to be considered as we make our energy choices. Our main emphasis is on the principles of energy analysis. For while the technology changes from year to year, or even from week to week, the principles remain the same. If people have a clear grasp of the principles—if they know how to ask the right questions—they will be able to deal with tomorrow's technology as well as today's.

Our purpose in this text is threefold. First, we hope to give readers a basic overview of the energy situation. The first four chapters are designed to provide much of the necessary fundamentals. Second, we hope to provide a set of analytical techniques to help readers scrutinize the many energy alternatives that will be offered to them. Chapters 5 through 9 are devoted to an examination of analytical methods. Finally, we hope to suggest some ways for readers to develop their own energy solutions. The last two chapters focus on specific applications of the analytical principles.

Each chapter of the book is intended to stand on its own. Readers should be able to deal with the analytical chapters without having finished the survey of energy resources. This means that the book can be read in nearly any order. There are two exceptions to this principle: Chapter 2 must be mastered before any later chapters are undertaken, and Chapter 11 should be saved for last.

One special feature of the book is the series of boxed inserts containing supplementary information. Some of the inserts present detailed technical information, while others look at energy principles from unusual angles. The text can be read and understood without reference to the boxed material, but we hope readers will find them interesting and illuminating.

Each chapter is followed by a selection of study questions and, where appropriate, by quantitative problems. Some of these questions and problems will require readers to go beyond the text for information. In this way, readers can gain some experience in doing basic energy-related research.

The authors welcome comments and suggestions from readers. Most of the material has been used, in one form or another, at Vermont Technical College as part of a course called Energy and Society. But we are sure that some errors and inconsistencies remain. No doubt readers will find us biased in some places, inadequate in others, and sometimes simply boring. We take full responsibility for any flaws in the text. But we trust that the book's merits will outweigh its defects, and that it will help readers to explore the intricate and stimulating discipline of energy analysis.

Arun Narayan Toké, Russell Mills

Acknowledgments

The authors would like to acknowledge the assistance of the following persons and organizations:

John Kilcullen, for getting the project off the ground.

Dean George Strout of Vermont Technical College, for personal and institutional support.

Dewey Patterson and the library staff at Vermont Technical College, for their assistance and forbearance.

Dr. Carl Brandon and *Ken Stevens,* for assistance with word processing.

Steve Bravar of the Vermont Yankee Corporation, for assistance with illustrations.

Carolyn B. for flexibility and forbearance.

Roger Hard, for commentary and desk space.

The offices of *U.S. Representative James Jeffords* and *U.S. Senator Patrick Leahy,* for assistance in obtaining government publications.

We would also like to extend our sincere thanks to the many authors and scholars whose work has made ours much easier.

Finally, the following groups deserve special thanks for supplying us with information, photographs, or illustrations: The National Center for Appropriate Technology, the U.S. Bureau of Reclamation, the U.S. Geological Survey, the

U.S. Department of Energy, The Atomic Industrial Forum, Central Vermont Public Service, The Vermont Yankee Corporation, The General Electric Corporation, General Motors, Bethlehem Steel, E.G. & G. of Idaho, Sandia Labs, Friends of the Earth, The Worldwatch Institute, the Johns Hopkins University Press, the American Association for the Advancement of Science, Rodale Press, and The Pennsylvania State University.

1

Historical Survey of Energy-Using Societies

The complex issues of energy, economics, and the environment cannot be adequately understood without an awareness of the historical perspective. In an important sense, the history of humanity is the history of our energy use. We seem to be unique among all organisms in having the power, will, and inclination to change the face of the planet. And we seem to be almost compulsive about changing our lot. We have proved ourselves to be one of the most constructive and also one of the most destructive forces that shape the destiny of our planet. In this pattern of dominance we can find the roots of our civilization, and of our current problems as well.

1.1 EVOLUTIONARY PREHISTORY

The first appearance of *Homo sapiens* is believed to have occurred some 2.5 million years ago. During the early periods of human prehistory, the environment changed drastically, and so did the climate. Many plant and animal species could not adapt to these changes and did not survive. Humans were among the species that made it through four glacial and three interglacial periods: times when the glaciers covered as much as 30% and as little as 10% of the earth's surface. As far as anthropologists have been able to determine, prehistoric human society survived because of social organizations, language, the use of weapons and tools, and the use of fire.

Early humans appear to have been hunters and gatherers of wild food.

They ate fruits, roots, and other plant parts, together with small amounts of meat obtained from hunting. Early prey were most likely small, unprotected animals. But as human skill and weapons became more sophisticated, the human diet began to include larger game animals. The new weapons allowed the hunters to apply increased amounts of *power,* and thus to hunt larger animals.

At the same time, the human brain was developing as well: storing information, learning to make and use more powerful tools and weapons, and most important, developing more sophisticated techniques of interpersonal communication. Although other animals have possessed each of these capabilities to a certain extent, only in humans have these three skills been so fully developed and so closely related.

Despite their intellectual capacities, early humans were still basically prey rather than predators, lacking claws, fangs, protective scales, great size, or exceptional speed. However, the reasoning capabilities of humans and their tool-making experience led to two very important developments in the history of the species: fire and the lever.

Fire is generally considered the single most important step in the proliferation of the human species. Before humans could control fire, they were limited to raw food. But with the development of fire, previously unpalatable or indigestible foods could be cooked. Food could also be smoked or dried for longer preservation, enabling humans to store food in times of plenty and eat well during periods of scarcity.

Fire also offered protection against predators. Humans no longer feared the darkness, and they had leisure time to socialize in the evenings. This led to great advances in communication, as people planned the next day's activities, developed religious and social rituals, and began to keep oral records of their achievements.

Finally, with fire for warmth, it was no longer necessary for humans to confine themselves to warm climates. Human rangelands and hunting grounds expanded to the northern temperate zones, to the very edge of the glaciers. Thus the use of fire was instrumental in the spread of humanity to all parts of the earth.

Although the mastery of fire (which probably took many thousands of years) gave humans a vast energy resource for cooking, warmth, and personal protection, the limits of human strength still presented an obstacle. But the invention of the lever provided early humans with a method for increasing the amount of power available. A single person, using a simple lever, could lift several hundred kilograms of mass, and groups of people could move immense objects. Just as the development of fire transformed the human diet, so the development of the lever transformed their living arrangements. Instead of living in caves and other natural dwellings, humans were able to build their own habitations using trees, stones, and other massive building materials. So this second major energy development also contributed greatly to the range, comfort, and influence of the human species.

Gradually, prehistoric people domesticated other animal species. Small animals such as the dog and the pig were probably the first to be domesticated. Then people learned how to herd larger animals such as caribou. These larger animals were not fully domesticated, in the sense of sharing people's work, but they certainly did not remain in a totally wild state either. In extreme northern Europe, nomadic groups still follow the reindeer herds from season to season, preserving this relationship of semidomestication.

Finally, large animals, such as oxen, horses, and elephants, were domesticated and used for tasks requiring large amounts of power. For the first time, humans had access to a dependable external source of mechanical power.

Some time after the domestication of animals, humanity began to domesticate the plant kingdom as well. Rather than gathering wild plants, people began to cultivate some of the more desirable strains of plants, selecting the best specimens to improve the breed. Current research suggests that agriculture was practiced in Southeast Asia and Mexico as early as 10,000 years ago.

The invention of agriculture transformed human living patterns. No longer did each family or clan require several square miles of living space to supply their food energy needs. People were able to remain in the same place for several years rather than migrating with the herds of semidomesticated animals. Domesticated animals helped with the chores and provided an additional food supply. Crops provided for seasonal needs and could also be stored for periods of inclement weather or poor hunting. Most important, the development of agriculture created large amounts of leisure time. For the first time, food production could be done by a fraction of the population, freeing the rest for other tasks such as manufacturing, crafts, and—unfortunately—warfare.

1.2 HISTORICAL DEVELOPMENT OF ENERGY RESOURCES

After the rise of civilization, humanity continued to develop new and more powerful ways of using energy. The wheel was invented and used more than 3000 years ago. The power of wind and water was harnessed to power ships and drive mills for grinding grain. Wood, charcoal, and eventually coal were used to produce metals: bronze, iron, and steel. Earthenware pottery began to be used alongside baskets; then stoneware and porcelain joined earthenware as higher temperatures were achieved in the ceramic kilns.

This level of energy use remained relatively constant for several hundred years until the development of industrial machines in the seventeenth and eighteenth centuries. But then, especially after the invention of the steam engine, human energy use developed a new and totally different character. Now, instead of depending on natural energy-conversion processes such as plants and animal power, human industry could artificially concentrate unprecedented amounts of power at a single point, and thus do tremendous amounts of work. Gradually, railroads and steamships linked all parts of the world. Steam-powered machinery made possible the mass production of manufactured goods on a scale that would have been unimaginable a few years earlier. Standards of living rose

Figure 1.2.1 First White steam car, built 1901 in Cleveland, Ohio by the White Sewing Machine Co. (*Source*: National Archives.)

steadily in the industrialized world as the benefits of energy-conversion technology became available to wider segments of societies.

By the late nineteenth century, approximately 100 years ago, the Industrial Revolution had transformed the Western world. Steam locomotives and steamships were the main modes of long-distance transportation, although horses were still used for short-distance transportation. Streetcars and steam trams ran on city streets, and factories produced consumer goods in large quantities. Communication links included telegraphs for public messages and letters for private correspondence. The telephone, which was first introduced commercially in 1877, was still something of a novelty.

Although industrialization was widespread in Europe and the United States by the late nineteenth century, many comforts which we take for granted were not widely available. Central heating was extremely rare and limited to city dwellers. During the cold season, the family kitchen served as the family room, dining room, living room, and bath room (for bathing purposes). A non-airtight wood-fired cook stove was the usual cooking appliance. This stove sometimes had a coil in the firebox for heating water; otherwise, water had to be heated on top of the stove. Electric lighting was unknown; candles, kerosene lanterns, and coal oil or coal gas lamps provided light in the evenings. These fuels were very expensive; in fact, fuel was a major expense for a typical family in the nineteenth century.

Besides lacking these modern home comforts, people in the nineteenth century lacked other amenities which are part of our everyday life. Foods were seasonal. The canning industry was still young, and only the rich could afford

Figure 1.2.2 Coal-mining operations of the early twentieth century: Production from a plant this size averaged 500 tons a day. (*Source*: National Archives.)

canned foods. Potatoes, carrots, cabbages, and other vegetables were stored in root cellars. Ice boxes were used in the cities, but food spoilage and poisoning were still serious problems.

As we look back at the late nineteenth century, we may be struck by the relatively primitive conditions compared to our own times. However, the people of that time compared their standard of living to previous times, where, as we have seen, even fewer amenities were available. It is easy for us, from our prosperous vantage point, to feel that our way of life is the norm. But in historical terms, we are living in a condition of unprecedented luxury.

1.3 ENERGY IN THE THIRD WORLD

For an even more illuminating comparison, we can compare our level of energy use with that of the less developed countries—the so-called "Third World." These are not civilizations of the past, but our own twentieth-century neighbors, comprising nearly three-fourths of the world's population.

The Third World countries, located generally in tropical climates, get plenty of sunshine and warmth. Space heating is not a problem. The main energy problem is a lack of food. Roughly half the population of the Third World—about 2 billion people—are undernourished.

Most Third World countries have been heavily populated for centuries, and resource management has been inadequate. Excessive cultivation has exhausted the soil of once-fertile farmland, and overgrazing and wood harvest have caused soil erosion. Animal dung, instead of being returned to the soil

BOX 1.3.1 LIFE IN DEVELOPING AFRICAN COUNTRIES

Consider these startling figures representative of the life in African developing countries.

Life expectancy of 47 years
Literacy rate of 35%
Infant mortality rate of 20%
Safe drinking water available to only 25% of the population
Per capita annual income of $350
Between 1960 and 1980:
- Food production dropped by about 20% on a per capita basis
- Real income dropped in 15 countries in the African continent

Source: National Public Radio.

as fertilizer, is used as cooking fuel where wood is scarce. Inadequate irrigation further reduces agricultural productivity. These problems are compounded by lower literacy rates and an accompanying lack of knowledge concerning farming techniques, health, and nutrition.

Inequitable distribution of farmland further contributes to the hunger problem. Few people own or have access to farmland. In India and Costa Rica, for instance, 50% of the rural families own no land. In Java, Guatemala, and Bolivia, 85% are landless. And the available land is not high in quality.

In response to the agricultural crisis, many Third World countries have opted for high-technology, high-energy solutions. Petroleum-fueled agricultural machinery and synthetic fertilizers have been imported, often with the assistance of development grants. But only large landholders can apply these large-scale agricultural technologies, so small landowners do not benefit. Also, many large farms, especially in Central America, produce food for export rather than for domestic consumption. So in many cases, the imported agricultural technology only aggravates the hunger problem.

Other energy resources are also scarce in the Third World. Wood gathering continues to deplete the already scarce forest resources. Oil and natural gas are available in limited supply and at high prices, so only the well-to-do can afford them. Hydro potential is fairly well developed, but rural and remote areas get none of the benefits due to the lack of power lines.

To compensate for their energy shortages, many Third World countries have opted for nuclear power. Unfortunately, the safety features of nuclear installations in less developed countries are not as good as in the developed countries. Also, many Third World countries lack the central power grids necessary to make efficient use of nuclear-generated electricity.

Solar, wind, and other renewable energy resources are still not widely used in Third World countries, largely because the technology for using these resources has been neglected in the more developed countries. However, the governments of Third World countries are now promoting the use of solar technologies for irrigation and other purposes.

From our own perspective in the more developed countries, we see an abundance of natural resources allowing us to maintain current levels of consumption. But what will happen as standards of living rise in these underdeveloped countries with their enormous populations? If the entire world used energy at the same rate as the United States, all the world's proven energy reserves would be exhausted within 10 years. So it is not enough to export Western-style energy technology to the less developed regions of the world. Energy solutions must be global in scope. Equitable and efficient utilization of energy resources in all countries of the world is essential if we hope to prevent an energy crisis of unprecedented proportions.

BOX 1.3.2 LIFE IN THE THIRD WORLD

Imagine living without:

Electricity	Bottled gas
Natural gas	Canned or frozen foods
Automobiles	Television sets
Refrigerators	Central heating
Telephones	Air conditioning
Stereo sets	Microwave ovens
Food processors	Gas or electric stoves
Mixers and blenders	Dishwashers
Washers and dryers	Attached bathrooms

You have just imagined what life is like for most people in the world.

1.4 THE WORLD ENERGY PICTURE

Energy resources are not distributed evenly across the world. Some regions have been blessed with an abundance of energy resources, whereas other regions and areas of the world are energy hungry. Human settlement patterns were based on climatic and soil conditions, so some of the most heavily populated regions are especially poor in energy resources, whereas sparsely settled desert areas, such as the Middle East, contain vast energy resources.

Human ingenuity has come up with one solution to this unequal distribution of resources: exports and imports of resources. We transport goods and services across continents and oceans. But this, too, creates problems: those countries with plentiful resources have an economic advantage over the rest of the world. This has translated into more power for the energy-rich nations. Let us briefly look at the world distribution of various energy resources.

Oil. The major world producers of oil are Saudi Arabia, Iran, Nigeria, Iraq, Mexico, Great Britain, Venezuela, the United Arab Emirates, Kuwait, Libya, the USSR, and the United States. The countries of the Middle East produce about 35% of the world's total oil production. The Iron Curtain countries (the USSR and its allies) produce about 20%, and North America produces 16%. Of these oil-producing countries, only the United States imports petroleum. This is because our oil consumption is disproportionately higher than the levels in other countries of the world.

Coal. North America, the USSR, China, Great Britain, Poland, Australia, India, West Germany, France, and South Africa are blessed with large coal reserves. More than 90% of the world's coal reserves are believed to be in these countries.

There is more coal in the ground than we can produce economically. Much of it is not economically recoverable because it is too far underground, the deposit is not rich enough, or it is located too far from industrial centers where the coal can be marketed. World coal reserves are estimated at 1300 billion metric tons, of which about 750 billion tons are economically recoverable. This supply could last for approximately 200 years, assuming that coal consumption will continue to increase as it has in the recent past.

Natural gas. Canada, Mexico, the Netherlands, Norway, the USSR, and the Middle East have abundant natural gas supplies. They account for more than 70% of the world's natural gas exports. The United States also has large supplies of natural gas, but almost the entire production is used domestically. Natural gas is produced commercially in more than 60 countries of the world.

Uranium. The world's recoverable deposits of uranium are located in the United States, Canada, the USSR, Australia, Gabon, and South Africa. But only a handful of countries have the technology to convert uranium to electricity. The United States, France, Canada, and Great Britain are the free world's major exporters of nuclear technology. Japan, Sweden, Germany, Belgium, and the USSR also have substantial nuclear industries. No reliable figures are available on nuclear power production in the communist countries. In 1981, the noncommunist countries of the world produced 731 billion kilowatthours of electricity using nuclear reactors.

Unconventional fossil fuels. The use of unconventional fossil fuels is still in the experimental stage. Oil shale is available in the western United States and Canada, together with the Estonian region of the USSR. Major tar sand deposits occur in Canada, Siberia, and Venezuela. But because oil shale and tar sands are not yet commercially available energy resources, the importance of these deposits is not yet known.

Renewable energy resources. The renewable energy resources (solar, hydro, wind, etc.) are decentralized and widely dispersed, although the renewable energy resources are available in somewhat different combinations in different regions of the world.

All tropical areas are abundantly supplied with direct solar and ocean thermal power. Good wind sites are somewhat harder to find, but ocean fronts, trade-wind paths, mountaintops, and some special geographical situations provide good wind sites for power production.

Hydro sites are not widely distributed, but some countries have hydro power in significant proportions. Canada, the United States, the USSR, India, Japan, Brazil, and other "mighty river" regions of the world can produce hydroelectricity in large quantities. In general, regions without hydro power tend to have abundant sunshine or wind resources.

Forests and commercial "agribusiness" enterprises produce biomass, including wood, for the energy needs of many countries. Most rural and developing regions depend heavily on biomass conversion.

Geothermal energy resources are currently significant in Iceland, the western United States, Canada, and Italy. Other volcanically active regions of the world present some potential for geothermal energy development.

Tidal sites are obviously limited to coastal regions; but even there, only a few dozen tidal sites have been identified as potentially profitable. Prime sites are limited to temperate regions in the northern and southern hemispheres.

1.5 LEVELS OF ENERGY USE: PAST, PRESENT, AND FUTURE

Looking at human history, especially in the Western world, we can see a continuous escalation of energy use. At a survival level, humans require approximately 2000 kilocalories (kcal) per day of food energy. (See Section 2.5 for an explanation of energy units.) But food energy represents only a small fraction of personal energy consumption in the developed world. Total energy use for the average citizen of the United States amounts to 250,000 kcal (approximately 1,000,000 Btu) per day. This represents a 100-fold increase over the subsistence rate. In the rest of the industrial world, energy consumption is much lower—about 130,000 kcal (520,000 Btu) per day—but still considerable. The greatest part of this tremendous increase in energy use has occurred during the past few decades, as fossil fuels and electricity have increasingly been substituted for human labor in nearly all areas of life.

During recent years, however, energy consumption in the industrialized world has not increased as rapidly as in previous years. In the United States, energy consumption has actually declined somewhat since 1979. This appears to be the result of several factors:

A conscious effort at conservation by the general public

Significant increases in energy prices

Government efficiency regulations and guidelines

Saturation of markets (when a household has one dishwasher, it is unlikely to purchase a second one)

Frequent recessions and economic downturns in the late 1970s and early 1980s.

The consequences of this slowdown are widespread. The nuclear power industry appears to have reached its peak: no new reactors have been ordered since 1978, and dozens of reactor orders have been canceled. Oil cartels seem incapable of raising prices or even of maintaining them at their current high levels.

Some energy analysts believe that energy use has reached its peak in the developed countries. They claim that energy-efficiency improvements will continue to reduce energy demand and that the use of energy-intensive consumer products has reached the saturation point. Other analysts believe that the low growth in energy use results from slow economic growth. They believe that when economic conditions improve, energy growth will resume along with it, leading to renewed shortages and ever-higher prices.

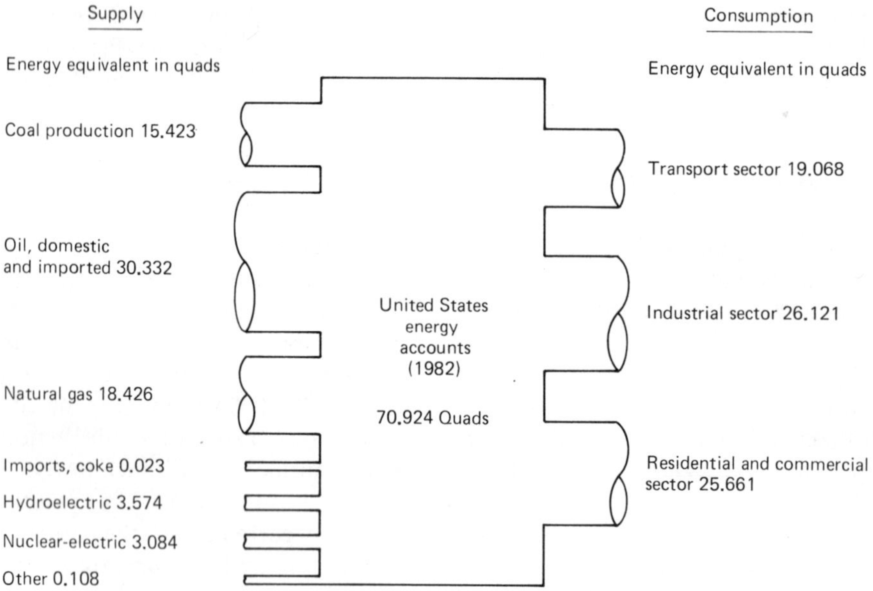

Figure 1.5.1 Energy survey for the United States, 1982, energy equivalent in quads. (*Source: Monthly Energy Review*, Energy Information Administration, Department of Energy, March 1, 1983.)

TABLE 1.5.1 Canadian Energy Consumption and Projections for the Year 2000

Energy Form	Current Use (quads)	Projected Use for the Year 2000 (quads)
Coal and peat	0.18	0.36
Natural gas	1.4	1.4
Petroleum	3.25	1.4
Electricity	1.01	1.35
Biomass	0.21	1.6
Solar	0.0	0.12
	6.05	6.23

Source: Soft Energy Notes, Vol. 5, No. 6 (February/March 1983).

1.6 WORLD ENERGY OUTLOOK FOR THE NEXT CENTURY: THREE PROJECTIONS

1. According to a 1982 report of the International Energy Agencies (IEA), there is no reason to believe that the world's energy problems are over, despite the oil glut that began in 1982. The report claims that the oil glut was caused primarily by a slowdown in world economic activity, not by growing supplies of oil. Assuming a resumption of economic growth, major oil market disruptions could occur again before 1990. Here are some of the highlights of the report:

a. Total world energy demand for the year 2000 is projected at the equivalent of 121 million barrels of oil per day (MBD). This is a significant reduction from an earlier (1977) projection of 185 MBD.

b. Projected world oil supply for the year 1990 is estimated at 50 to 52 million barrels per day (MBD), whereas demand could be slightly higher: 50 to 56 MBD. The situation may get worse by the year 2000, when projections call for a supply of only 49 to 53 MBD and a demand of 58 to 74 MBD.

c. The nuclear power contribution for the year 2000 is projected at approximately 136 to 179 gigawatts (GW; 1 gigawatt = 1 million kilowatts) for the United States. This is approximately 1.9 to 2.5 million barrels per day. Projected nuclear power production for the European Common Market countries is estimated at 394 to 489 GW, or 5.5 to 6.9 million barrels per day.

d. Renewable energy resources and technologies are expected to provide only 0.2% of the world's energy supply by 1990 and 1% for the year 2000. (These figures seem quite low; 1983 levels were already much higher than this.) Solar technology is regarded as an immature technology which is not economically competitive with other energy supply technologies.

e. Reliance on the free market system will not be able to handle the energy situation. Government actions are necessary to "strengthen and supplement" free market decisions.

f. Other recommendations include the following:
   (1) Greater energy efficiency
   (2) Energy conservation to curb demand
   (3) Energy planning policies
   (4) Increased use of mass transportation systems
   (5) Switching to a non-oil-based energy economy

2. According to the Bicentennial Project of the U.S. Energy Research and Development Administration (1976), the energy future of the United States in 2001, its 225th birthday, might look something like this:

> People will live in outer space and beneath the sea.
>
> Transportation will be the largest energy-consuming sector, with passenger-carrying rockets, ocean-spanning hydrofoils, underground jet-turbine trains, electric vehicles, and moving sidewalks.

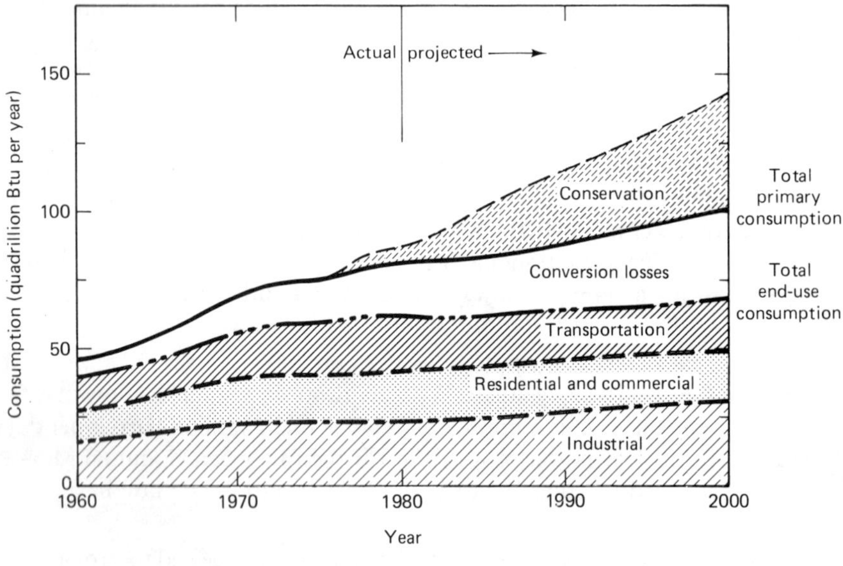

Figure 1.6.1 U.S. energy consumption by sector (quadrillion Btu per year). Conversion loss includes energy consumed to produce synthetic fuels and to produce, transmit, and distribute electricity. (*Source:* U.S. Department of Energy.)

Huge chains of suburbs and supercities will stretch the length of the East coast and all along the shores of the Great Lakes.

Industries will use round-the-clock work schedules for efficiency.

Recycling of energy and used materials will be a common and necessary occurrence.

Strict conservation measures will be mandated by law.

Energy shortages will be a permanent fact of life.

New electric power generating techniques will be in use, including magnetohydrodynamics, thermionics, and breeder and fusion reactors.

New fuels, including synthetic fuels, solar energy, and geothermal resources, will be in widespread use.

A totally new source of energy may be discovered.

The indispensable energy resource will still be in use: human intellect, creativity, curiosity, and wisdom.

3. *Pathway to Energy Sufficiency: The 2050 Study,* published by Friends of the Earth, presents a "lowest energy use" scenario requiring no major technological advances, no political revolutions, and no basic changes in human behavior. The study, which was conducted at the University of Wisconsin Institute for Environmental Studies, concludes that a 60% per capita reduction and a 53% overall reduction from 1975 levels of energy use can be accomplished by the middle of the twenty-first century. This scenario also projects other desirable results, such as a reduction in environmental pollution and decreased international tension.

To accomplish such a drastic reduction in energy use, many social and physical changes would be required, including the following:

Reduction in automobile use and highway construction

Sharply increased use of mass transit

A decline in food processing

Development of solar energy

Changes in patterns of consumption

The study suggests that these changes will occur naturally, as a result of the evolution of tastes, preferences, and social structures.

If these changes occur as suggested, then by the year 2050, *without any reduction in standards of living,* U.S. energy consumption in the various sectors can be projected as shown in Table 1.6.2. On a per capita basis, this would be approximately 120 million Btu (MBtu) per person per year, compared to the present U.S. average of 350 MBtu per year.

TABLE 1.6.2 Low-Use Energy Projections for the Year 2050

Sector	Energy Consumption (quads)	
Residential	6.35	
Commercial	4.64	
Transportation	6.16	
Personal		3.60
Freight		2.56
Industrial	16.11	
	33.26	

Source: John S. Steinhart et al., *Pathway to Energy Sufficiency: The 2050 Study* (San Francisco: Friends of the Earth, 1979).

1.7 WHERE DO WE GO FROM HERE?

Are the days of cheap energy finished? Will we reach a point in our lifetimes when we will simply run out of oil and natural gas? Are we headed for a no-frills energy diet? Are there limits to growth? Have we reached the end of economic development?

Or will we discover a new, limitless energy resource: nuclear fusion, perhaps, or efficient solar electricity? Will we be able to free ourselves from the need to do manual labor? Will we be able to offer abundant energy to underdeveloped countries and move into a future of unrestrained luxury?

Experience suggests that the truth lies somewhere between the two extremes. The laws of nature tell us that as long as the sun shines and the earth rotates, there is energy on the planet. If we can avoid eliminating ourselves through nuclear war, human civilization is likely to continue evolving, discovering new ways of using energy. Table 1.7.1 provides an overview of this evolution to date. Clearly, we have come a long way from our stone-age beginning.

However, we will not have a free ride. Formidable economic, environmental, political, and social problems will face us as we transform our patterns of energy production and use. History suggests that not all people will benefit equally from the new energy technology. But history also suggests that humanity will probably manage to survive the way we have always done: never quite achieving our highest expectations but never giving up in complete despair. We will probably muddle through.

Figure 1.7.1 The end of cheap oil? (*Source:* National Archives.)

TABLE 1.7.1 Brief History of Energy-Related Developments

Year	Event
2 million B.C.	First use of stones as tools
1 million B.C.	First use of fire by hunter-gatherer societies
15,000 B.C.	Beginnings of pottery
8000 B.C.	Agriculture begun in Near East
3500 B.C.	Rise of cities, first use of metal, invention of wheel
1500 B.C.	Invention of lever
1300 B.C.	First tools made from smelted iron
1000 B.C.	Chinese produced natural gas from 3000-ft wells and transported the gas in bamboo pipelines; gas was used for space heating and light
700 B.C.	First use of pulley
400 B.C.	First drilling for oil
27 B.C.	Invention of waterwheel
250 A.D.	Chinese first to use water mills and gunpowder
600	Windmills developed by Arabs
1118	First use of cannons in Europe

(Continued)

TABLE 1.7.1 Brief History of Energy-Related Developments *(Cont.)*

Year	Event
1200	Coal mining begun in Europe
1230	Rockets and hot-air balloons developed in China
1306	Coal use banned in London because of air pollution
1438	Invention of printing press
1600	Firewood and timber scarcity in England
1688	First distillation of coal gas
1694	Shale oil produced in England
1711	Invention of sewing machine
1712	Invention of steam engine by Thomas Newcomen
1743	Water turbine developed
1745	First coal mining in the United States
1747	Discovery of atmospheric electricity by Benjamin Franklin
1769	First automobile (three-wheeled steam-powered vehicle built by Cugnot in France)
1800	Electric battery built by Volta
1802	First steam locomotive
1831	Electric generator developed by Faraday in England
1834	Electric motor developed by Thomas Davenport
1837	First electric automobile
1859	Edwin Drake tapped first oil well in the United States
1860	First internal combustion engine (two cycle)
1869	Opening of transcontinental railway across the United States
1876	Bell invented telephone; Otto built first four-cycle internal combustion engine
1879	Edison invented incandescent light bulb
1881	First hydroelectric power plant, Godalming, England
1883	Steam turbine developed
1894	Diesel engine developed
1895	Discovery of X-rays
1903	Wright brothers flew first airplane
1905	Einstein discovered the relationship between energy and mass: $E = mc^2$
1913	Assembly line developed by Henry Ford
1927	First nonstop transatlantic flight from New York to Paris (Lindbergh)
1936	Hoover Dam built on Colorado River
1937	First regular television broadcast (Great Britain)
1942	First atomic fission reactor (Chicago)
1945	Atomic bombs used for the first time on human populations (Hiroshima and Nagasaki, Japan)
1951	First electronic computer built for commercial applications
1952	Hydrogen bomb exploded by the United States (first fusion bomb)
1957	First U.S. commercial nuclear power plant, Shippingport, Pa.; USSR launched first satellite

TABLE 1.7.1 Brief History of Energy-Related Developments *(Cont.)*

Year	Event
1960	First geothermal power plant in the United States
1969	First lunar landing
1973	Oil embargo by the Arab countries
1977	Opening of Alaskan oil pipeline
1979	Accident at Three Mile Island nuclear power plant
1981	United States launched first reusable space vehicle

STUDY QUESTIONS

1. Trace the development of early civilization, focusing on the development of energy technology.
2. How were the great pyramids built? What kinds of energy were used for such a massive construction project? How does this compare with modern construction projects?
3. Imagine that you are living in the jungles of Africa 50,000 years ago. Suddenly you discover how to control fire. How would this new discovery change your life? What could you do now that you could not do before?
4. Anthropologists believe that agriculture was the development that allowed people to establish cities. Why should this be so? How did the development of agriculture change people's energy consumption patterns to permit the development of civilization?
5. What are the five most important energy-related discoveries of human history? Why are these discoveries so important?
6. Have you heard people talk about the "good old days"? What energy resources did they use then? Do you agree that living conditions were better then, or are we better off? By studying those living conditions, what can we learn that could help us during the rest of the twentieth century?
7. What are the historical causes of the present energy crisis? What does this historical background suggest about ways of dealing with the crisis?
8. What steps can the developed countries take to improve the energy situation of the Third World? What obstacles do you see that might limit the effectiveness of your solution?
9. What are the current trends in energy consumption? Do you believe that energy use can continue to grow? What limiting factors might change current patterns?
10. Make your own energy projection for the year 2000. Do you expect serious shortages? Development of new resources? Increased economic well-being? Widespread economic hardship?
11. What new energy-related inventions do you see on the horizon? Do you think space colonies and underwater cities will be common in the next 25 years? What would be the energy requirements of these new inventions?

STUDY PROBLEMS

1. Per capita energy consumption in the United States is approximately 1 million Btu/day. Assuming that the total population of the United States is 225 million, what is the total U.S. energy consumption for 1 year? (You may want to convert your answers to quads: 1 quad = 10^{15} Btu.)
2. If the Third World countries raised their energy consumption to the level of the United States, what would be the total energy consumption in the Third World? (The combined population of Third World countries = 2.5 billion.)
3. The industrialized nations have an average per capita energy consumption of 500,000 Btu/day. If that were the worldwide average, what would be the total annual energy requirements of the world?
4. How much energy would the United States save each year if energy consumption were reduced to the average level of the industrialized countries (500,000 Btu per day)?
5. In 1979, U.S. energy consumption was 79 quads (1 quad = 10^{15} Btu). The 2050 study proposes a 60% reduction in per capita energy consumption for the year 2050. Assuming a 1979 population of 225 million and a 2050 population of 275 million, calculate the projected energy consumption in 2050.

FURTHER READINGS

BROWN, LESTER, and PAMELA SHAW. *Six Steps to a Sustainable Society,* Worldwatch Paper 48. Washington, DC: Worldwatch Institute, 1982.

COOK, EARL. *Man, Energy, Society.* San Francisco: W.H. Freeman, 1976.

DETWEILER, W.H., ET AL. *Environmental Decay in Its Historical Context.* Glenview, IL: Scott, Foresman, 1973.

ENERGY RESEARCH AND DEVELOPMENTAL ADMINISTRATION, BICENTENNIAL PROJECT. *Energy History of the United States.* Washington, DC: ERDA, 1975.

STEINHART, JOHN S., ET AL. *Pathway to Energy Sufficiency: The 2050 Study.* San Francisco: Friends of the Earth, 1979.

2

Fundamentals

Energy is the basis of life and of everything that happens in the universe. All movement is a form of energy, and as Einstein discovered, matter itself is a form of energy. A barrel of oil, a ray of sunshine, a cord of wood, a bucket of hot water, a turning wheel, a waterfall, a bubbling stream, an ear of corn, an apple, a horse, or a person all represent energy in one form or another. When we do some work, turn on a switch, or ask someone to do something for us, we initiate a flow of energy from one form to another. Routine activities such as breathing, eating, reading, sitting, seeing, sleeping, even doing nothing, all take energy—some more, some less. So exactly what do we mean by the word *energy*?

BOX 2.0.1 ENERGY IN DAILY LIFE

The amount of energy that we need for our daily existence depends on the kind of life we live. A blue-collar worker may require twice as much energy (kcal/day) as an average white-collar worker. The source of this energy is food intake. Thus the food we consume is the fuel for our body machine.

In a normal working day, an office worker consumes about 2000 kcal. Heavy activities such as mountain climbing, lumbering, coal mining, and so on, require upward of 4000 kcal/day. The following table lists human metabolic rates for sample activities.

BOX 2.0.1 *(Cont.)*

Metabolic Rates for Representative Human Activities

Activity	*Energy Consumed (kcal/hr)*	*Examples*
Minimal	Up to 25	Sleeping, sitting, reading
Light	20–100	Typing, housework, drawing, reading aloud, teaching
Moderate	80–250	Washing, carpentry, painting, walking, sweeping
Heavy	200–600	Blacksmithing, bicycling, rowing, swimming, polishing, sawing wood, carrying weight
Very heavy	500–1000	Bicycle racing, mountain climbing, running, climbing stairs

Sources: Earl Cook, *Man, Energy, Society* (San Francisco: W. H. Freeman, 1976); Amit Goswami, *The Concepts of Physics* (Lexington, MA: D. C. Heath, 1979).

In one sense, everything is energy. But energy is elusive and hard to see. Often we do not perceive energy until it is used, and then we cannot recover it to examine it again. But we *can* perceive the changes that energy causes in the universe. In this text we are principally concerned with energy as the ability to produce change. So to begin, let us define *energy* as whatever can be used to produce change.

2.1 BASIC CONCEPTS

To understand the nature of energy, we need to define a few physical principles. The physics of energy begins with the concept of *mass.* Mass is a measure of the quantity of matter in an object. It is measured in kilograms (kg) in the metric system and in pounds (lb) in the British system. (Units of measurement are discussed more fully in Section 2.5.) One pound equals approximately 0.45 kg. Another fundamental property is *length* or *distance*, which is measured in meters (m) in the metric system and in feet (ft) in the British system. One meter equals approximately 3.3 ft. A third fundamental property is *time*, which is measured in seconds (s or sec) in both systems.

From these basis units, we can define several other concepts. The first of these is *velocity*. Velocity is the distance traveled in any given direction per unit of time. In scientific terms,

$$\text{velocity} = \frac{\text{distance}}{\text{time}} \tag{2.1.1}$$

For example, if a car travels a distance of 1 kilometer (1000 meters) in 50 seconds, it has the following velocity:

$$\begin{aligned}\text{velocity} &= \frac{\text{distance}}{\text{time}} \\ &= \frac{1000\ \text{m}}{50\ \text{s}} \\ &= 20\ \text{m/s (meters per second)}\end{aligned}$$

Velocity can also be expressed in feet per second and miles per hour.

When the velocity of an object changes, there is an *acceleration*. Acceleration is defined as the change in velocity per unit of time:

$$\begin{aligned}\text{acceleration} &= \frac{\text{chance in velocity}}{\text{time}} \\ &= \frac{\text{distance/time}}{\text{time}} \\ &= \frac{\text{distance}}{\text{time}^2}\end{aligned} \tag{2.1.2}$$

Suppose that a car is traveling at a rate of 10 m/s [this is approximately 20 miles per hour (mph)]. Over a period of 10 s, it accelerates to 25 m/s (approximately 50 mph). The acceleration of the car is equal to the following:

$$\begin{aligned}\text{acceleration} &= \frac{\text{change in velocity}}{\text{time}} \\ &= \frac{15\ \text{m/s}}{10\ \text{s}} \\ &= 1.5\ \text{m/s}^2\end{aligned}$$

With these fundamentals, we are ready to define *force*, *work*, and *power*.

When a *force* is exerted against an object, the object accelerates. For example, when we push against a wheelbarrow, it increases in speed. The general expression for this relationship follows:

$$\text{force} = \text{mass} \times \text{acceleration} \tag{2.1.3}$$

This principle has two obvious consequences. A greater force will cause a greater acceleration (a harder shove makes the wheelbarrow go faster), and a greater mass requires a greater force to produce the same acceleration (we need a harder shove to get the wheelbarrow moving when it is full of wet concrete).

The acceleration due to gravity is called g. At the surface of the earth, $g = 9.8\ \text{m/s}^2$ or $32.2\ \text{ft/s}^2$. Any object, no matter how large or small, will accelerate downward at this rate, discounting the effects of air resistance. This means that the gravitational force, or weight, which pulls an object downward is equal to g times the mass of the object. This makes sense: we all know that gravity pulls harder at more massive objects.

2.1.1 Laws of Motion

The behavior of moving objects is described in Newton's laws of motion, developed over 200 years ago. Newton's first law states:

> A body at rest tends to stay at rest and a body in motion continues to be in motion, unless acted upon by an external force.

The existence of forces in the universe is the only possible way to make changes; otherwise, as this law tells us, everything will go exactly the same as it has been.

In the world of everyday experience, objects do not continue to move on their own, although they do tend to remain at rest. This is because moving objects in the universe are slowed down by dissipating forces that exist in nature. These forces consist of:

1. Friction between two surfaces
2. Air resistance (a form of friction)
3. Electrical resistance
4. Electromagnetic effect

To keep an object moving steadily, force must be applied to overcome these dissipative forces. The work done in overcoming these forces is converted to heat, which raises the temperature of the body and/or its surroundings.

Newton's third law tells us that all forces act in pairs:

> For every action, there is an equal and opposite reaction.

Let us consider the following examples to illustrate this law:

1. Water flows from a higher level to a lower level and eventually reaches the ocean. Why? Because just like everything else, water is constantly being acted upon by the force of gravity. Unless an obstruction is placed in its way, the water will continue to flow to sea level. When a dam is built, an obstruction is present and water accumulates behind the dam. The dam exerts a restraining force on the water, but the water exerts an equal force on the dam.
2. When an automobile engine is turned on, the fuel and air mixture ignites and expands very rapidly. This pushes against the piston, which in turn pushes against the crankshaft. The piston resists (pushes back against) the expanding gases just as hard as the gases push against the piston.

3. When you try to push a wheelbarrow full of concrete, it pushes back against you with an equal force. No wonder a person feels tired after pushing a few loads of concrete!

It should be clear by now that a change in the motion of an object is due to a force or forces acting on it. When this change occurs, *work* is said to be done. The work done by a force is equal to the energy lost by the forcing mechanism, and the energy gained by the body is equal to the work done on it. This means that energy can be transferred, but it can never be created or destroyed. Energy is conserved.

Work, then, is a measure of the amount of energy that is converted from one form to another, or from one body to another. Work is a term that we usually associate with an intentional energy conversion, presumably for some useful purpose. Thus work represents energy in transition or an energy transaction.

2.1.2 Force and Work

Work can be defined in terms of properties that we have already discussed: *force* and *distance*. A force must be exerted in order to do some work. There must also be some displacement or movement, and at least some of the displacement must be in the direction of the force. If all the displacement takes place at right angles to the line of force, the work done by the force is equal to zero. Figure 2.1.1 shows the concepts of physics involved in the process.

To simplify matters, let us assume that there is no friction between the two surfaces. The force acting on the block is indicated by an arrow which shows the strength and direction of the force. The force works against the force of gravity, which as we all know, tends to pull everything downward. Let us assume that the slave is able to move the block up the side of the pyramid for a distance

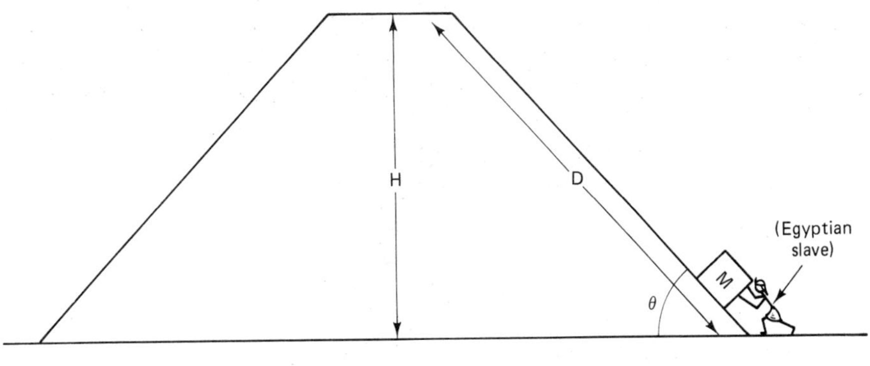

Figure 2.1.1 Eygptian slave pushing a block up the side of a pyramid.

of D meters. At the completion of this task, as shown in the diagram, the block (which has a mass of M) has been lifted by a height of H meters. By being lifted, the block has gained a certain amount of energy. However, this energy did not come out of nowhere. In this case, the slave has lost, or expended, an equivalent amount of energy.

When a force acts on an object, the work done is equal to the following:

$$\text{work} = \text{force} \times \text{distance} \tag{2.1.4}$$

If the slave is pushing against the block with a force of F, then the total amount of work done on the block is equal to $F \times D$, where D is the distance that the block has been pushed. This equation is useful if we know the exact strength of the force being exerted. In this case, though, it is difficult to measure the force directly. However, there is another way to look at the amount of energy transferred to the block.

When an object is raised above the ground, it gains energy according to the following formula:

$$\text{work} = \text{mass} \times g \times H \tag{2.1.5}$$

where mass = mass of the object

g = acceleration due to gravity

H = height through which the object is lifted

This formula is similar to the one we introduced earlier:

$$\text{work} = \text{force} \times \text{distance} \tag{2.1.4}$$

Earlier, we established that the gravitational force exerted on an object was equal to mass $\times$ g. Since this force acts vertically, it acts through distance H. Substituting mass $\times$ g for force and H for distance, we can see that Eq. (2.1.5) is identical to (2.1.4).

If the block has a mass of 250 kg and the height H is equal to 25 m, the total amount of energy gained by the block is equal to the following:

$$\begin{aligned} \text{work} &= \text{mass} \times g \times H \\ &= 250 \text{ kg} \times 9.8 \text{ m/s}^2 \times 25 \text{ m} \\ &= 612{,}500 \text{ newton-meters (joules)} \end{aligned}$$

Now we can use this formula to determine the force that the slave exerts in order to push the block to the top of the pyramid. We have already established that the amount of work done on the block is equal to $F \times D$, where F is the force exerted by the slave and D is the distance the block is moved up the side of the pyramid. Since this is equal to the energy gained by the block, we can write the following equation:

$$F \times D = \text{mass} \times g \times H \tag{2.1.6}$$

Using this equation, we can now calculate the amount of force required to move the block along the inclined surface.

since $$F \times D = \text{mass} \times g \times H, \text{ then}$$

$$F = \frac{\text{mass} \times g \times H}{D}$$

We can see by this equation that as D increases, F decreases; thus it is easier to move the block along an inclined surface than to lift it directly upward. We are all familiar with this principle. For example, mountain roads are constructed with switchbacks rather than going straight up the hill. This is so that automobiles can drive to the top without straining their engines.

If the distance D along the side of the pyramid is equal to 40 m, then

$$F = \frac{\text{mass} \times g \times H}{D}$$

$$= \frac{250 \text{ kg} \times 9.8 \text{ m/s}^2 \times 25\text{m}}{40 \text{ m}}$$

$$= 15{,}300 \text{ newtons}$$

BOX 2.1.1 TRIGONOMETRIC EXPRESSION OF EQUATION (2.1.6)

If we apply a simple trigonometric relation to the diagram in Figure 2.1.1, we find that $H/D = \sin\theta$, where θ equals the angle between the ground and the inclined surface of the pyramid. Applying this principle, we can rewrite Eq. (2.1.6) in the following terms:

$$F \times D = \text{mass} \times g \times H \qquad (2.1.6)$$

$$F = \text{mass} \times g \times H/D$$

and since $H/D = \sin\theta$, then

$$F = \text{mass} \times g \times \sin\theta \qquad (2.1.7)$$

2.1.3 Work and Power

Power is defined as the *rate* at which work is done. In mathematical terms, this is expressed as follows:

$$\text{power} = \frac{\text{work done}}{\text{time required}} \qquad (2.1.8)$$

For example, suppose that it took the slave half an hour (1800 s) to move the block to the top of the pyramid. We already know that the total amount of work done is equal to 612,500 joules (J). So, in order to determine the power output of the slave, we apply Eq. (2.1.8).

$$\begin{aligned} \text{power} &= \frac{\text{work done}}{\text{time required}} \\ &= \frac{612{,}500 \text{ J}}{1800 \text{ s}} \\ &= 340 \text{ J/s} \end{aligned}$$

Power is an important concept because it ties the dimension of *time* into the energy picture. As we look back in the history of energy use, we see a definite accelerated growth in power consumption, that is, in the rate at which energy is being utilized.

Energy and power are often confused with each other. But the distinction between the two is as clear as the difference between the odometer and the speedometer in an automobile. One gives you the total distance traveled by the car in kilometers or miles, and the other gives you the velocity or the rate at which that distance is being traveled. Similarly, *energy* refers to the total amount of work that is being done, whereas *power* refers to the rate at which that work is being done, in work per unit of time. A jet engine, obviously, has more power than an automobile engine, which in turn has more power than a lawn-mower engine.

Power plants, as the term suggests, are rated by their power output, or the amount of energy they can convert per unit of time. This conversion is usually from thermal (heat) to kinetic (motion) form. In general, the higher the power consumption, the higher the energy consumption. But it ought to be made clear that power and energy requirements are two different things. Consider the most common example in the United States, the automobile.

An automobile, like any other physical entity, has a mass (generally between 900 and 2500 kg, or 2000 and 5500 lb). When traveling on a freeway, the car averages between 80 km per hour (km/h) and 90 km/h, or between 50 and 55 mph. Any object moving at a certain velocity has motion or kinetic energy, K.E., equal to half the product of the mass and velocity squared:

$$\text{K.E.} = 1/2 \times M \times V^2 \tag{2.1.9}$$

For the sake of computation, let us assume that our automobile has a mass of 900 kg and that the driver and luggage have an additional mass of 100 kg. Further, let us say that the driver is in no rush and is driving at a velocity of 80 km/h. The velocity must be converted into meters per second in order to apply our formulas, using the conversion factor 1 km/h = 0.28 m/s. This means that 80 km/h equals 80 × 0.28 or 24.4 m/s. The kinetic energy of the moving vehicle equals:

$$\begin{aligned} \text{K.E.} &= 1/2 \times M \times V^2 \\ &= 1/2 \times 1000 \text{ kg} \times (24.4 \text{ m/s})^2 \\ &= 288{,}000 \text{ J} \end{aligned}$$

Suppose that at the entrance ramp of a freeway, the driver of this vehicle has to wait at a stoplight (i.e., the car is completely stopped). When the light turns green, the driver steps on the gas pedal (the accelerator) and accelerates to 80 km/h (24.4 m/s) in 12 s. As a result, starting at rest with a kinetic energy of zero, the car acquires a K.E. of 288,000 J in 12 s. Let us now calculate the power required to achieve this kinetic energy. We know that

$$\begin{aligned} \text{power} &= \frac{\text{work done}}{\text{time required}} \\ &= \frac{288{,}000 \text{ J}}{12 \text{ s}} \\ &= 24{,}000 \text{ J/s} \end{aligned}$$

Power is commonly measured in horsepower, not in joules per second. To convert, we use the conversion factor, 1 horsepower (hp) = 747 J/s. Therefore, 24,000 J/s = 32.7 hp. We see, then, that the engine must deliver 32.7 hp to achieve this acceleration of 0 to 80 km/h in 12 s. (Actually, the horsepower would have to be much greater to account for the losses in the system. Not all the power generated by the engine is transferred to the motion of the automobile.)

To many drivers, it makes a big difference whether their automobiles can reach 80 km/h in 12 or 6 s. It should be obvious that it will take twice as much power to accelerate in 6 s as in 12 s. But the energy gained by the vehicle is the same in both cases as long as the final speed is the same. In other words, the power requirement is doubled if the acceleration time is reduced to 6 s and halved if the time is increased to 24 s. But in all cases, the kinetic energy of the car running at 80 km/h is 288,000 J.

By the same token, it must now be clear that a jet engine, which accelerates a much larger mass than the mass of an automobile, and at a much greater rate, must have a power rating many times greater than that of an automobile engine. This means that the energy consumption of a jet engine for a given length of time will be much higher than the energy consumption of an auto engine in the same amount of time.

A power plant converts energy into electrical form continuously, unless it is shut down for repairs or for some other reason. But there is a limit to the rate at which it can convert energy. Therefore, all power plants are rated by their power output: the maximum energy conversion that can be safely accomplished by the power plant in a given unit of time.

If the power rating of an energy conversion device is known, it is possible to calculate the energy consumption of the device over a given period of time using the following formula:

$$\text{energy} = \text{power} \times \text{time}$$

Thus, if a power plant has an output of 200 kilowatts (kW) for a period of 10 hours, the total amount of energy converted by the system is equal to 200 kW × 100 h, or 20,000 kilowatthours (kWh).

A wood stove, a gas furnace, an electric heater, or an electric amplifier is also rated by its power output. In general, the higher the power rating of a device, the higher the cost, both when we purchase the device and when we use it. When we purchase such a device, we are paying in part for its power-handling capability. When the device is used, be it a vacuum cleaner or an oil furnace, more fuel (in the form of oil, coal, natural gas, wood, or electricity) is consumed by the device with the higher power rating; thus more energy is converted and used.

2.2 BEHAVIOR OF ENERGY

Although we cannot create or destroy energy, we can convert energy from one form to another. Some conversions are quite easy, whereas others can be very difficult, if not almost impossible. For example, the chemical energy in gasoline is easily converted to mechanical motion in an automobile, whereas nuclear fusion is very hard to convert to a useful form such as electricity. Just about any form of energy can be converted easily to heat energy, and with relatively high efficiency. But going the other way is both difficult and inefficient, as we shall see.

2.2.1 Laws of Thermodynamics

There are two important, well-established laws that put limits on our activities dealing with energy—and that means on everything we do. First, energy can be neither created nor destroyed. However, it can be converted from one form to another, and it is always converted when work is done. The law of conservation of energy, also known as the *first law of thermodynamics*, tells us:

> There is a constant amount of energy in the universe. Energy can be neither created nor destroyed.

No matter what anyone does, the total amount of energy will remain unchanged.

This law alone might lead us to believe that we can use as much energy as we want without running out of it. So why worry? What is this energy crisis? Well, this is where the second law of thermodynamics comes in.

The second law tells us that no conversion process conserves the full value of the energy at the beginning of the process. There is always some loss of useful energy. More formally, the *second law of thermodynamics* states:

In any energy transaction, there is a decrease in the amount of useful energy available for future transactions.

One consequence of this principle is that very few energy-conversion processes are 100% efficient. Some energy is always lost. For instance, when fuel oil is burned to provide heat to a dwelling, at most only 80% of the oil's stored chemical energy is converted to useful space heat. The rest is lost.

In some cases, the loss of useful energy is caused by a change in *form* rather than simple waste of energy. In the case of electric resistance heating, 100% of the electrical energy is converted to useful heat. Thus it seems that no energy is lost. But in this conversion, essentially all the electricity is converted from a highly versatile and useful form (electricity) to a less useful form (heat). Thus there has still been a decrease in the total amount of useful energy.

It makes sense that the amount of useful energy decreases when electricity is converted to heat energy. Why? Because heat energy is a form of kinetic energy due to *random* motion of molecules or atoms, whereas other forms of energy, such as chemical, mechanical, electrical, nuclear, or gravitational energy, are due to a *systematic* motion or arrangement. It takes energy to organize! In fact, this is the most important point. There is plenty of energy out there. But the organization of this energy causes the shortage of energy for useful work. In other words, although we are not using up the energy, we are continuously making it less available for future use. Every time we use energy, there is a net reduction in the amount of useful energy left for our future. Where does this energy go? It goes to heat the universe by a tiny fraction of a degree.

2.2.2 Grades of Energy

As one consequence of the second law of thermodynamics, we find that there are various *grades* of energy. Some forms of energy are very easy to work with: easy to use, easy to transport, easy to convert to other forms, and so on. These are called high-grade forms of energy. For example, electricity is one of the highest grades of energy. It can be converted to many other forms with relative ease and high efficiency. It can be controlled very accurately and transported easily.

Other forms are not so easy to work with. These are called low-grade energy forms. For example, heat energy is very difficult to store, transport, and convert to other forms; and when heat energy is converted to other forms, the efficiency of the process is usually quite low.

Very few high-grade energy resources are available in nature. This means that we must create high-grade energy from lower-grade resources. If we begin with a medium-grade energy resource such as coal, we can convert it to a low-grade energy form such as heat very easily and with little energy loss. But if we try to convert it to a high-grade form such as electricity, the process is very difficult and more energy is lost. This is one of the inevitable limits imposed on us by the nature of energy.

2.2.3 Efficiency

Another consequence of the second law of thermodynamics is that we cannot have a 100% conversion of energy from heat to mechanical forms. When we drive an automobile, not all the energy released by a gallon of gasoline is converted to equivalent mechanical motion of the vehicle. Some energy is lost in overcoming friction between the road and the tires. (That is a necessary evil—you need a little friction in order to have traction. Otherwise, the tires would simply spin in one place, the way they do on icy roads.) Some energy is lost in the power train of the auto (the driveshaft, transmission, and differential), and a large amount is lost as waste heat from the auto's combustion chamber. The heat carried away by the engine coolant (water with antifreeze, or air, as in a Volkswagen Bug) is a very high percentage of the energy supplied to the engine by the automotive fuel.

To slow down or stop, we apply the brakes. This represents an energy conversion from mechanical motion (kinetic energy) to heat energy because of the friction between the brake pads and the discs or drums. This heat cannot be converted again to mechanical motion with a high degree of efficiency; and some energy has been lost for future use.

Thus we see that there are many avenues for energy loss in a given system. All of them will, eventually, end up as waste heat. We can represent this general principle of energy transactions by a simple block diagram, as shown in Figure 2.2.1. Since practically all systems behave in this way, we define a concept called *efficiency* as the percentage ratio of useful energy output to input energy. In mathematical terms,

$$\text{efficiency} = \frac{\text{energy output}}{\text{energy input}} \times 100\% \qquad (2.2.1a)$$

The efficiency of a system must be greater than or equal to zero and less than or equal to 100; that is,

$$0 \leq \text{efficiency} \leq 100\% \qquad (2.2.2)$$

It is not necessary to express efficiency as a percentage. It can be expressed as a simple ratio:

$$\text{efficiency} = \frac{\text{energy output}}{\text{energy input}} \qquad (2.2.1b)$$

This figure must range between 0 and 1. This is clear because the output cannot be greater than the input. If the device has an efficiency of zero, it wastes all the energy that is put into it. Nothing can be less efficient than that.

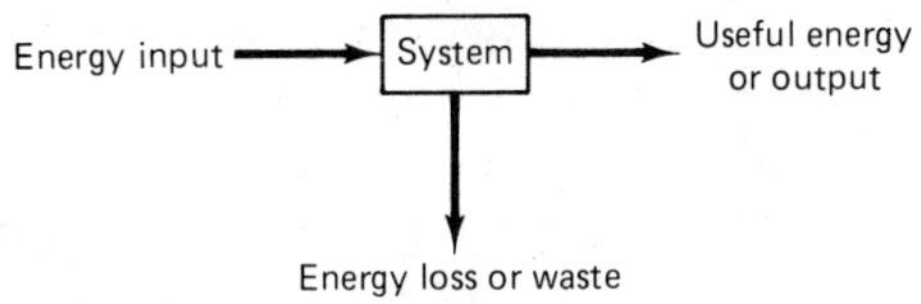

Figure 2.2.1 Block diagram of an energy transaction.

Since power = time × energy, these efficiency equations can also be written in terms of power. For example, suppose that an industrial furnace burns 600 cubic feet (ft^3) of natural gas per hour and has a rated output of 500,000 Btu/h. Each cubic foot of natural gas produces 1000 Btu of heat energy when burned.

$$\begin{aligned}
\text{power input} &= 600\ \text{ft}^3/\text{h} \times 1000\ \text{Btu/ft}^3 \\
&= 600{,}000\ \text{Btu/h} \\
\text{power output} &= 500{,}000\ \text{Btu/h} \\
\text{efficiency} &= \frac{\text{power output}}{\text{power input}} \qquad (2.2.3) \\
&= \frac{500{,}000\ \text{Btu/h}}{600{,}000\ \text{Btu/h}} \\
&= 0.8333 \quad \text{or} \quad 83.3\%
\end{aligned}$$

BOX 2.2.1 ILLUSTRATIVE EXAMPLE

Suppose that an oil furnace produces 73,000 Btu of useful space heat from 100,000 Btu worth of heating oil. To calculate the efficiency and the energy loss of the system, we use the following methods:

$$\begin{aligned}
\text{energy output} &= 73{,}000\ \text{Btu} \\
\text{energy input} &= 100{,}000\ \text{Btu} \\
\text{energy efficiency} &= \frac{\text{energy output}}{\text{energy input}} \qquad (2.2.1b) \\
&= \frac{73{,}000\ \text{Btu}}{100{,}000\ \text{Btu}} \\
&= 0.73 \quad \text{or} \quad 73\% \\
\text{heat loss} &= \text{energy input} - \text{energy output} \\
&= 100{,}000\ \text{Btu} - 73{,}000\ \text{Btu} \\
&= 27{,}000\ \text{Btu}
\end{aligned}$$

Most processes consist of two or more steps. When this is the case, the total efficiency of the process is *not* simply the sum of the step efficiencies. Instead, it is equal to the product of the step efficiencies. (To perform these calculations, it is best to work with ratios rather than percentages.)

$$\text{total process efficiency} = E_1 \times E_2 \times \cdots \times E_n \qquad (2.2.4)$$

Consider how this formula works in the following example, illustrated in Figure 2.2.2. Suppose that we start with 100 watts (W) of water power. The useful output of the hydro plant is 85% of this, or 85 W. This is the input to the transmission lines. The output of the transmission lines is 85% of the output of the hydro plant. This equals the following:

$$\text{transmission-line output} = 0.85 \times 85 \text{ W} = 72.25 \text{ W}$$

This 72.25 W is sent to the record player, where 60% of it emerges as sound. This means that the total sound output is equal to the following:

$$\text{sound output} = 72.25 \text{ W} \times 0.60 = 43.35 \text{ W}$$

The total efficiency of the entire system equals the output divided by the input:

$$\text{system efficiency} = \frac{43.35 \text{ W}}{100 \text{ W}} = 0.4335 = 43.35\%$$

As this example shows, the efficiency of this total process is equal to the product of the ratios for the individual steps. Efficiency considerations are discussed in greater detail in Chapter 5.

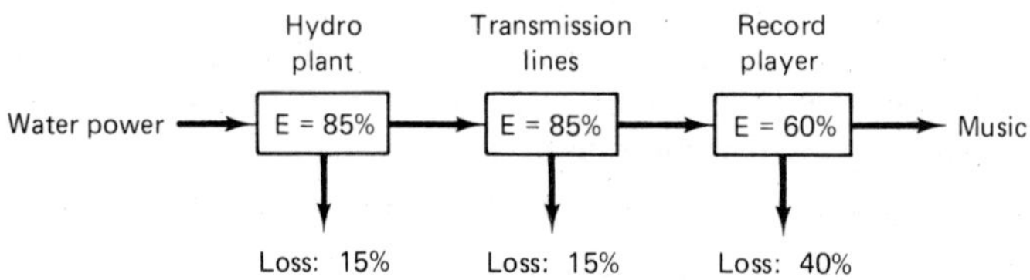

Figure 2.2.2 Energy conversions from water power to stereo sound.

2.3 ENERGY CONVERSIONS

It is well known that with some exceptions, one has to transform energy from one form to another to obtain a desired effect: *useful* work. For example, to heat a building, fuel is burned, in a furnace or a stove, and thus the stored chemical energy in the fuel is transformed to thermal energy by the process of rapid oxidation. In a hydroelectric plant, the mechanical energy of rushing water is changed to the mechanical energy of the turbine, which is, in turn, used to power the generators. There are cases where energy can be used directly without going through any conversion process. A simple example is the use of windows to illuminate a room with sunlight. But in general, energy must be converted in order to be used.

Another point that can be made about energy conversions is that when more steps are involved in an energy-conversion process, the overall efficiency of the process will suffer. For example, we may choose to heat a house by burning coal in a stove or furnace. This is relatively efficient, although not altogether clean or convenient. We may also choose to heat a house with electricity produced by a coal power plant. In the plant, the heat from coal is used to boil water and produce kinetic energy in the steam lines. This kinetic energy is converted into electricity, which is transmitted to the house. Finally, the electricity is converted back into heat. Clearly, it takes fewer steps to burn the coal directly; and as we shall see in Chapter 5, it is also more efficient. This is true because, as has been mentioned earlier, energy-conversion steps generally have efficiencies of less than 100%. In other words, there is energy loss at each step, so that the net output energy of the system will be less at each step. Therefore, the efficiency will be reduced.

Basically, then, there are two types of energy conversions: direct energy conversions and indirect, or multistep, energy conversions. Direct energy conversions achieve the desired result in a single step; indirect conversion involves at least one intermediate step. Direct conversion is generally more efficient than indirect. However, due to technical limitations, it is not always possible to achieve higher efficiencies with direct conversions. For example, photovoltaic cells, which convert sunlight directly to electricity, generally have operating efficiencies below 20% (see Chapter 4).

The total efficiency of an energy-conversion system can be improved greatly by applying the principle of *cogeneration* to the system. The principle is simple and very practical in most cases. Here the energy wasted in the conversion process is used for another secondary purpose. For example, the waste heat or steam from a thermal power plant can be used to heat nearby buildings (district heating) or to provide steam for manufacturing processes (process steam). This principle was widely used early in the twentieth century, but it was discontinued because of utility policies. Today, energy shortages are causing industries and utility companies to take a second look at cogeneration. We discuss this principle more fully in Chapter 11.

Yet another point must be emphasized: it is inefficient to convert a low-grade energy form such as heat to a high-grade form such as electricity. A high-grade form can be converted to another high-grade form, or to a low-grade form, with high efficiency. This means that a small amount of energy is lost when we convert kinetic energy to electricity, but more is lost when we convert heat to electricity. For example, a hydro power plant is generally more than 80% efficient, but the efficiency of a thermo-electric power plant rarely exceeds 40%.

BOX 2.3.1 THEORETICAL EFFICIENCY

The maximum efficiency of thermal to mechanical conversion is given by the following equation, which describes the ideal thermal-to-mechanical, or Carnot,[a] cycle:

$$\text{efficiency} = \frac{T_1 - T_2}{T_1} \tag{2.3.1}$$

where T_1 = maximum (intake) temperature, degrees Rankine (°R)

T_2 = minimum (exhaust) temperature, °R

°*R* = degrees Fahrenheit + 460

The expression will also apply if the temperature is expressed in degrees Kelvin, the unit of absolute temperature in the meter–kilogram–second (MKS) system. Degrees Kelvin = degrees Celsius + 273.

[a]Named after Sadi Carnot (1796–1832), a French scientist who studied thermomechanical conversions. He established, among other things, the relationship between ideal efficiency and the temperature of the medium used. It needs to be made clear that the limit noted in Eq. 2.3.1 is seldom achieved in practical thermal conversions.

Next we will look at block diagrams illustrating typical energy conversions. These show diagrammatically how some energy is lost as waste heat at nearly every stage of the conversion process.

1. *Direct conversion:* photovoltaics

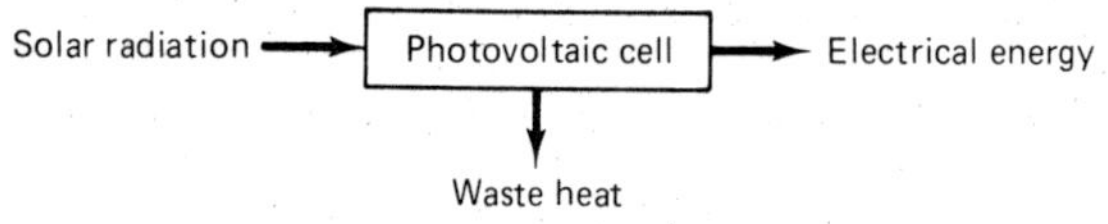

2. *Direct conversion:* passive solar

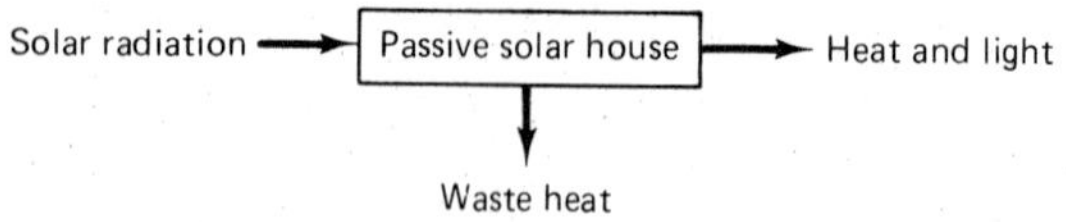

3. *Indirect conversion:* solar thermal

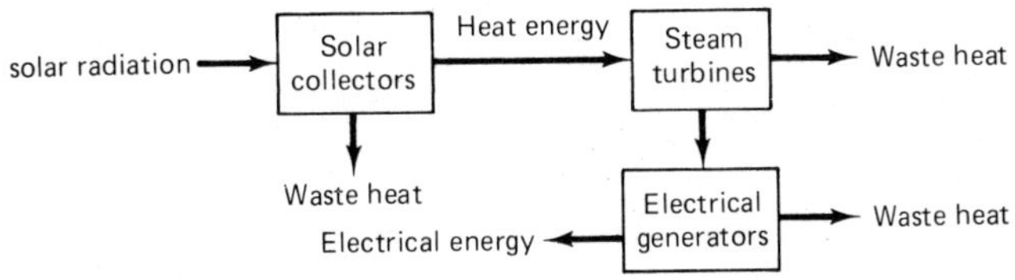

4. *Indirect conversion:* oil heat

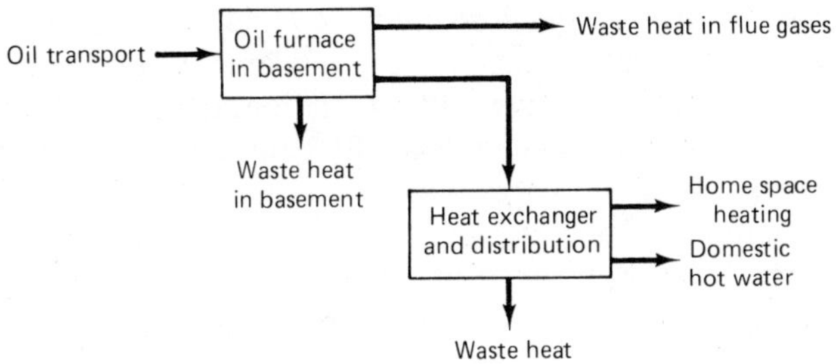

5. *Indirect conversion:* automobile

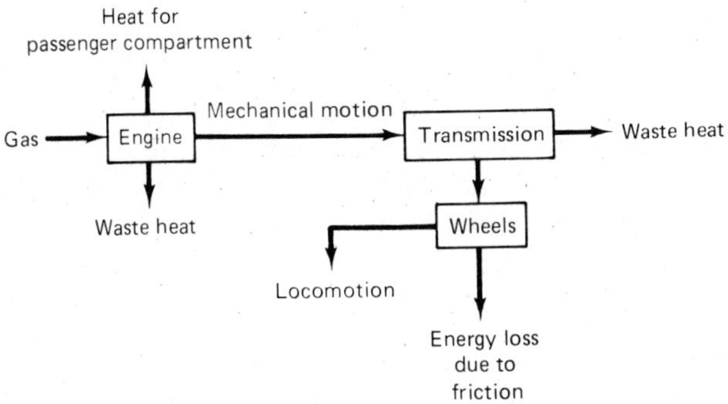

These simplified block diagrams give us a general idea of the process of energy conversion. Some of the important considerations in an analysis of a given energy-conversion system are:

The overall efficiency of energy conversion
The cost of the technology involved
The economics of financing the project
The environmental and health risks and impacts
The social and political impacts of the project
The lifetime of the project

The operation and maintenance of the system
The benefits derived from the conversion
The exploration of feasible alternatives

We will discuss all of these matters in subsequent chapters.

2.3.1 Energy Consumption

Human societies use energy for various purposes. Energy is used more in industrialized nations than in developing and underdeveloped countries. The United States, for example, uses an enormous amount of energy per capita. With only about 5% of the total world population, the United States accounts for about 30% of the world's energy use. By contrast, the two most populous countries, China and India, together are responsible for less than 10% of the world's energy consumption although they have 39% of the world's population.

The sources of energy vary from country to country and from region to region depending on resource availability, economics, politics, and many other factors, such as life-styles. In Israel, only solar water heaters can be sold legally at present. In France, nuclear power is heavily relied upon. In India, it is common to see animal refuse being used for fuel, either directly or by conversion to methane gas. In the Middle East there is such an abundance of natural gas and oil that natural gas is allowed to burn without any useful purpose.

In this country, natural gas is the main resource for space heating, hot water, and cooking in the Midwest. Wood stoves are very common in northern rural parts of the country. Heat pumps are getting more attention in the milder climate of the South, and solar is gaining ground in the sunny Southwest. It is hardly necessary to mention our heavy reliance on gasoline for transportation.

U.S. energy use can be classified under the headings listed in Table 2.3.1, shown in pie chart form in Fig. 2.3.1.

As indicated in Figure 2.3.1, residential-sector consumption of energy ac-

TABLE 2.3.1 Uses of Energy

Energy Use			
Residential	Industrial	Commercial	Transportation
Space heating	Process steam	Space heating	Automobile
Water heating	Direct heating	Air conditioning	Trucks
Cooking	Mechanical work	Feedstock	Mass transit
Refrigeration	Agriculture	Refrigeration	Comforts
Air conditioning	Electrolytic processes	Communications	Air conditioning
Lighting	Communications	Entertainment	Lighting
Communications	Maintenance	Controls	Other
Controls	Other	Maintenance	
Entertainment		Other	
Other			

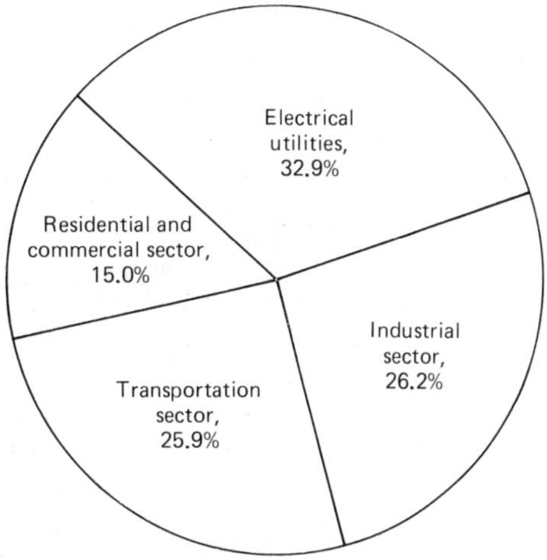

Figure 2.3.1 Primary energy use in the United States, 1982. Total U.S. energy use: 73.6 quads (includes contribution of wood energy).

counts for approximately 19% of the total, the industrial sector uses about 41%, the commercial sector approximately 15%, and the transportation sector a whopping 25%. The end uses are listed in Table 2.3.2 in order of their relative importance in terms of energy use and shown in pie chart form in Fig. 2.3.2. For example, space heating accounts for about 11% in the residential sector and roughly 7% in the commercial sector, accounting for 18% of total energy consumption. Process steam uses more than 16% of the total energy use in the country, although it is used almost exclusively in the industrial sector.

TABLE 2.3.2 Approximate Energy Consumption by End Use

Transportation	25%
Space heating	18%
Industrial process steam	16.5%
Industrial process heating	11.5%
Motor work (mechanical)	8%
Agriculture	5%
Water heating	4%
Other (lighting, air conditioning, controls, communications, etc.)	12%

Source: U.S. Department of Energy.

According to government figures, the U.S. population was estimated at 223 million and U.S. energy consumption was 79 quads for the year 1979. In 1960, the U.S. population was 181 million and its energy use stood at 44 quads. This amounts to an increase of 80% in energy use and a per capita increase of 37% over a period of 19 years. By 1982,* energy consumption had dropped to 74 quads, but this still represents a 70% increase in energy consumption.

*U.S. energy consumption (excluding wood energy) in the first half of 1983 amounted to 34.8 quads compared to 36.6 quads in the first half of 1982.

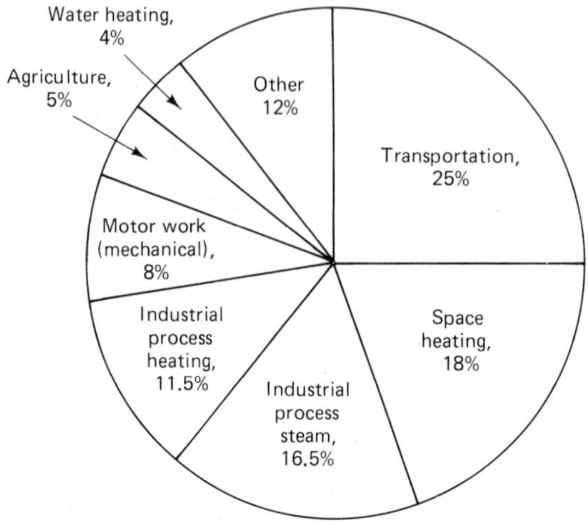

Figure 2.3.2 Domestic energy consumption by end use.

Of the 74 quads of energy used in 1982, 42 quads were lost in the process. In other words, only 32 quads represent useful energy out of a total of 74 quads, an energy efficiency of 43%. Much of this low efficiency can be attributed to the transportation sector and the process of electrical energy production. For example, in 1982, an input of 24.163 quads to the electrical industry produced an electrical output of 7.12 quads, for an industry-wide efficiency of 29.5%. As we will see in Chapter 5, most electrical energy conversion processes, except hydro conversion, are low-efficiency processes.

To conclude the discussion of energy consumption in this country, in Table 2.3.3 we give a breakdown of the total U.S. energy consumption for a few recent periods, classified by energy resource. Figure 2.3.3 shows the relative percentages represented by these resources in 1982.

TABLE 2.3.3 U.S. Consumption of Energy Resources

	U.S. Consumption (quads)		
Resource	1979	1982	1983 (January to June)
Petroleum	37.12	30.33	14.68
Natural gas	20.66	18.43	9.10
Coal	15.11	15.42	7.50
Hydro	3.14	3.57	2.05
Nuclear	2.71	3.18	1.52
Wood (est.)[a]	2.1	2.75	1.40
	80.84	73.68	36.25 (6 months)

[a]The Department of Energy figures do not report wood consumption data.

Source: U.S. Department of Energy.

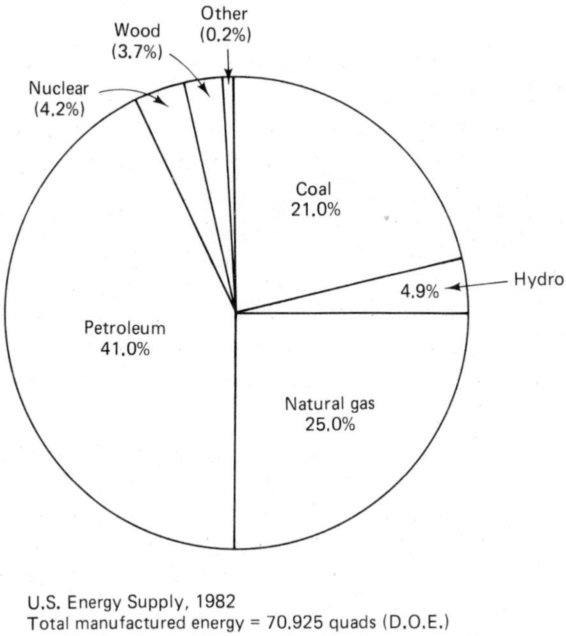

Figure 2.3.3 U.S. energy resource consumption for 1982: 73.6 quads. Wood energy consumption is estimated, and the solar contribution does not appear.

2.3.2 Energy Transfer

Energy does not move in a mysterious way. All energy can be transported from place to place in order to reach the point of use. Sometimes a significant amount of energy is lost in the process of transfer (see Fig. 2.3.4); and sometimes we waste energy because we choose an inappropriate means of transfer. It will be useful to know the different means of transferring energy from one place to another, especially in connection with the study of energy systems and conservation. For example, a familiarity with the mechanisms of heat transfer is essential if we want to understand the processes of heat transfer and heat loss in residential space heating. Here is a partial list of energy-transfer mechanisms.

1. Thermal energy
   a. Conduction
   b. Convection (includes infiltration)
   c. Radiation
2. Electrical energy
   a. Conduction
   b. Discharge

   c. Electromagnetic radiation
   d. Magnetic effect
3. Mechanical energy
   a. Gear trains
   b. Pulleys and ropes
   c. Belts and chains
   d. Friction couplings
   e. Fluid pressure (hydraulics)
4. Chemical and nuclear energy
   a. Land transport
   b. Pipelines
   c. Barges and ships
   d. Air and space transport
5. Light energy
   a. Reflection
   b. Refraction
   c. Diffusion
   d. Radiation

As we shall see in subsequent chapters, the transfer of energy can be a major concern in energy policymaking.

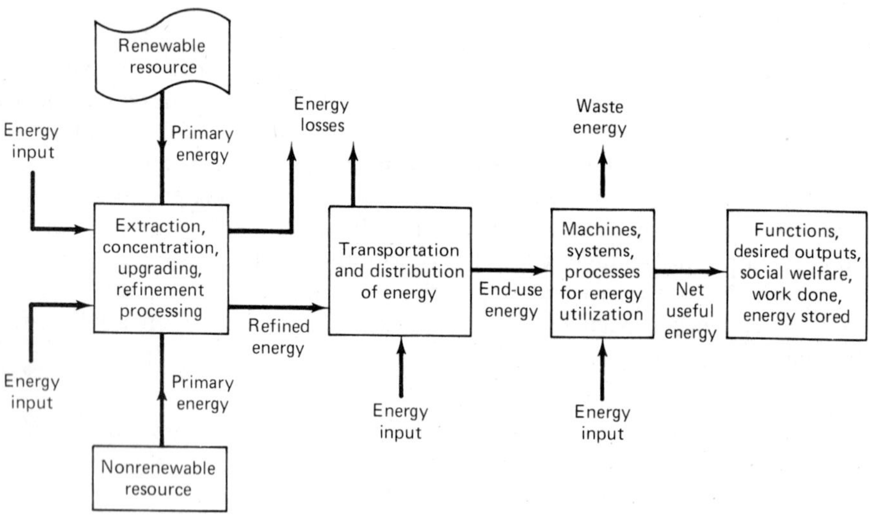

Figure 2.3.4 Energy flow diagram.

2.4 CLASSIFICATION OF ENERGY RESOURCES

There are several different ways to classify energy. The most basic way is to classify energy according to the form in which it is found. Basically, energy occurs in two forms: kinetic and potential.

Energy due to motion is called *kinetic energy*. Kinetic energy can be found in a rolling stone, blowing wind, moving water, or a turning flywheel.

Energy due to position is called *potential energy*. It is energy that has been stored for later use. Water behind a dam, nuclear fuel, coal, and oil are all examples of energy stored in various forms.

Table 2.3.4 presents one classification based on the forms in which energy can be found.

TABLE 2.3.4 Kinetic and Potential Energy

Energy	
Kinetic Energy	Potential Energy
Wind	Nuclear
Mechanical	Gravitational
Heat, geothermal	Electric (potential)
Electric (current)	Chemical (food and fuels)
Light, solar	Hydro dam (stored gravity)
Hydro turbine	
Tidal	

Another classification, based on the physical properties and characteristics of energy types, can be found in most science and engineering texts. Here are the basic physical types of energy:

Mechanical energy	Magnetic energy
Electric energy	Light energy (optics)
Chemical energy	Nuclear energy
Thermal energy	Gravitational energy
	Electromagnetic radiation

A third approach is to classify energy resources as renewable or nonrenewable. This system of classification is becoming increasingly important as we deplete our stock of fossil fuels and are faced with decisions about the economic, physical, and political implications of energy use. The basic categories of renewable and nonrenewable resources are presented in Table 2.3.5.

TABLE 2.3.5 Renewable and Nonrenewable Resources

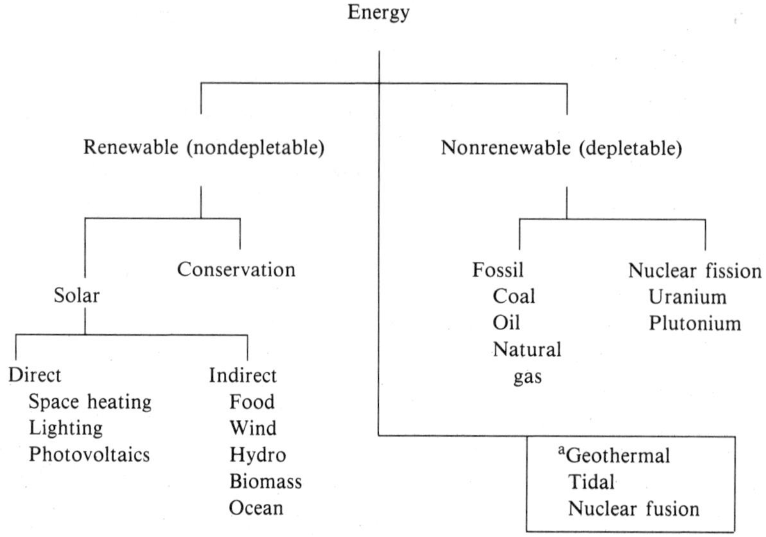

[a]The resources in the box can be considered as renewable or nonrenewable, depending on the system of classification.

Renewable energy resources are presumed to be ever-present as far as the human time scale is concerned. Most of them are related to solar energy. For example, wood and biomass (animal and vegetable wastes) are stored forms of solar energy—energy which was converted to chemical energy by photosynthesis. Similarly, winds are caused by the uneven solar heating of the surface of the earth. So wind energy is also derived from solar energy. (As we shall see later, solar energy is essentially a form of nuclear fusion energy.) Renewable energy resources are usually found close to where we live: in the air, on the land, and in the seas. Some, like biomass, have been used by human beings since prehistoric times.

Nonrenewable energy resources are the storehouses of energy, from prehistoric times. And where are these warehouses found? Naturally, underground! In prehistoric times, solar energy was trapped in the form of wood and other vegetable and animal matter (in short, biomass), and was converted to fossil fuels under extreme pressure. Thus coal, oil, and natural gas deposits represent huge storehouses of energy from the past. Nuclear and geothermal energy are also resources that come from beneath the surface of the earth.

There are two exceptions to the principle that nonrenewable resources are generally found underground. Heavy water, a source of deuterium and tritium used in nuclear fusion reactions, occurs naturally in seawater. Also, in a few isolated cases, fuel for nuclear fission reactions (primarily uranium) occurs naturally in exposed sand deposits on the surface of the earth. But most of the nuclear fuel is found underground.

2.4.1 Resources and Economics

Yet another way of looking at the same classification is from an economic point of view. Here, renewable resources are defined as "income" resources, and nonrenewable resources are treated as "capital" energy resources. Renewable resources can be replenished or renewed over a short period of time, much like income that one receives weekly or monthly. Renewable resources can therefore be called *income* energy resources: energy incoming continuously.

Similarly, nonrenewable resources are the ones that are stored, and which, once used, will never be seen again, as far as the human time scale is concerned. (We know that these resources are still being formed underneath the surface of the earth, but at a very slow rate: a rate that is not at all meaningful to human activities.) Thus these resources can be likened to *capital,* or funds that have been saved up over a period of time but can be spent quickly. This is the origin of the term "capital energy resources."

Once we define these two types of resources from an economic standpoint, it becomes clear that we should try to live within our energy income and use energy capital as sparingly as possible. But it is also clear that intelligent use of capital energy resources can enable us to increase our supply of energy income. For example, we can use nonrenewable petroleum products to make glazing for solar collectors. We consider the economics of energy resources and systems in Chapter 6.

2.4.2 Resources and the Environment

Another way of looking at the classification is from an environmental point of view. Here, renewable resources are defined as the ones found in the biosphere (the region of living things). For the most part, an environmentalist will tend to advocate the use of only those energy resources available in the biosphere. The environmentalist will reason that adding things, such as coal and oil, from outside the biosphere may disturb the delicate ecological balance. For this reason, he or she may be reluctant to use nonrenewable resources. We discuss the environmental analysis of various energy resources and systems in Chapter 7.

2.4.3 Resources and Power

Renewable resources. Now let us tie the concept of *power* to the classification of renewable resources. If a resource can be renewed, such as wood, or replenished, such as water behind a dam, at a rate comparable to the rate at which the resource is being used to produce useful energy, then we call that resource renewable.

This sets a limit on the rate at which renewable resources can be utilized to obtain energy. For example, a hydro power plant cannot produce an unlimited supply of electricity because the size of its electric output is limited by the following factors:

1. The rate of water flow
2. The available head for power production (the difference between the water level behind the dam and the level where the turbines are located)
3. The average rainfall in the river's watershed
4. The available land for dam construction and for the reservoir that will be created behind the dam
5. The maximum rate at which water can be released from the dam without flooding useful agricultural and residential land downstream

A similar set of limits can be drawn on the use of all renewable resources. Therefore, we can say that, in general, renewable resources can be rated by their power output.

Nonrenewable resources. On the other side of the coin, there is no theoretical limit to the rate at which nonrenewable resources can be converted to useful energy. For example, if Saudi Arabia wanted to speed up its production of oil, it could increase the rate of extraction either by increasing the number of oil wells or by increasing the output per well. However, there is a limit on the total amount of energy that can be obtained from a nonrenewable resource. The Saudis can produce oil only until the oil wells run dry; and we know that they will run dry eventually.

A savings account contains only a finite amount of capital. Unless money is deposited as rapidly as it is drawn out, we cannot continue to draw it out forever. It is the same with our use of energy resources. It is basically our choice how long we wish the supply to last, and we make that choice when we decide what our rate of consumption will be. (For example, in 1979, the United States consumed roughly 10% of its proven reserves of natural gas. If that rate of consumption continues and no more gas is discovered, the U.S. reserves will be depleted in 10 years. If U.S. consumption increased tenfold, the entire reserve would be exhausted in one year.)

In practice, however, it must be realized that a nonrenewable resource can become economically unavailable if we have to spend more money or energy on the extraction and refinement than we can get back from the refined product. This can happen easily when a resource gets scarce or when the easily accessible part of the resource is used up. There may be plenty of the resource left, but it will not be extracted because of the amount of work or expense involved in the process of extraction.

Here are some of the factors that determine the rate at which a nonrenewable resource is used:

Needs and desires of the society
Available technology
Supply and demand economics
Availability of renewable resources
National and local energy policy
Environmental and social impacts

In the past, technological innovations have steadily increased the rate at which we consume energy resources. An automobile consumes much more energy than a horse, and an airplane consumes much more than either. But the rate of consumption cannot increase indefinitely. There are limits on the total amount of energy available.

Sometimes the limits on a resource are meaningless. For example, there are plenty of heavy-water (deuterium) molecules in seawater. Deuterium occurs at a frequency of approximately one atom in 6500, or about half a gram of deuterium per gallon of seawater. Just 1 gram of deuterium can potentially release the energy equivalent of 600 gallons of oil. So we have, theoretically, enough energy to last for hundreds of thousands of years. But commercial fusion power is still not feasible, economically or technologically. In fact, we are at least 25 years away from any commercial production of fusion power for peaceful purposes. (It is interesting to note that we do have all the fusion power we need—and then some—for military purposes. Thousands of nuclear weapons, based on an application of an uncontrolled fusion reaction, have been manufactured by the military-industrial complexes of the superpowers. The first fusion bomb was made over 30 years ago! But control of this immense force, like control of the weapons that use it, has not been easy to achieve.)

To conclude this section, we might add that the system of classification presented here is not exhaustive. As we mentioned before, for example, since fusion can potentially supply energy for many thousands of years, we might wish to classify it as a renewable resource because this kind of time span is beyond human comprehension. What can we predict about, say, the year 40,000? Not much, except that the sun will still shine on earth, and the universe will still be there. Yet, theoretically, we have more than enough fusion power to last until the year 40,000 once the technology is perfected. Because of these kinds of arguments, we have classified fusion, tidal, and geothermal resources with the other energy resources which they resemble. Fusion is classified as a nonrenewable resource, together with nuclear fission; geothermal is classified with nonrenewable resources because it comes from underground; and tidal energy is classified as a renewable resource together with hydro.

It should also be pointed out that even solar energy will not be available forever. The sun will exhaust all its fuel in a matter of several billion years. But these time scales are totally beyond our comprehension, so solar energy is always classified as a renewable energy resource.

2.5 UNITS OF MEASUREMENT

Since the purpose of this chapter is to lay the groundwork for a discussion of energy and power issues, policies, and forecasts, it goes without saying that we need to discuss the units in which these quantities are measured. Proper units are essential for any intelligent dealing with physical quantities and qualities, and many social or nonphysical questions as well. There are many units in use: no shortage here! In fact, most scientists feel that there are too many units. They would prefer worldwide adoption of the metric system and abandonment of the cumbersome British system.

Of the two systems of measurement currently in use, the most widely used is the metric or MKS system (MKS stands for meter–kilogram–second). A less widely used system is the British system, also called the FPS (foot-pound-second) system. Because of the existence of these two systems, there are at least two varieties of units for each quantity with the exception of time. Oddly enough, Great Britain has abandoned the British system; today it is used in only a few countries, including the United States.

No unit is too large or too small in the whole energy picture. We must be able to describe the energy binding the parts of an atom, and also, at the other end of the scale, the energy consumption of the whole world. Therefore, we have multiples and submultiples of units. Finally, because energy appears in several forms, we have energy units specifically devoted to a particular form of energy. But as long as we realize that "energy is energy is energy," there should be no crisis; all we need is a conversion table for various units. Conversion tables and other useful tables are presented in Tables 2.5.1 through 2.5.7 and in Appendix A at the end of the book.

TABLE 2.5.1 Basic Units of Measurement

Quantity	System	Unit	Symbol	Conversion
Mass	Metric	kilogram	kg	1 kg = 2.2 pounds = 0.069 slug
	British	slug		1 slug = 14.5 kg
Length	Metric	meter	m	1 meter = 3.25 ft
	British	foot	ft	1 foot = 0.304 m
		inch	in.	1 inch = 2.54 centimeters
Time	All	second	sec, s	
Temperature	Metric	degree Celsius	°C	°C = (°F − 32) × 0.556
		degree Kelvin	K	K = °C + 273
	British	degree Fahrenheit	°F	°F = °C × 1.8 + 32
		degree Rankine	°R	°R = °F + 460

TABLE 2.5.1 Basic Units of Measurement *(Cont.)*

Quantity	System	Unit	Symbol	Conversion
Force	Metric	newton	N	1 newton = 0.225 pound
	British	pound[a]	lb	1 pound = 4.45 N

[a]In the British system, there are two units called *pounds*. One is a measure of mass and the other is a unit of force. The two should not be confused. Strictly speaking, we should distinguish between the units *pound-mass* and *pound-force* (also called *pound-weight*).

TABLE 2.5.2 Units of Energy

Btu (British thermal unit): British system unit of heat energy. One Btu is the energy required to raise the temperature of 1 pound of water by 1 degree Fahrenheit. When a wooden kitchen match is completely burned, the amount of energy released is approximately 1 Btu.

Barrel (bbl): A liquid measure of oil; approximately 42 U.S. gallons. 7 bbl = 1 ton. This unit is used extensively to express large-scale energy use.

Calorie (kcal or Cal): Metric system unit of thermal energy. One kilocalorie is the amount of heat required to raise the temperature of 1 kilogram of water by 1 degree Celsius. 1 kcal = 4 Btu.

Foot-pound (ft-lb): British system unit of mechanical work. One foot-pound is equal to the amount of mechanical energy required to lift a weight of 1 pound through a height of 1 foot.

Joule (J): Metric system unit of mechanical energy. One joule is the amount of energy transferred when a force of 1 newton acts on a body for a distance of 1 meter. (J = N-meter.)

Kilowatthour (kWh): Système International (extension of metric system). One kilowatthour is the electrical energy consumed when an electrical power of 1 kilowatt is maintained for 1 hour. A 100-W bulb burning for 10 hours uses 1 kWh of electrical energy.

Quadrillion Btu (quad): A quad is the energy equivalent of 10^{15} Btu. The energy consumption of the United States in 1982 was approximately 74 quads.

TABLE 2.5.3 Units of Power

Horsepower (hp): British system unit of mechanical power. The horsepower is a measure of the rate at which mechanical work is done. One horsepower is equivalent to 550 ft-lb per second. It is more than the power capability of an average horse.

Watt (W): Système International unit of electric power. One watt is the electrical power consumption equivalent of 1 joule per second. It is also equal to the power supplied by an electrical circuit when 1 volt potential difference is applied and a current of 1 ampere flows in a resistive circuit. 1 volt $\times$ 1 ampere = 1 watt.

TABLE 2.5.4 Scientific Notation

When quantitites under consideration are very large or small, multiples and submultiples of these units are used. Here are some of the common names and symbols of these multiples.

Name	Symbol	Meaning	
micro	μ	1/1,000,000	One millionth of a unit
milli	m	1/1,000	One thousandth of a unit
centi	c	1/100	One hundredth of a unit
deci	d	1/10	One tenth of a unit
kilo	k	1,000	One thousand units
mega	M	1,000,000	One million units
giga	G	1,000,000,000	One billion units
tera	T	1,000,000,000,000	One trillion units
quad	q	1,000,000,000,000,000	One quadrillion units
quint	Q	1,000,000,000,000,000,000	One thousand quads

TABLE 2.5.5 Equivalence Table for Energy Units

Unit	Btu	kcal	ft-lb	hp-hr	kWh	MJ
Btu	1	0.25	780	0.0039	0.0029	0.00105
kcal	4	1	3100	0.0016	0.0012	0.00418
ft-lb	0.0013	0.0032	1	5.1×10^{-7}	3.8×10^{-7}	1.35×10^{-6}
hp-hr	2500	640	2×10^{6}	1	0.746	2.07
kWh	3400	860	2.7×10^{6}	1.34	1	2.65
megajoule (MJ)	948	239	7.4×10^{5}	0.50	0.37	1

TABLE 2.5.6 Equivalence Table for Power Conversions

1 hp =	746 W	1 kW =	1000 W
	550 ft-lb/s		740 ft-lb/s
	0.746 kW		1.34 hp
	0.18 kcal/s		0.24 kcal/s
	0.71 Btu/s		0.96 Btu/s

TABLE 2.5.7 Approximate Heat Contents of Some Common Fuels[a]

Fuel	Heat content
Liquid fuels	
Crude oil	5.80 mBtu/Bbl
Gasoline	5.25
Kerosene	5.67
Distillate fuel oil	5.82
Residual fuel oil	6.29
Natural gas liquids	4.03
Gaseous fuels	
Natural gas, dry	1020 Btu/cubic foot
Natural gas, wet	1093 Btu/cubic foot
Solid Fuels	
Coal, lignite	6000-7500 Btu/pound
Bituminous	12,000-15,000 Btu/pound
Anthracite	13,500-15,000 Btu/pound
Wood[b]	20-35 mBtu/cord
Peat	5000 Btu/pound

[a]See Chapters 3 and 4 for details.

[b]Depends upon the type of wood; see Table 4.6.1 for details.

Source: U.S. Department of Energy.

STUDY QUESTIONS

1. What are energy, work, power, and force? How are they interrelated?
2. Circle the units of power and underline the units of energy in the following list:

   Btu kcal/h kWh quad hp kW hp-hr MW kJ bbl/day

3. Explain the difference between energy and power requirements using examples from your own experience.
4. What forms (electrical, mechanical, etc.) can energy take? How are they converted from one form to another?
5. State and explain the two laws of thermodynamics.
6. Explain the concept of efficiency. Why can the efficiency of a system not exceed 100%?
7. Explain renewable and nonrenewable resources. Which resources do you use most in your daily life? Draw up a chart showing your daily consumption of renewable and nonrenewable energy resources.

8. Draw a block diagram for each of the following energy conversions:
   (a) Wind energy-conversion system (wind to electricity).
   (b) Active solar system (sunlight to space heat).
   (c) Coal power plant (coal to electricity).
   (d) Hydro power plant (water to electricity).
   (e) Food on your plate (starting with solar energy).
   (f) Electric lighting (starting from the coal mine).
9. Write a short essay on the energy crisis, explaining why it occurred and how it can be solved.
10. Would you classify nuclear fusion, geothermal, tidal, and wood energy resources as renewable or nonrenewable? Give your reasoning.

PROBLEMS

1. **(a)** Calculate the energy consumption of a 2-kW space heater that is run an average of 8 hours a day for a complete heating season of 6 months.
   (b) At 10 cents/kWh, calculate the cost of energy consumed.
2. Calculate the efficiency and input power for the energy conversion symbolized by the following block diagram:

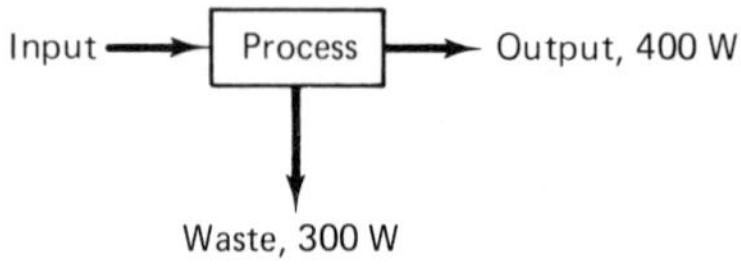

3. Suppose that you own a European car with a speedometer measuring only km/h, from 0 to 150 in increments of 10. Where would you locate the following speeds:
   (a) 20 mph,
   (b) 30 mph,
   (c) 45 mph,
   (d) 55 mph, and
   (e) 70 mph?
4. Alfred is carrying his baby up a flight of stairs. How much energy does he spend in climbing to the upper level of his house if his baby has a mass of 10 kg and the height of the staircase is 4 m? Assume that there are no friction losses and that all the energy that he spends is converted to potential energy. Alfred has a mass of 55 kg.
5. Calculate the kinetic energy of:
   (a) A football player of mass 75 kg moving at a speed of 10 m/s for a touchdown.
   (b) An automobile of mass 1500 kg traveling at 85 km/h.
   (c) A person of mass 60 kg walking at 6 km/h.

   Express your answers in ft-lb, joules, and kWh.
6. Elaine goes downhill skiing on an expert trail. At one point on her adventure, on the steepest part of the trail she reaches a speed of 50 km/h.

(a) What is her kinetic energy at that point?

(b) Where does she get this kinetic energy?

(c) What was her potential energy at the very start of the run?

(d) What was her kinetic energy at the start?

Assume that Elaine and her ski equipment together have a mass of 65 kg and the vertical height of this ski slope is 500 m.

7. A 60-kg skier is being chair-lifted up a mountain slope. The lift goes at an angle of 30 degrees, so that for every 2 m traveled by the lift, 1 m of height is gained. The total distance along the slope is 1 km. Calculate the potential energy gained by the skier.

FURTHER READINGS

COOK, EARL. *Man, Energy, Society.* San Francsico: W. H. Freeman, 1976.

DORF, RICHARD. *Technology, Society, and Man.* San Francisco: Boyd & Fraser, 1974.

Energy. A special report. Washington, DC: National Geographic Society, February 1981.

Energy and Power. A special issue on energy and power. *Scientific American*, September 1971.

FOWLER, JOHN. *Energy and Environment.* New York: McGraw-Hill, 1975.

GOSWAMY, AMIT. *The Concepts of Physics.* Lexington, MA: D. C. Heath, 1979.

NEBEL, BERNARD. *Environmental Science.* Englewood Cliffs, NJ: Prentice-Hall, 1981.

SINGER, S. FRED, ed. *Energy: Readings from Scientific American.* San Francisco: W. H. Freeman, 1979.

3

Nonrenewable Energy Resources

At present, human beings rely on nonrenewable energy resources for nearly all of their energy needs. Coal, oil, natural gas, and nuclear energy account for the vast majority of our energy supply. Automobiles run almost exclusively on gasoline, most electricity is produced in fossil or nuclear plants, and most residential heating systems use natural gas, fuel oil, or electricity as their primary fuel. Industry also consumes enormous amounts of coal, oil, and natural gas. In the United States, of 71 quads of primary energy use accounted for by the U.S. Department of Energy (1982), only 3.6 quads came from renewable resources. (2.75 quads of wood energy use were not included in the DOE figures.)

As our energy use has increased in recent years, we have begun to recognize the finite nature of nonrenewable energy resources. The "energy crisis" has primarily been a crisis in the supply of petroleum and natural gas. As we look to the future, we can anticipate further crises as the world's supply of nonrenewable resources decreases.

In order to understand the nature of the energy problems we face, it is essential to understand where and how we are getting the energy we use today. Even the most outspoken advocates of renewable energy resources will agree that a thorough understanding of our present patterns of energy supply and consumption is essential if we are to work toward constructive change.

We have included nuclear fusion and geothermal energy among the nonrenewable energy resources. It can be argued that the supplies of these energy resources are so vast that they cannot be depleted at any time in the foreseeable

Figure 3.0.1 Large steam turbine-generator. This 500,000-kW 3600-rpm turbine-generator set is capable of providing a rapid change in the generator field voltage in response to power system needs. (*Source:* General Electric Company of the United States.

future, so in principle they should be classified as renewable, or nondepletable resources. We have included them in this chapter because of similarities between their production and the production of other nonrenewable energy resources. Readers are encouraged to develop their own classifications, based on the criteria that seem most appropriate to them.

3.1 FOSSIL FUELS: COAL

Coal is our most abundant fossil fuel. We have roughly five to six times as much as oil in terms of total energy content (Tables 3.1.1 and 3.1.2). The potentially recoverable reserves are roughly equivalent to 12,000 billion barrels of oil, while the economically recoverable reserves are equivalent to about 3000 billion barrels. The total world production of coal in 1977 was 3400 million metric tons. Major producers of coal today are the United States, USSR, China, Poland, Great Britain, West Germany, and India. In 1982, U.S. coal energy production was 18.65 quads.

Because of its abundance, coal is often considered a "bridging fuel" from today's oil-based economy to a future economy based on renewable resources. Indeed, coal reserves appear to be sufficient to sustain present rates of consumption for 200 to 500 years. But to achieve this transition, according to some studies the United States will have to mine 50% more coal in the year 2000 than it does today. For this to happen, serious economic, social, and environmental issues related to coal production and use will have to be resolved. These problems are explored in some depth in subsequent chapters.

TABLE 3.1.1 Estimates of U.S. Fossil Energy Resources: Known Recoverable Reserves[a]

Item	Standard Units	Quads (10^{15} Btu)	Percent of World Reserves
Petroleum	26.4 billion barrels	153	4.7
Natural gas	191.0 trillion cubic feet	196	8.5
Natural gas liquids	6.0 trillion cubic feet	35	2.4
Bitumens	2.5 billion barrels	14	12.9
Coal	218.4 billion short tons	4980	33.5
Shale oil	74.0 billion barrels equivalent	429	4.1
Tar Sands[b]	NA	NA	NA
		5807	17.07

[a]Known recoverable reserves are approximately half of demonstrated recoverable reserves.

[b]Not available.

Source: Adapted from Phillip G. LeBel, *Energy Economics and Technology* (Baltimore: Johns Hopkins University Press, 1982).

TABLE 3.1.2 Estimates of World Fossil Energy Reserves: Known Recoverable Reserves

Item	Standard Units	Quads (10^{15} Btu)
Petroleum	566 billion barrels	3,225.0
Natural gas	2230 trillion cubic feet	2,299.0
Natural gas liquids	363 trillion cubic feet	1,456.0
Bitumens	26.7 billion barrels	107.9
Coal	651.7 billion short tons	14,858.0
Shale oil	1784 billion barrels equivalent	10,352.0
Tar sands	300 billion barrels equivalent	1,740.0
		34,037.9

Source: Adapted from Phillip G. LeBel, *Energy Economics and Technology* (Baltimore: Johns Hopkins University Press, 1982).

3.1.1 History of Coal

Coal was used in China and Britain as early as 2000 years ago. Hopi Indians used coal in the thirteenth century. More recently, European cities were (and some still are) heated by coal. In the eighteenth and nineteenth centuries, the London smog caused by coal burning was notoriously unhealthy. After the invention of the steam engine, coal consumption increased dramatically. In the year 1900, coal supplied about 70% of U.S. energy requirements. But the availability of easy-to-use petroleum and natural gas caused a decline in coal use between 1950 and 1973. Since then, however, coal consumption has been on the rise in the United States, particularly since the late 1970s. Currently, the United States uses more than 15 quads of coal energy every year. We are also a net exporter, sending abroad between 10 and 12% of our total coal production.

3.1.2 Physical Characteristics of Coal

Coal is a form of stored solar energy, as are all fossil fuels. During the Pliocene period of prehistory, approximately 300 million years ago, tremendous amounts of vegetation, in the form of giant ferns and other herbaceous plants, grew around the shores of ancient seas. This vegetable matter was buried under sediment (overburden), which compressed the fossil matter and transformed it into a variety of substances.

Of the various coal-related fossil fuels, peat is the closest to the original vegetable matter. Peat is brown, contains vegetable plant remains, and has a high moisture content (see Section 3.2).

Under the pressure of the earth's overburden, together with the passage of time, large quantities of peat were transformed into lignite, the softest form of coal. Lignite is black, dense, and drier than peat, although it still contains substantial amounts of carbohydrates.

Under greater pressure and the passage of more time, some lignite was further converted into bituminous coal. In this process, most of the carbohydrates in lignite were converted into hydrocarbons and then into elemental carbon. In some deposits, bituminous coal has been further converted into semi-bituminous or anthracite coal, which have even less moisture, fewer carbohydrates and hydrocarbons, and a higher percentage of elemental carbon.

3.1.3 Grades of Coal

Because of the great variety in the chemical composition of coal, a number of different categories of coal have been established. These categories are determined by moisture content (which causes a loss of net energy during combustion), presence of volatile impurities, percentage of elemental carbon, and energy content per unit of weight. In general, the higher grades of coal are purer and have a higher energy content, although the highest grade—anthracite—has a somewhat lower energy content than the next-lower grade. Higher grades of coal generally have less clay, sand, and sulfur contamination than the lower grades.

Although the system of classification shown in Table 3.1.3 is the most widely used, coal is also classified in other ways. For example, coal is often broadly classified into *hard* and *soft* coals. Anthracite and the better grades of bituminous coal are generally considered to be hard coals, while lower grades of bituminous coal, lignite, and brown coal are considered soft coals. These categories are useful when a larger number of categories is confusing: for example, in comparing overall coal production of two countries, or of two regions in a single country.

Yet another system of classification is based on the ability of coal to form *coke.* Coke is coal that has been processed to form a highly pure substance consisting almost entirely of elemental carbon. Coke is very important in the manufacture of steel and other metallic alloys. Thus it is also possible to classify coals

TABLE 3.1.3 Grades of Coal and Their Characteristics

Grade	Moisture Content (%)	Volatiles (%)	Carbon Content (%)	Energy Content (kcal/kg)	Combustion Characteristics
Anthracite	2–6	Up to 13	More than 90	8000	High temperature, short, blue flame, nonsmoking
Semibituminous	2–6	15–20	80–90	8600	Short, luminous flame, high temperature, little smoke
Medium bituminous	2–8	20–30	65–85	8400	Long flame, medium temperature, smoky combustion
Low-rank bituminous	10–20	30–40	40–75	7500	Smoky, long flame
Lignite	35–45	35–55	25–50	3800	Low temperature, smoky, high moisture emissions

according to their uses. This gives rise to two categories: (1) *metallurgical* or coking coals, that are used for chemical processes, and (2) *thermal* coals, that are used to provide heat and/or steam.

3.1.4 Coal Resources in the United States

The United States is blessed with tremendous quantities of coal; in fact, the United States is the world's largest producer of coal (Fig. 3.1.1). The United States accounts for more than 50% of all coal mined and consumed in the Western world. In 1981, a total of 824 million short tons of coal was produced in the United States, of which 81.5% was used to produce electrical energy. (In July 1983, a total of 56.5 million tons of coal was produced in the United States, of which 50 million tons was consumed by the electric utilities.)

According to estimates by the U.S. Geological Survey, the United States has between 1 and 7 *trillion* tons of coal. Of course, not all of that is accessible. More realistic figures are 255 billion metric tons of economically recoverable reserves and 410 billion metric tons of measured reserves. ("Measured reserves" are those deposits which are well surveyed and whose extent and quality have been established by adequate sampling.) The U.S. share of total world coal reserves is approximately 30%.

Bituminous coal is the most abundant and widespread variety found in the United States (see Table 3.1.4). It is used generally for industrial, electrical, and heating energy requirements. It is also exported, primarily to Japan, Canada, and Europe. Lignite also comprises a significant fraction of U.S. reserves. However, lignite has a low heating value because of its high moisture content. This makes transportation and storage of lignite prohibitively expensive. Therefore, lignite is generally used near the mining site in so-called "mine-mouth" electric power generating stations.

TABLE 3.1.4 U.S. Recoverable Coal Reserves

Grade	Reserves (billions of tons)	Percentage of Total
Anthracite	4.3	2
Bituminous	113.6	52
Subbituminous	83.0	38
Lignite	17.5	8
	218.4	100

Sources: Energy: A Special Report (Washington, D.C.: National Geographic Society, 1981); U.S. Geological Survey.

Coal is found in several regions of the United States. In the eastern United States, coal is found in the Appalachian region, especially in Pennsylvania, Ohio, West Virginia, and Kentucky. The Midwest also has large coal deposits, especially in Illinois. The Rocky Mountain region (Colorado, New Mexico, Wyom-

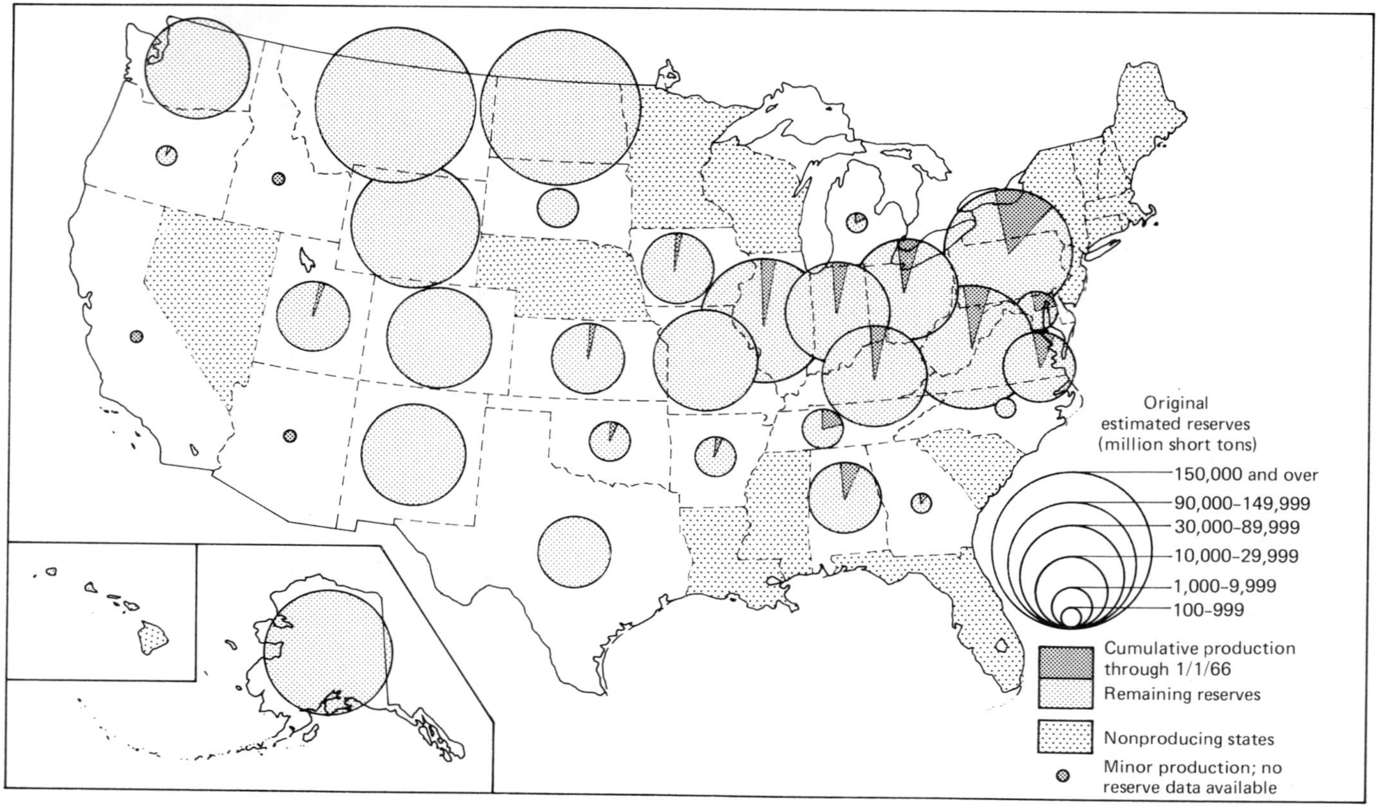

Figure 3.1.1 Coal production and reserves: anthracite, bituminous, and lignite, 1966. (*Source:* U.S. Geological Survey.)

ico, Wyoming, and Montana) has very large coal deposits. Other regions of the country also have locally or regionally significant deposits of coal.

3.1.5 Coal Production

Coal is mined in three different ways:

1. *Surface mining* accounts for 60% of coal production in the United States. In the eastern United States, *strip mining* is the main type of surface mining. In the West, *open-pit* mines are more common. In hilly regions, hillside *contour mining* is practiced. The methods used in these three types of mining are quite similar, so we will treat them all as forms of surface mining.

Surface mining is used when the coal deposits are relatively close to the surface. First, large mechanical shovels or draglines are used to remove the surface material (overburden) to expose the coal deposits. The surface material is stored, usually on a previously mined surface nearby, so that it will be available when the mined land is to be reclaimed. Then the coal bed is blasted to break it up for shoveling. The broken-up coal is loaded onto "unit trains," which may have as many as 100 coal hoppers and carry up to 10,000 tons of coal in a trip. The trains transport the coal to power plants, where it is converted into electric energy. A large surface mine can be up to 200 ft deep and can produce more than 10,000 tons of coal a day, employing 500 or more people.

Figure 3.1.2 Open-pit mine. Reclamation out West is a special problem because of inadequate topsoil and rainfall. (*Source:* Office of Surface Mining, U.S. Department of the Interior.)

Surface mining is a very efficient method of producing coal. It allows for more complete recovery of coal deposits than either of the other two methods, and it is more economical. However, surface mining requires land reclamation to avoid environmental damage. In the United States, 23,000 km^2 (almost 5.8 million acres) of land have been surface-mined and approximately half the land has not been reclaimed. All too often, huge scars have been left on the land where coal has been mined. Erosion has ruined the land for cultivation, and acidic runoff has damaged nearby water tables. At present, however, mining companies are required to restore the land close to its previous condition, preserving the original contours of the land as well as its previous vegetation and soil condition.

2. *Underground mining* accounts for most of the remaining 40% of coal production in the United States. This method is used when the coal deposits are deep underground (up to 1000 m deep) or when the deposits are located near or under heavily populated or heavily used areas. As in surface mines, the mining is done by large machines, but the machines (called *continuous miners*) are used inside deep tunnels rather than out on the surface. The coal is transported on railroads laid within the mines and then transferred to other railroads or barges when it reaches the surface.

Underground mining is not as efficient as surface mining. It is economical only when the coal deposits are at least a few feet thick. Even then, only half the available coal can be extracted from an underground mine; the rest must be left to support the weight of the overburden. The deep mines need to be ventilated and pumped to prevent the buildup of water and toxic gases and to

Figure 3.1.3 Underground continuous miner. This machine can bite into the face of a coal seam and rip out as much as 12 tons of coal per minute. (*Source:* Bethlehem Steel Corporation.)

reduce the fire hazard from methane gas. Despite the progress that has been made in recent years, deep mining is still one of the most dangerous jobs in the world in terms of health effects and safety conditions on the job. In 1980, 145 miners lost their lives on the job, and many others died from work-related disease such as black lung (see Chapter 8). But at present, deep mines are the only way to extract coal from deposits far below the surface.

3. *Auger mining* is relatively unimportant, but in special situations it can be the best method for extracting coal. When a thick horizontal coal bed is exposed on the sides (on steep hillsides, for example), large horizontal augers are used to shear the coal from the exposed face. The auger drills into the coal seam and as the auger rotates forward, the broken coal is passed backward and loaded.

3.1.6 Transportation of Coal

Because coal is a heavy, solid fuel, transportation is one of the most difficult aspects of coal production. Unlike oil or natural gas, coal cannot simply be pumped to the point of use. This has made the coal industry vulnerable to breakdowns in transportation and railroad strikes, as well as to problems in coal production.

As mentioned earlier, rail transport is widely used to transport coal. Large trucks and barges are also used, but in the United States these two methods are less common. A newer method of transporting coal is *slurry pipelines*. The

Figure 3.1.4 Surface mining: A gigantic drag-line shovel almost dwarfs the huge coal hauler truck on a strip mining site. (*Source:* The Office of Surface Mining, U.S. Department of the Interior.)

coal is mixed with water to make a slurry that can be pumped through a pipeline in the same manner as oil or natural gas. This is a much more efficient method of transportation than surface methods. Many new slurry lines have been proposed since the renewal of interest in coal use in the aftermath of the world oil crunch. However, slurry pipelines require large amounts of water to be transported together with the coal. Much of the coal mined in the western portions of the country cannot benefit from this new technology because of water supply shortages. In these regions, which are significant producers of coal, rail transport remains the basic means of transporting coal.

3.1.7 Use of Coal

Coal accounts for 21% of primary energy use in the United States. In other words, one-fifth of the energy consumed in the United States comes from coal. Other noncommunist countries are equally dependent on coal, with the exception of Japan, which uses coal for only 16% of its energy needs. By contrast, the communist countries depend on coal for approximately 30% of their primary energy use.

Not all the coal produced in this country is consumed at home. Japan is the largest importer of coal in the world, and 14.5% of U.S. coal exports go to Japan. (The largest supplier of coal to Japan is Australia, which accounts for 50% of Japan's coal imports.) The United States also exports coal to Canada (21% of U.S. exports) and Europe (57.5%).

Within the United States, 84% of all domestically consumed coal is used for electric power generation (see Fig. 3.1.5). To minimize coal transport, much western coal is burned in power plants located at the mine site (these are called "mine-mouth" plants). The balance is used mainly for industrial purposes, including iron and steel and chemical production. The coke industry accounts for 5.8% of domestic coal use, and other industries use 9.05%. A relatively

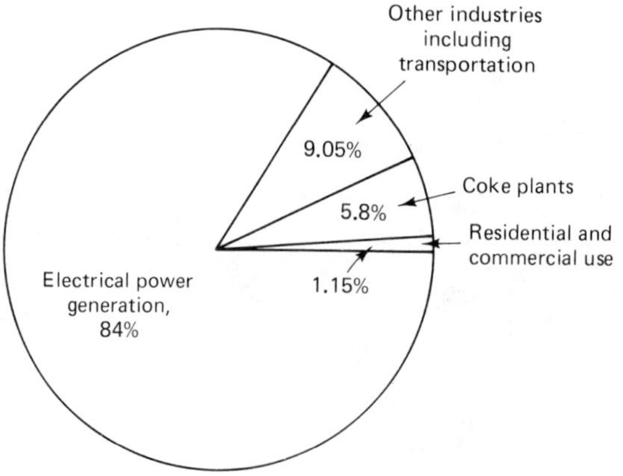

Figure 3.1.5 Use of coal in the United States, 1982. Total coal consumption; 15.423 quads = 21.75% of total U.S. Energy use. (*Source:* Based on data from the U.S. Department of Energy.)

small portion (1.15%) is used for residential and commercial heating. This is a far cry from the nineteenth century, when coal was the primary fuel for all industrial and residential energy uses.

3.1.8 Future Prospects for Coal

In today's energy-intensive economy, coal appears to be an attractive energy source. However, increased production and consumption of coal will create problems on many fronts. Capital shortages, together with high interest rates, limit the rate of investment in heavy coal-mining equipment. Job safety issues and environmental concerns will make coal mining an increasingly expensive proposition. Transportation problems can be expected, especially in the United States, where the railroad system has fallen into a serious state of disrepair. And the environmental consequences of coal burning will require the development of expensive and technically sophisticated antipollution equipment. In short, coal will probably continue to be a major energy source in the future, but it is hardly a cure-all for the world's energy problems.

3.2 FOSSIL FUELS: PEAT

Peat is a brownish-black, highly porous, spongy organic material usually found in shallow water on the surface of the earth. It has a naturally high moisture content and the plant remains in peat are sometimes clearly visible. Even after drying (or *seasoning*), the moisture content can be as high as 40 to 50% moisture by weight. What this means is that a substantial amount of energy is required just to drive out the moisture during the combustion process. Thus peat has a relatively low heat yield of about 12,200 Btu/kg or about 950 kcal/kg.

The sources of peat and coal are one and the same. Actually, coal is a metamorphism of peat; that is, coal is simply peat which has been transformed by high pressures and temperatures into a drier, more compact substance. The pressure and heat drive off the moisture content of buried peat and slowly convert the carbohydrates into hydrocarbons and elemental carbon (see Section 3.1).

3.2.1 Peat as an Energy Resource

Major peat deposits occur in the USSR, Canada, the United States, northern Europe, and the Asian countries of Indonesia and Japan. The USSR claims to have most of the world's peat resources. Recently, Canada has discovered vast quantities of peat in the northern regions of the country. In the United States, a major peat bog with deposits almost 2 m deep can be found in the Dismal Swamp region of Virginia and North Carolina.

Peat is not a recognized worldwide energy resource as coal and oil are, even though enormous reserves can be found. Regionally, however, peat does have considerable importance as an energy source because of the unavailability of other fuels or the ease with which peat can be obtained in the area. Some countries, like Ireland, are highly dependent on peat. There, peat-burning electric power plants produce over one-fourth of the country's electricity. In other northern regions, peat is burned in much the same way as firewood to provide residential heating.

The total amount of peat in the world is unknown, but estimates range between 20×10^{10} and 33×10^{10} tons dry weight. This is a considerable amount, representing a potential heating value of approximately 5000 quads. (By comparison, annual world energy consumption amounts to approximately 225 quads.) But most peat is not economically recoverable because peat is bulky and expensive to transport, and because deposits are too shallow or too distant from major energy markets to be developed economically.

3.2.2 Is Peat Renewable?

It can be argued that because new peat deposits are continually being created (in the Dismal Swamp region, for example), peat should be considered a renewable resource. In fact, peat is renewed at the rate of about 3 tons per hectare per year, a relatively rapid rate of renewal. However, the renewability of a resource depends on the rate of consumption compared to the rate at which the resource is renewed. Currently, world consumption of peat amounts to some 90 billion tons per year, which is not appreciably greater than the rate of renewal. But many of the peat deposits currently being created are in very remote areas and will never become available for commercial use. Therefore, for the present, peat must be classified as a nonrenewable resource.

3.3 FOSSIL FUELS: OIL

Petroleum is the largest single energy source for the Western world. Oil is used to power automobiles, railroads, ships, and planes. It can heat buildings, generate electricity, and provide raw material (feedstocks) for a number of synthetic materials. Unfortunately, supplies of this vital fuel are drying up rapidly. At the present rates of production and consumption (Table 3.3.1), some experts fear that only a few decades' worth of oil reserves remain in the world.

TABLE 3.3.1 U.S. Uses of Petroleum (January 1983)

Use	Amount (MBD)	Percentage of Total Use
Fuel for transportation	6.0	39.2
Distillate fuel oil	3.1	20.2
Residual fuel oil	1.8	11.8
Liquefied petroleum gases and ethane	1.6	10.5
Industrial feedstocks and other products	2.8	18.3
	15.3	100.0

Source: U.S. Department of Energy.

As of July 1983, the United States used approximately 14.8 million barrels of petroleum per day (1 bbl = 42 U.S. gallons). This is approximately 30% of the world's total consumption of petroleum, by a country with only 5% of the world's population. Because total domestic oil production only amounts to 8.67 MBD, the U.S. imports roughly 6 MBD, approximately 40% of its total petroleum consumption (Table 3.3.2).

TABLE 3.3.2 U.S. Oil Production and Imports (MBD)

Year	Domestic Production	Imports	Total Use
1973	10.97	6.26	17.30
1975	10.04	6.06	16.32
1977	9.91	8.80	18.43
1979	10.18	8.45	18.51
1980	10.21	6.90	17.06
1981	10.23	6.00	16.06
1982	10.28	5.00	15.29
1985[a]	7.6	6.3	13.9
1990[a]	7.0	4.8	11.8
2000[a]	6.0	1.2	7.2

[a]Midrange projections.

Source: U.S. Department of Energy.

Other Western countries are even more dependent on imported oil. Japan, for instance, must import almost all of its oil (4.4 MBD). Most western European countries are also heavily dependent on imported oil, although some countries on the North Sea are net exporters of petroleum. Altogether, the oil-importing countries must buy approximately 12 MBD from foreign sources.

Most imported oil comes from the Organization of Petroleum Exporting Countries (OPEC), which is dominated by countries in the Mideast, especially in the area of the Persian Gulf. (See Chapter 9 for a fuller discussion of this organization.) In the first half of 1983, the United States imported an average of 3.065 MBD of crude oil, of which 1.414 MBD came from OPEC. OPEC produced an average of 19 MBD in 1982, of which 11.5 MBD were produced by Arab members. This organization, called an *oil cartel,* succeeded in raising the price of crude oil from under $2.00 a barrel in 1970 to over $34.00 per barrel in 1981. By 1982, the price of OPEC oil had declined to a level of $29.50 per barrel, but any decline is generally regarded as temporary. Although the OPEC countries were able to manipulate world oil prices in the late 1970s, and may be able to do so again as oil supplies dwindle, the main causes of high oil prices can be traced to the traditional market forces of supply and demand (see Chapter 6).

Whatever the cause of the shortages and the astronomical increases in the price of oil, the effects have been devastating on world economics. Enormous amounts of capital have been flowing from the Western industrial countries to the oil-exporting countries. The increases in oil prices probably led to massive

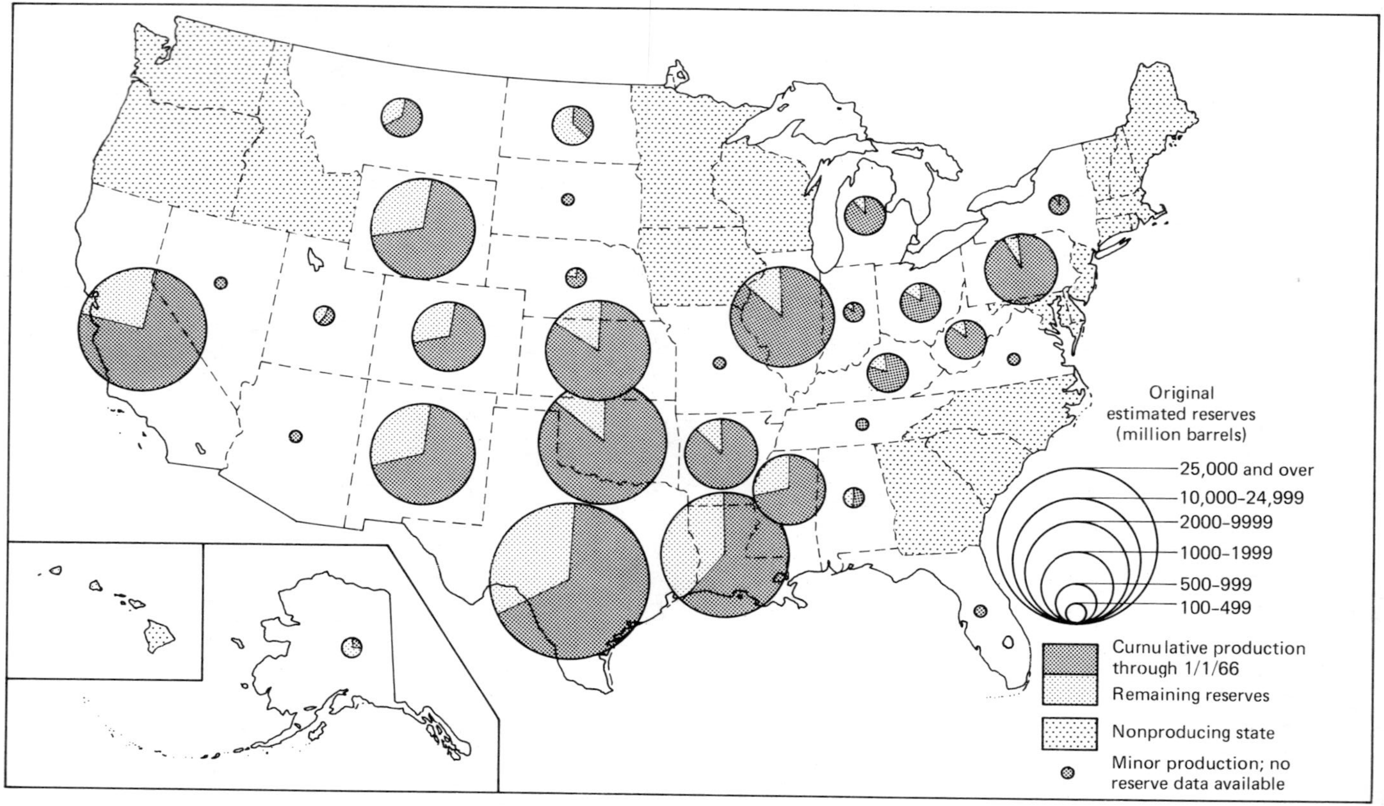

Figure 3.3.1 U.S. oil production and reserves, 1966. (*Source:* U.S. Geological Survey.)

increases in the prices of other commodities. (This view is supported by the fact that world inflation began to cool down in 1981–1982, at the same time that a temporary world oversupply of oil, called an "oil glut," produced falling oil prices all over the world.) Some analysts believe that oil shortages and the resulting price increases were primarily responsible for the worldwide recessions of the late 1970s and early 1980s.

Politically, the effects of the oil situation have been no less significant. Perhaps the best index of this importance came in 1979, when President Carter declared that the United States would take "any necessary steps" to maintain the independence of the Persian Gulf region against threats from outside forces. These "necessary steps" were generally understood to include nuclear war.

No single natural resource in the history of the world—with the possible exception of gold—has ever had such wide-ranging political and economic effects. For this reason, oil is sometimes referred to as "black gold." The questions we must ask ourselves are these: What is it about oil in particular that gives it such great importance in world affairs? Perhaps more important, how should we respond to the serious economic and political crises raised by the current problems of oil supply and demand?

3.3.1 History of Oil Use

The discovery of oil dates back at least to 400 B.C. By A.D. 1000, oil was being produced in Burma and Mexico. Italy began producing oil in the seventeenth century. The first oil well in the United States was drilled in 1829 in the state of Kentucky. But it was not until 1858–1859, when the Drake well was drilled in western Pennsylvania, that oil became an important industrial commodity in the United States, mainly for kerosene production.

During the late nineteenth and early twentieth century, world oil production was dominated by the Standard Oil Trust, which had a virtual monopoly on the production and distribution of oil in the United States until 1911, when the Trust was broken into a number of smaller firms. More recently, a group of very large corporations—firms such as Exxon, Shell, British Petroleum, and Mobil—have begun to dominate the energy industry. These companies are called "multinational" corporations because their power and influence stretches throughout the world. Using the enormous profits from their oil operations, these corporations have bought smaller companies in all fields of energy production and distribution, including coal, nuclear, and even solar energy. Because of the massive size and international scope of these corporations, oil production and consumption can no longer be regarded in nationalistic terms but must be considered on the world level.

The discovery of uses for petroleum has paralleled, and to a large extent been responsible for, the tremendous growth in oil production. During the later part of the nineteenth century, petroleum was a relatively unimportant commodity used primarily for lighting and lubrication. Then, in 1894, Gottlieb Daimler developed the first successful commercially produced internal-combustion automobile. (The first automobile was developed 31 years earlier by Etienne Lenoir of Belgium.) From that time on, oil use began to increase dramatically. The twentieth century saw a tremendous increase in the use of oil for a variety of purposes. With the development of the automobile and diesel locomotives, oil replaced coal as the principal fuel for transportation. Soon, fuel oil replaced coal as the primary home heating fuel because of oil's greater convenience. As concern with air quality increased, coal-burning electric power plants were converted to burn fuel oil. By the middle of the twentieth century, oil had become the world's primary energy source.

As the use of oil has changed, so have the methods of producing and refining oil. In the early days of oil production, oil could be extracted from surface pools. Under tremendous pressure from beneath the earth, oil from underground deposits seeped to the surface, where it formed isolated shallow pools. These surface deposits helped guide early drillers in their search for underground oil. Once a natural oil deposit was found, wells were drilled at random, often near established oil production wells.

As surface pools dried up, this method of exploration was replaced by the "educated guess" method of oil exploration, the main method in use today. Geological and geophysical data are combined with information based on years of oil exploration. Researchers analyze the geological and geophysical characteristics of a possible drilling site and compare them with the characteristics observed at successful production sites. Test drillings are carried out only if the research findings are positive.

Most oil and gas is found in sedimentary rocks, usually organic shale. The formation of petroleum took place over millions of years under conditions that prevented oxidation and promoted concentration of organic matter (remains of plants and animals). The accumulation of organic matter was followed by burial and subsidence under the pressure of a layer of sedimentary "overburden" rocks which was formed with the passage of time. The result of this process is a highly volatile fluid composed of a variety of hydrocarbons. Unlike coal deposits (see Section 3.1), petroleum deposits contain only negligible amounts of elemental carbon.

3.3.2 Oil Exploration Technology

There are several available methods for determining the structure and composition of the underlying rock strata beneath the earth's surface. All of these methods are useful in oil exploration.

1. The earth's magnetic and gravitational fields change slightly depending on the density of the rock structure beneath the earth. A dense rock structure will display a greater gravitational attraction than a less dense structure, and oil-bearing rocks display different electromagnetic qualities than those of other geological formations. These comparatively small variations in gravity and magnetic field can be detected by sensitive meters.

2. Seismic measurements are often more useful than gravitational measurements. A small explosion is set off underground, and measurements are taken of the time required for the shock wave to reach the upper surface of the rock bed and reflect back to the point of measurement. When the data from a number of such explosions are combined, geologists can begin to understand the structure of the underlying rock.

3. The third and most direct method of exploring underground structures is to drill for a rock sample. A circular bit cuts a cylindrical section of the rock, which is stored inside a core barrel mounted just above the bit. When the sample is brought to the surface, it can be examined for electrical properties, radioactive characteristics, density, and other physical properties. This method of exploration is more expensive than gravitational or seismic measurements, but it is also much more informative. It is generally used if the other two methods indicate promising results.

3.3.3 Oil-Drilling Technology

In earlier days, oil drilling took place on land near surface deposits or other known sources of petroleum. When drillers struck a primary oil deposit for the first time, the underground pressure sometimes forced oil up through the bore hole and sprayed it into the air. Now, however, these "gushers" are rare indeed.

As land deposits became less abundant, oil drilling moved offshore. The first offshore shallow oil well was drilled off southern California in 1912. The first deep-water offshore oil well was drilled along the Gulf coast of Louisiana in 1947. Today, the search for oil has expanded beyond these relatively temperate climates to such harsh regions as the North Sea and the northern coast of Alaska.

As the search for oil has moved to more remote regions, the cost and effort involved in drilling a well have increased. In 1980, the average depth of a U.S. oil well was 1425 m (nearly nine-tenths of a mile), and the average cost was $130,000, measured in constant 1972 dollars.

Oil-drilling operations use huge rotary assemblies to penetrate the rock formations overlying an oil deposit. A simplified sketch of a drilling assembly is shown in Figure 3.3.2. The drill-bit tip is made of diamond, and it rotates continuously while the shaft is being lowered. This maintains steady progress in drilling. The drill shaft is made up of 10-m (30-ft) segments linked together to make a shaft of the desired length. During drilling, the drill bit requires constant cooling and lubrication. This is achieved by a stream of specially prepared "mud," which also brings the drill cuttings to the surface. To avoid groundwater contamination, the hole is lined with a casing of steel pipe and cement.

Of course, not all drillings are successful. In 1982, 66,000 wells were drilled in the United States, of which 40,300 were successful. When the drilling reaches a certain predetermined depth without striking oil, the whole drill assembly is withdrawn and uncoupled piece by piece. (This must also be done when the drill bit wears out.) An unsuccessful well is called a "dry hole" or a "duster,"

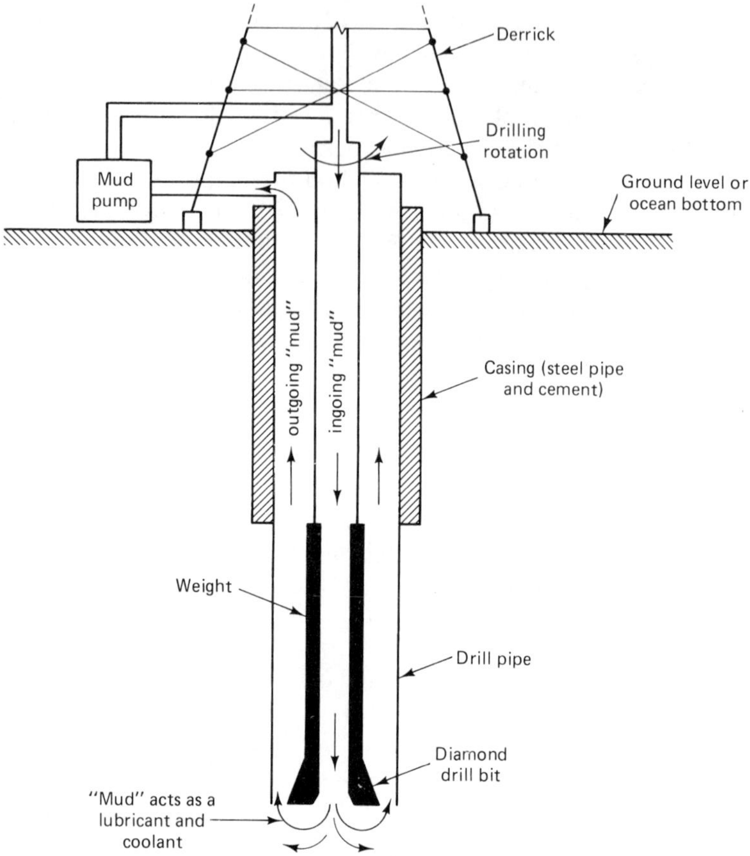

Figure 3.3.2 Drill assembly

even if the drill has struck groundwater. The unsuccessful well is plugged, and drilling operations move on to another site.

In a minority of cases, the drillers strike oil. When the bit strikes an oil or gas pool, there is usually an increase in pressure, sometimes enough to cause a "blowout," which can endanger both workers and equipment and cause environmental damage. A special valve, appropriately called a "blowout preventer," is usually attached to the wellhead as a safety precaution. Once the oil has been found, the drill is extracted and disassembled, and oil recovery equipment is installed to extract the oil that has been tapped.

3.3.4 Oil Recovery Techniques

Oil wells produce varying amounts of oil. In the United States, an average well produces around 20 bbl/day, but some wells can produce as much 100,000 bbl/day. Over its entire lifetime, the average oil well produces upwards of 125,000 bbl of oil. To recover this oil, oil producers use a variety of techniques.

Sometimes, the pressure from the oil pool is sufficient to bring the oil to the surface naturally, occasionally even sending the oil dozens of feet into the air. However, when the pressure drops, pumping is usually necessary to bring the oil to the surface.

If simple pumping does not produce oil at a sufficient rate, oil recovery can be enhanced by breaking up the rock formations that contain the oil. One common method is *hydraulic fracturing,* which consists of forcing water at high pressure into underground cracks. Another technique, called *acidization,* consists of injecting acid in order to dissolve alkaline rocks underground. Both these methods make it easier for the oil to migrate through the underground rock formations to the pumping site.

To increase the production of petroleum further, a "secondary recovery system" may be employed. This consists of injecting water or gas at high pressure into the underground reservoir. Injection works in three ways: first, it flushes out oil that otherwise adheres to the surface of the reservoir rock; second, it increases the pressure in the oil formation and thus helps oil recovery; third, it raises the level of the oil, which floats on top of the water and can be pumped out normally.

The primary and secondary extraction processes described here can recover only about 30% of all the oil in the underground deposit. To recover more oil, a process called "enhanced oil recovery" can be employed. In this new and expensive technique, detergents and carbon dioxide are injected to mix with the oil and dislodge it from the rock. After about one or two years, the oil–deter-

Figure 3.3.3 Enhanced oil recovery. (*Source:* U.S. Department of Energy.)

gent mixture can be pumped out and then separated and refined at the surface.

Other enhanced recovery schemes are also being studied. One promising possibility is mining: digging shafts along the bottom of the rock formations in order to drain the residual oil into an artificial reservoir, from which it can be pumped to the surface. This technique is not currently feasible, but if oil prices rise significantly in the future, it may be used together with other methods of enhanced oil recovery.

3.3.5 Transportation of Oil

Because oil reservoirs are unevenly distributed and often far away from the place where the oil is in demand, the crude oil must be transported, sometimes thousands of miles. The methods of transporting oil vary considerably in speed and efficiency.

When oil must be transported across the ocean (e.g., from the Persian Gulf to the United States or Japan), it is carried in specialized oil tankers. However, ocean transport of oil can cause environmental problems. Some oil tankers are enormous and have significant maneuverability problems. It can take more than 2 miles to stop a loaded "supertanker." If a tanker is damaged in an accident, the consequences can be very serious. For example, in 1978, the oil supertanker *Amoco Cadiz* broke up off the French coast, spilling 67 million barrels of crude oil. This oil washed up along 325 km (200 miles) of the French coastline, causing serious damage to fish and waterfowl and disrupting the region's lucrative tourist industry.

Accidents are not the only source of oil contamination of the oceans. Some operators routinely flush out their empty tankers with seawater before picking up a new shipment of oil. This practice causes ocean pollution as well. In fact, it accounts for roughly 30% of the oil discharged to the environment. These sources of ocean pollution are discussed in greater detail in Chapter 7.

New technologies are being developed to prevent oil spills at sea and to cope with them when they occur. Double-bottomed tankers, deep-water "superports," new detergents for cleaning up oil spills, oil-absorbent sponges, and even oil-eating bacteria have been developed in an attempt to cope with the environmental problems caused by ocean transportation of petroleum.

Whereas tank trucks are used for distributing petroleum products to gas stations and residential users, pipelines and railroad tank cars are the preferred methods of long-distance land transport. Oil travels from the north slope of Alaska to the port of Valdez (about 800 miles) in a pipeline, constructed in the 1970s. When it reaches Valdez, the oil is pumped into oil tankers which transport it to the lower 48 states. Once the oil reaches the West coast of the United States and is refined, it may be pumped into another pipeline for distribution, or it may be transported overland by railroad to reach its destination.

Pipelines are expensive to build and maintain. The Alaskan Oil Pipeline cost $7.7 billion to construct. Breaks along a pipeline can cause significant oil spills. Because of the great length of a pipeline, locating and repairing leaks can be quite expensive, often requiring repair crews to travel large distances over difficult terrain. However, pipelines are the most energy-efficient means of transporting petroleum overland.

3.3.6 Refining of Petroleum

Crude oil is a complex mixture of various hydrocarbons. Typically, it is composed of 85 to 90% carbon and 10 to 14% hydrogen, together with a small amount of sulfur and traces of other elements. The basic hydrocarbons occur in a bewildering variety of structures, including paraffins, cycloparaffins, and aromatics. (The intricacies of hydrocarbon structures are well beyond the scope of this textbook, and interested readers are advised to consult a chemistry handbook for further particulars.) In the refining process, this complex mixture is separated out into a wide variety of fuels and other useful petrochemicals, which are then shipped to the appropriate consumers.

The first stage of refining is *distillation.* This process consists of gradually raising the temperature of the crude oil and drawing off the various fractions of hydrocarbons as they boil off individually. Some of the most important distillation products are listed in Table 3.3.3, together with their boiling points and some of their uses.

Distillation products are further blended, cracked, reformed, polymerized, and hydrogenated to produce desired petrochemicals. Not all of the petroleum products are used for their energy value. About 18% of crude oil is transformed into products such as lubricants, detergents, solvents, waxes, insulation, insecticides, herbicides, paving asphalt, synthetic clothing, plastics, fertilizers, and much more. Antifreeze, preservatives, adhesives, vinegar, explosives, and automobile tires can also be made from petroleum.

TABLE 3.3.3 Crude Oil Distillation Products

Product	Boiling Range (°C)	Applications
Methane, ethane	−175 to −45	Petrochemical feedstock, fuel gas
Propane and butane	−40 to 0	Liquefied petroleum gas for domestic cooking, industrial fuel, and feedstock
Gasoline	20 to 80	Automobile fuel, small engines
Naphtha	80 to 150	Catalytic cracking feedstock, jet fuel, blending purposes
Kerosene	150 to 250	Fuel for heating, lighting, aviation; solvent and thinning agent
Diesel oil	250 to 350	Fuel for locomotives and heavy engines, heating systems, feedstock
Residual oil	350 and above	Large power and heating plants, lubricants, asphalts, waxes

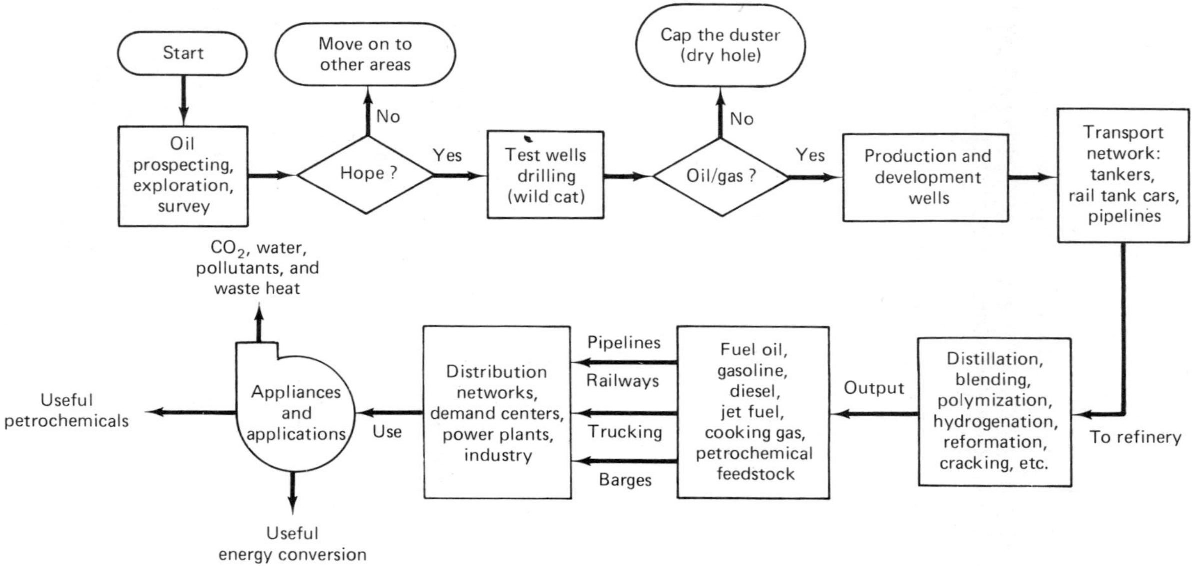

Figure 3.3.4 The petroleum connection.

3.3.7 World Oil Production and Oil Reserves

The world's total supply of oil is limited. There can be no doubt of that, because the earth itself is finite. Estimates of total world oil supply have changed over the years, and large new oil deposits continue to be found. There is probably no way to know exactly how much oil there is in the world. Some oil reserves may never be found. Other areas, currently believed to contain oil deposits, may prove unproductive. However, projections of near-term oil production can reasonably be based on estimates of *measured* or *proven reserves.*

Proven reserves represent an informed estimate of the amount of oil that can be recovered at a profit using available technology. Even this relatively precise estimate has been found to be off by as much as 25% in some areas. The criterion of profitability can change as market conditions change. In 1970, when oil prices stood at approximately $2.00 per barrel, many oil recovery techniques were not profitable. In 1984, with oil prices near $30.00 per barrel, much oil that was formerly considered "unrecoverable" can be profitably recovered. Also, with new technologies constantly being developed, more oil may become available in the future.

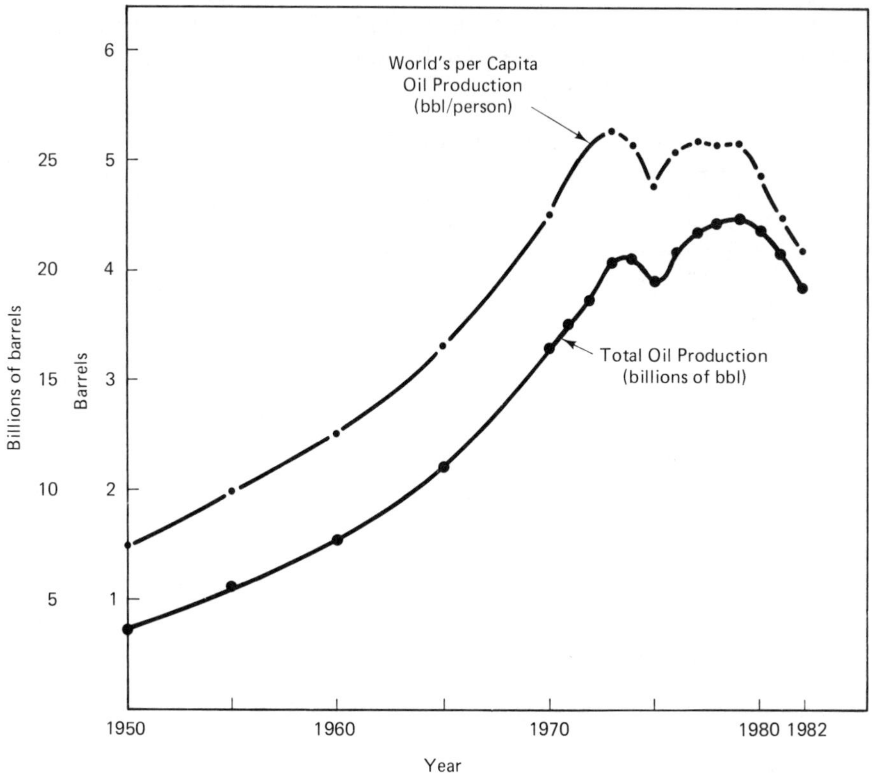

Figure 3.3.5 World oil production. (*Source:* U.S. Department of Energy.)

Currently, the proven conventional oil reserves in the United States amount to 26.4 billion barrels. Potential reserves amount to another 150 billion barrels. To put these figures in context, it should be pointed out that total U.S. petroleum production up to 1982 consisted of 155 billion barrels. Thus the United States may have already used up nearly half of its proven and potential oil reserves. Current U.S. production is 8.66 MBD (July 1983). At this rate of production, the total U.S. proven and potential reserves will last for less than 50 years. If production is increased to the level of U.S. oil consumption (currently about 15 MBD), the reserves will not even last that long.

The situation is less serious in other oil-producing countries, but the pattern remains the same. The world's known petroleum reserves are estimated at 566 billion barrels and the ultimately recoverable reserves (including deposits that are believed to exist but have not yet been discovered) are estimated at around 2000 billion barrels. World oil production is equal to 52 MBD (see Table 3.3.4). At these rates, the world's oil supply will last approximately 100 years. Increases in world oil consumption could reduce this figure significantly. The message of these figures is clear: we are using up our oil reserves at a measurable rate, and thus we cannot continue to increase world consumption of petroleum indefinitely. But for the present, the world's dependence on petroleum has not shown any sign of decreasing significantly. This imbalance between high demand and dwindling supplies is certainly one of the most urgent energy problems facing the world today.

TABLE 3.3.4 World Oil Production (May 1983)

Nation or Nation Group	Production (MBD)	Percentage of World Total
OPEC Arab members	9.6	19.0
OPEC non-Arab members	7.4	14.7
United States	8.7	17.3
USSR	12.4	24.7
Canada	1.3	2.6
Mexico	2.8	5.6
China	2.1	4.2
Other countries	6.0	11.9
	50.3	100.0

Source: U.S. Department of Energy.

3.4 FOSSIL FUELS: NATURAL GAS

Natural gas and crude oil (petroleum) are usually found in the same geographic areas. Often, the same well will produce both oil and gas. Natural gas is usually found floating above an oil pool, although there are cases where an oil well contains no natural gas, and some natural gas wells produce no oil. The geological requirements for natural gas formation are the same as for oil, and so are the prospecting, exploration, drilling, and production procedures.

Natural gas also shares some of the advantages of petroleum: both are easy to transport and control. In fact, natural gas is somewhat easier to produce and distribute, and its clean-burning properties make it preferable to oil for many purposes. But like the world's oil supplies, the world's supplies of natural gas are being used up, although not quite as rapidly as the supply of petroleum. Some energy policy analysts believe that biogas can gradually be introduced into the natural gas distribution system—in fact, this is already happening in some parts of the country. They contend that the nation should switch to a *greater* reliance on natural gas in order to facilitate the introduction of alternative fuels. Other critics believe that the necessary development of alternative fuels will not be completed in time, and that we should not place so much dependence on a fuel whose reserves may be exhausted in the near future. Still other analysts believe that huge reserves of natural gas remain to be tapped, and that natural gas presents fewer supply problems than most other forms of energy.

3.4.1 Production of Natural Gas

Natural gas used to be regarded as an undesirable by-product of petroleum production, and it was burned, or "flared," at the drilling site. This is still done in the Middle East, where no nearby markets exist for the natural gas and transportation to more distant markets is not economically feasible. In other areas where natural gas is produced as a by-product of oil drilling, it is used in the oil fields to drive pumps and engines. Most commercially sold natural gas comes from wells which are used solely for natural gas production.

Although natural gas is considered to be the cleanest fossil fuel, sometimes a wellhead produces high-sulfur natural gas, containing as much as 5.5% sulfur by weight. This sulfur must be removed not only because of pollution problems but also because of the highly corrosive effect of sulfur on gas pipelines. High-sulfur gas is called "sour gas" because of its foul odor: a powerful smell of rotten eggs. (Interestingly enough, although the foul-smelling hydrogen sulfide is removed from natural gas, another bad-smelling compound is added to the gas for safety reasons before it is sold.)

When natural gas is to be used commercially rather than flared or used for oilfield operations, it is shipped by pipeline from the wellhead to a nearby refinery. The water vapor, petroleum liquids, and hydrogen sulfide are removed from the "wet gas" using a device called a separator. This is necessary to prevent pipeline corrosion and to reduce operating costs.

3.4.2 Transportation of Natural Gas

Transportation of natural gas can be a difficult and costly process because in its gaseous state, natural gas occupies an enormous volume compared to liquid or solid fuels. However, natural gas flows through pipelines readily, so most overload transportation of natural gas is carried out by means of pipelines. In many parts of the country, distribution of natural gas to individual customers is also done through pipelines. This is highly convenient, since the customer controls the flow of gas and is not dependent on a company employee to deliver the fuel. However, it can also be a dangerous arrangement. Explosions and fires can occur due to improper installation or operation of gas-fired appliances, breaks in gas lines, or even sabotage.

When natural gas must be transported over bodies of water, pipelines are not always appropriate. In narrow, shallow seas, the gas can be transported in submerged pipelines. But in order to be transported across major oceans (e.g., from Saudi Arabia to North America), natural gas is usually liquefied. When the natural gas is liquefied, its volume decreases substantially: 1 ft^3 of LNG is the equivalent of 625 ft^3 of natural gas. The liquefied natural gas (LNG) is then shipped in specially designed tankers which keep the gas at or below the boiling point of natural gas: $-160\,°C$ ($-260\,°F$).

LNG transport has several drawbacks. For one thing, LNG tankers are highly expensive to build. Three designs are in common use:

1. *Freestanding tanks.* Each tank is insulated from the outside. These tanks are not incorporated into the ship's structure.
2. *Membrane tanks.* These are connected to the hull by load-bearing insulation.
3. *Spherical tanks.* These are fabricated separately and then integrated into the structure of the ship during final construction.

LNG can also be dangerous. In 1944, a LNG tanker exploded in the harbor at Cleveland, Ohio, killing 135 people and injuring several hundred. Fears of other such accidents have slowed the construction of LNG terminals in major cities.

One promising alternative to LNG is the idea of converting natural gas to methanol. Since natural gas consists primarily of methane, the technology is relatively straightfoward. Methanol is a liquid at normal outdoor temperatures and thus does not require refrigeration, and it can be burned directly in its liquid form or converted back to methane gas.

3.4.3 Natural Gas Reserves and Production

The world's top 10 producers of natural gas are Saudi Arabia, the USSR, Iran, the United States, Algeria, the Netherlands, Canada, Nigeria, Venezuela, and Australia. In the United States, natural-gas-producing areas are essentially the same as oil-producing regions: the Gulf coast, the Appalachian region, the Rocky Mountain states, Alaska, California, and the northern Midwest (see Figs. 3.4.1 and 3.4.2).

New natural gas fields are still being discovered. In 1982, 19,000 new wells were drilled in the United States for natural gas exploration and development.

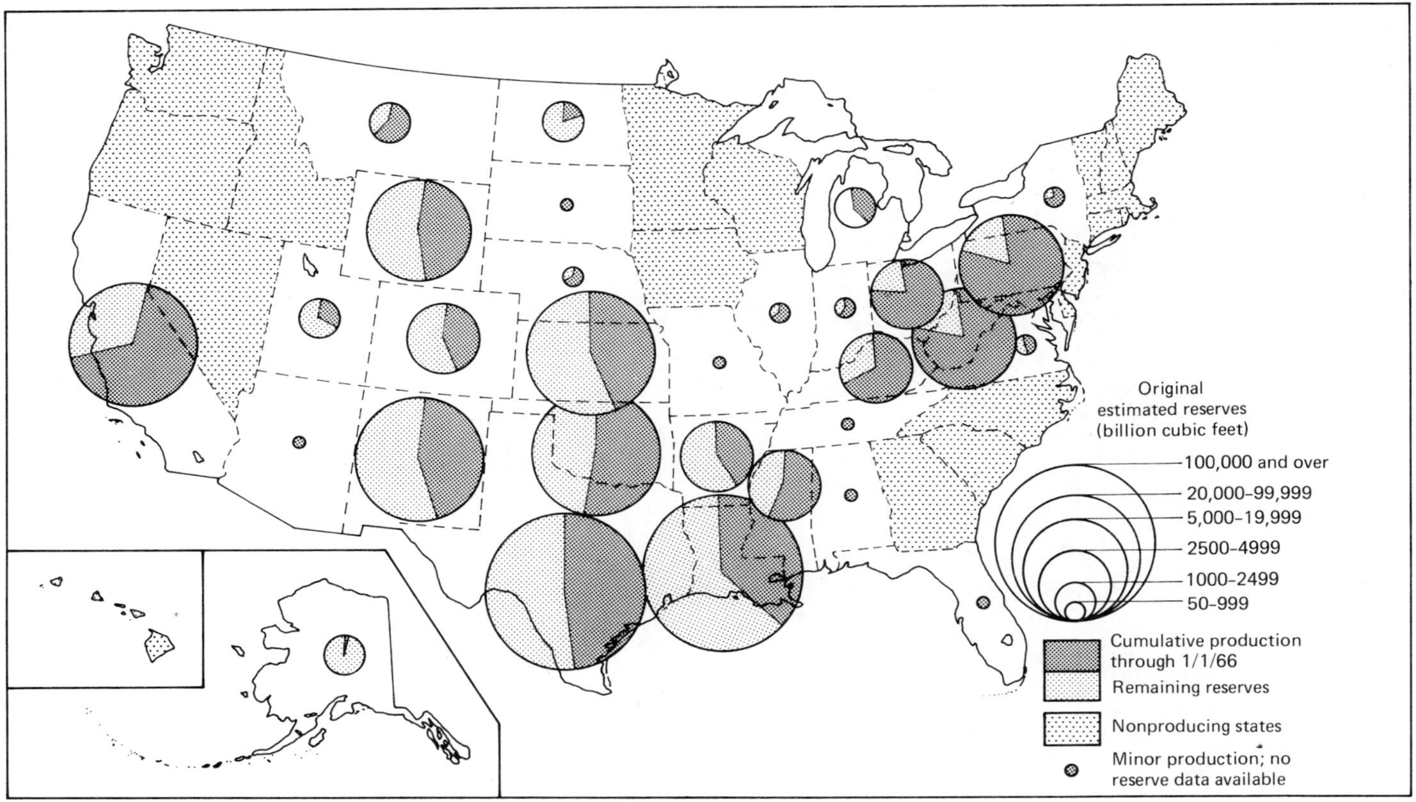

Figure 3.4.1 Natural gas production and reserves, 1966. (*Source:* U.S. Geological Survey.)

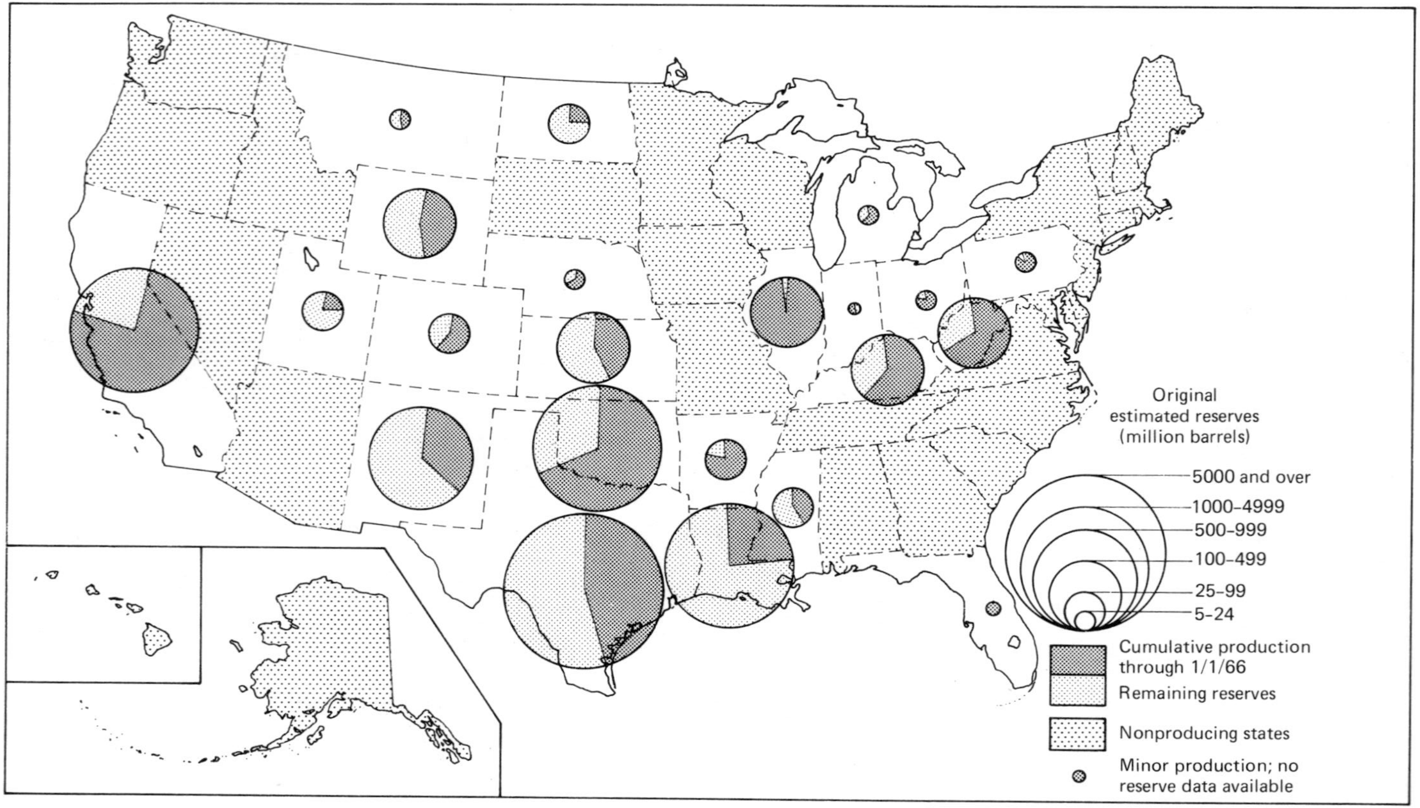

Figure 3.4.2 Natural gas liquids production and reserves, 1966. (*Source:* U.S. Geological Survey.)

As of 1980, U.S. proven reserves consisted of approximately 200 trillion cubic feet (TCF) of natural gas, with an energy content of approximately 200 quads (total U.S. energy consumption in 1982 was 74 quads). Total recoverable reserves are close to 1000 quads. "Unconventional" sources of natural gas, such as tight sands, Devonian shale, coal seams, and geopressurized zones, may provide additional reserves. At the moment, these unconventional sources cannot be developed economically. But when gas prices are fully deregulated, they are expected to increase substantially—perhaps by a factor of 3—and this may make some of the unconventional sources more feasible.

World reserves of natural gas are estimated at 2230 TCF, equivalent to 2290 quads of primary energy. Natural gas liquid reserves are estimated at 363 TCF, yielding 1456 quads of energy. Ultimately recoverable reserves (assuming improved technology and favorable economic conditions) are put at 13,000 TCF of natural gas and 2100 of liquid reserves, equivalent to 13,400 and 8500 quads, respectively.

Projections for the future of natural gas production are based on many uncertain and variable factors, including the following:

The resource base (availability of proven reserves)
Economics of production
Retail and wholesale prices
Governmental regulations
Technological developments
The priorities of the group making the projections

Additional discoveries, more successful drilling, greater leasing activity, and higher prices at the wholesale level all tend to increase the rate at which natural gas is extracted, thus increasing gas supplies for the short term but accelerating the rate at which overall reserves are depleted.

Recent U.S. Department of Energy studies have tried to project natural gas supplies over the next few decades. One study assumes a price of \$5.40 per thousand cubic feet in 1985, \$6.60 in 1990, and \$8.30 in the year 2000, not considering the effects of inflation. At these price levels, conventional gas production is projected as shown in Table 3.4.1.

TABLE 3.4.1 U.S. Gas Supplies (TCF/Yr)

Year	Conventional Sources	Unconventional Sources[a]	Synthetics	Imports	Total
1985	17.4	—	0.2	2.0	19.6
1990	17.1	1.3	0.5	2.0	20.9
2000	14.4	3.2	1.4	2.0	21.0

[a]See Section 3.5.

Source: U.S. Department of Energy.

3.4.4 Unconventional Gas Sources

In addition to conventional reserves, natural gas is available from a number of unconventional sources (see Table 3.4.2). These sources are discussed in somewhat greater depth in Section 3.5. We list them here because the supplies from these unconventional sources may have an impact on world supplies of natural gas by the year 2000.

Tight sands are extremely hard geological formations which can be fractured underground to allow the trapped gas to come to the surface. High-pressure steam or fluids are injected into the underground area near the well to produce this fracturing effect. This unconventional source produces close to 1 TCF of gas per year. Analysts believe that there may be up to 800 TCF of potentially recoverable natural gas locked up in tight sands.

Just as oil can be extracted from tar sands, oil shales, and coal deposits, natural gas and synthetic gas can also be manufactured. In fact, the processes of synthetic oil production also produce natural gases as by-products, very much like oil extraction. These gases can be refined and distributed. Changes in temperature and pressure during the refining process can change the mix of liquid and gaseous by-products. If the process is managed for maximum gas production, 100 to 300 cubic meters (3500 to 10,000 ft^3) of gas can be extracted from each ton of raw material (coal, tar sands, or oil shale).

TABLE 3.4.2 Unconventional Sources of Natural Gas (TCF)

Resource	Resource Base	Projected Production 1990	Projected Production 2000
Tight sands	150	1.6	3.1
Devonian shale	25	0.5	0.7
Coal seams	25	0.1	0.3
	200	2.2	4.1

Source: U.S. Department of Energy.

3.4.5 Economics of Natural Gas

Natural gas production is much less costly than the production of most other fossil fuels. For this reason, costs of natural gas have traditionally been lower than the costs of other nonrenewable resources. As long as natural gas supplies were abundant and oil prices were low, this presented no serious problems for producers, consumers, or government. However, as oil prices increased, natural gas prices began to rise along with them.

In an effort to protect the public against unwarranted price increases, the U.S. government regulated natural gas prices. However, the price restrictions applied only to gas that was shipped across state lines. If gas was used in the state where it was produced, it was exempt from government price controls.

In an effort to encourage production, the government also exempted natural gas from newly drilled wells.

As a result of the "two-tiered" pricing structure, natural gas shortages soon developed in non-gas-producing states. Producers preferred to sell their gas in-state at higher prices. Soon the gas pricing system was out of control. Unscrupulous operators bought "old" gas at controlled prices, and then, through a series of questionable maneuvers, reclassified it as "new" gas, free from price restrictions. Partly in response to these and other abuses, in 1978 the government extended price controls to eliminate the two-tiered pricing structure. At the same time, price controls began to be phased out, scheduled to expire on or before January 1, 1985.

In 1973, natural gas had an average wellhead value of $0.22 per thousand cubic feet. By 1980 this had increased to $1.60, and by May 1983, it was $2.63. In May 1983, an average family paid $6.63 per 1000 ft^3 for natural gas, compared to $1.08 in 1973. This created serious hardships for low income families who used natural gas for home heating. Many were threatened with cutoff of their gas supplies. State and local governments placed restrictions on utility cutoffs, and fuel assistance programs were developed to help customers pay their gas bills. As decontrol continued, there were fears of continued sharp price increases and further hardships for the poor.

The history of natural gas has been a changeable one. From a worthless by-product of oil production, natural gas developed into a major energy resource—the second most widely used fuel in the United States. The future appears equally changeable. Natural gas reserves are not large enough to support a continued shift to natural gas from coal and competing fuels. But some proponents of renewable energy resources believe that natural gas is a perfect "transitional" fuel because of its similarity to biogas (see Section 4.7). We cannot predict the future of natural gas; however, there is no doubt that for the immediate future, it will continue to be one of the world's most important energy resources.

3.5 SYNTHETIC FUELS: SHALE OIL, TAR SANDS, COAL CONVERSION

Synthetic fuels, often called *synfuels* for short, are viewed in the United States as an alternative to imported petroleum, which currently accounts for 40% of U.S. oil consumption. As we have seen, the Middle Eastern countries have been blessed with more than their share of oil. But when it comes to synfuels, the balance tilts strongly in favor of the United States. Many, including some governmental officials, regard synfuels as a way of achieving energy self-sufficiency without sacrificing conveniences like the automobile.

Basically, there are four promising sources of synthetic alternatives to petroleum and natural gas: (1) oil shale, (2) tar sands, (3) coal gasification, and (4) coal liquefaction. Production of methane, methanol, and ethanol from woodland or farm by-products is sometimes considered a synthetic fuel technology, but because of their renewable nature, these energy resources are discussed in Chapter 4.

Synfuel technology holds out the promise of plentiful, clean-burning fuel produced by a variety of methods and using a variety of natural resources. However, it must be emphasized that all synthetic fuel technologies are still essentially in the experimental or demonstration stages. The manufacture of synfuels is an expensive, dirty, energy-intensive process with possibly adverse environmental impacts. Its future is far from certain. For example, in May 1982, the world's largest energy company (Exxon) and the second-largest U.S. refiner of gasoline (Tosco) pulled out of their joint multibillion dollar venture, the Colony Shale Oil Project in Colorado. This had been the nation's most ambitious synfuel undertaking, but it was abandoned because of economic constraints. With oil selling below $30.00 per barrel, shale oil does not appear to be an economically feasible energy resource. However, when oil prices reach $50 to $60 per barrel, the pendulum may swing toward synfuels once again.

Despite these setbacks, the involvement of the U. S. government and major energy corporations makes synfuels a significant component of the energy resource picture. Projections by the Department of Energy indicate that if technical developments proceed smoothly, synthetic fuel production will amount to the equivalent of 50,000 barrels of oil per day by 1985 and 100,000 bbl/day by 1990. For the year 2000, total synfuels production is projected at 2.8 million bbl/day.

3.5.1 Shale Oil

Actually, oil shale does not contain oil or petroleum, per se. What it does contain is kerogen, a solid bituminous material that can be used as a substitute for crude oil. When kerogen is heated to a high temperature, it gives off liquid hydrocarbons which can be refined like petroleum distillates. If controlled heating is carried out in a hydrogen-rich atmosphere, the result is methane gas rather than liquid hydrocarbons.

In small quantities, oil shale has been extracted for many decades, particularly in Europe. In Great Britain, the use of shale oil dates back to 1695, and in Canada it was used as early as 1815. By 1850, the United States imported Canadian oil shale and processed it into oil at more than 40 locations. But the large-scale operations envisioned for the turn of the twentieth century have never been carried out on a commercial basis in the United States. American companies currently involved in shale oil ventures have required substantial government subsidies and guarantees before agreeing to proceed with shale oil development. However, commercial operations are known to exist in Brazil, the USSR, and China.

Basic production processes. Shale oil processing can be done in two basic ways: mining the shale and then processing it above ground, or process-

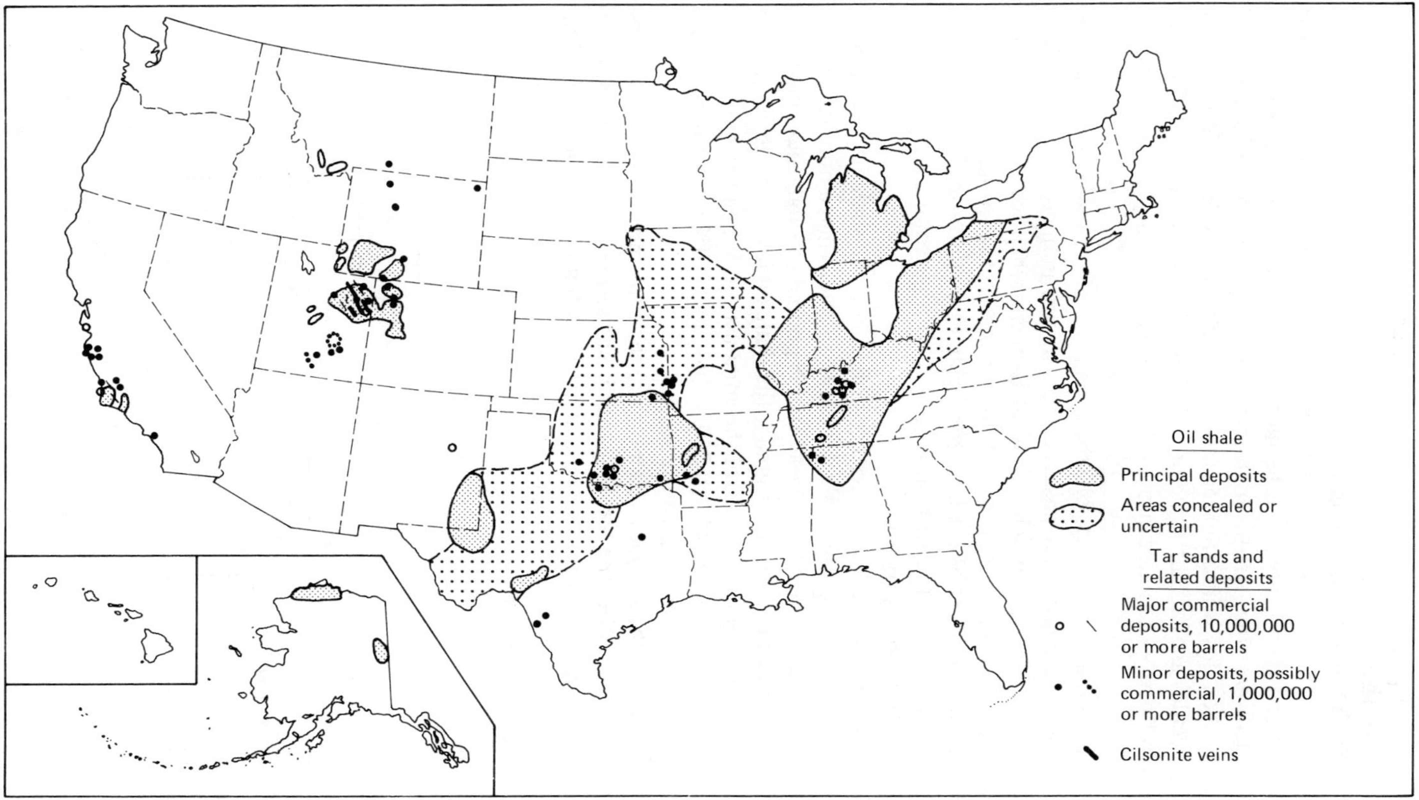

Figure 3.5.1 Oil shale and tar sands, 1965. (*Source:* U.S. Geological Survey.)

ing the shale without extracting it from the ground (the in situ method). Both methods have their advantages and drawbacks.

1. Shale can be mined in much the same way as coal, with strip mining the most common technique for extracting the shale. Once the shale is mined, it is crushed and then heated to a temperature of 550 to 600 °C (1000 °F) in a process called *retorting*. At this high temperature, the heavy molecules of oil shale are broken down into lighter molecules and some solid residues. These lighter molecules are called "shale oil" and are similar to crude oil. They are shipped to refineries and made into products such as gasoline, jet fuel, and diesel fuel.

In the production of shale oil, the expansion of the oil-bearing rock and the release of high-temperature gases causes a significant increase in the volume of the shale. (This is very much like what happens when we make popcorn.) The resulting large volume of processed shale creates storage problems. To maintain the original contours of the land for reclamation purposes, it is necessary to crush the processed shale to restore it to its former compactness. This is a highly energy-intensive process.

2. In the *in situ* method, holes are drilled into the shale deposit and explosives are used to reduce the shale deposit to rubble. Then a number of pipes are introduced into the deposit. Diesel fuel is poured through these pipes and ignited. As the diesel fuel burns, it ignites the shale oil in the ore deposit. Fresh air and steam are supplied to control the combustion. When the burning rubble reaches a temperature of 425 to 550 °C (800 to 1000 °F), oil emerges from the shale and is then pumped to the surface.

The in situ method is more economical than the more laborious process of mining and processing the oil shale, but it has serious environmental problems. Groundwater pollution can be caused by pollutants leached out of the burned deposits by water percolating through them. Also, it appears that a smaller percentage of the available shale oil is actually recovered by the in situ method.

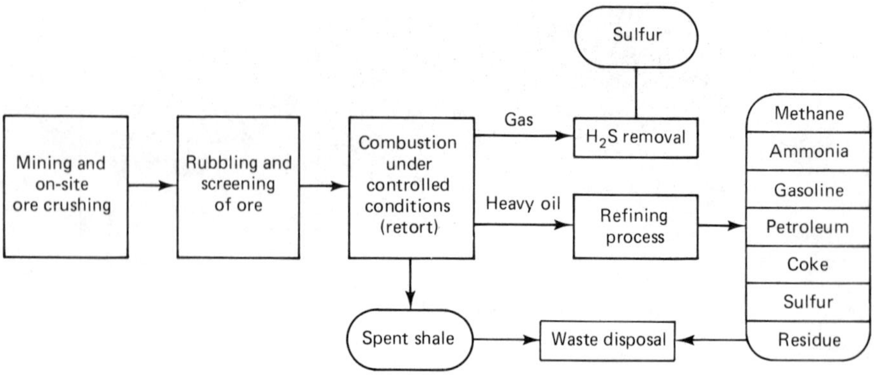

Figure 3.5.2 Shale oil project block diagram.

Oil shale reserves. There is an abundance of oil-bearing shale underground. But only a small fraction of that shale can be processed economically. One obstacle is the *quality* of the ore, or the amount of oil that can be recovered from a volume of shale. Very rich shale oil deposits can be found in Estonia (in the eastern USSR), where the shale can deliver up to 80 gallons per ton. In Utah and Colorado, the shale oil deposits are estimated to contain at least 25 gallons of oil per ton. Rich shale oil deposits such as these can be developed economically. But in areas where the shale is less rich in oil, production of shale oil is not economically feasible.

Location is another variable affecting the feasibility of shale oil production. Some areas are simply too remote, or the deposits are too far underground, for economical processing of shale oil. Considering these two factors of ore quality and location, current estimates place the total U.S. reserves of recoverable shale oil at 600 billion to 1 trillion barrels.

Current status of shale oil projects. With the cancellation of the Colony Shale Oil project in Colorado, there remains only one major oil shale project in construction in the United States. This one is also in Colorado. This project was undertaken by Union Oil and holds out the promise of 10,000 bbl/day. The U.S. government has guaranteed to pay $42.50 per barrel for this oil, compared to a 1984 price of $29.50 per barrel for OPEC petroleum. Thirty percent of this oil will be processed into jet fuel and 70% into diesel fuel.

Figure 3.5.3 Oil shale mountain in the state of Colorado. (*Source:* Bureau of Land Management, U.S. Department of the Interior.)

There are three major reasons oil shale projects have not developed more rapidly in the United States.

1. *Economics.* The process of shale oil production is highly capital intensive. Because kerogen is widely dispersed, more energy and greater expense is required to mine, retort, and refine the shale oil. This results in a low net output of energy and low profitability.
2. *Water supply.* Large amounts of water are required to process the oil shale. Water is scarce in the regions of the United States where oil-bearing shale is found in large quantities.
3. *Environment.* Waste disposal problems, ground- and surface-water pollution, and air pollution from particulate matter are all serious problems in shale oil development.

If these problems can be solved, shale oil has the potential of contributing significantly to the energy supply of the United States and other industrial countries. So far, however, even generous government subsidies and initiatives have not been sufficient to produce a thriving shale oil program in the United States.

3.5.2 Tar Sands

Tar sands, also called oil sands or bituminous sands, represent another source of fossil fuel in a form that cannot be pumped out like gas or oil from drilled wells. Tar sands, as the name suggests, are deposits of sands containing a high-sulfur tar known as bitumen, or, less frequently, a very heavy, viscous crude oil. This "tar" can be more than 80% carbon by chemical composition. The oil in tar sands is believed to have the same origin as the crude oil produced from drilled wells.

There are a few very large tar sand deposits and several smaller deposits located in different parts of the world.

1. The Athabasca tar sand deposits in Alberta, Canada, contain approximately 300 billion barrels of economically recoverable oil. Not all the Athabasca deposits are shallow; in fact, only about 10% can be strip-mined. The rest are deep underground.
2. The Orinoco tar belt and Maracaibo sedimentary basin in Venezuela are believed to contain potentially as much oil as the Athabasca deposits, although this region is virtually unexplored.
3. Northeast Siberia, in the USSR, contains large subsurface deposits of tar sands in comparable proportions.
4. Other deposits occur in Albania, Romania, Colombia, Trinidad, Nigeria, Argentina, Ecuador, Peru, and the United States.

In the United States, the major tar sand deposits occur in Utah, California, Colorado, Kentucky, New Mexico, and Texas. Eighty percent of U.S. tar sands, as much as 29 billion barrels of oil equivalent, are buried in Utah.

Processing of tar sands. Tar sands are similar to oil shale except that the hydrocarbons are trapped within the pores of sandstone instead of in shale. Like coal or shale oil, tar sands are first mined, usually by strip-mining techniques. The mined deposit aggregate is then heated with steam. This makes the viscous tar flow out and float on the surface, where it can be extracted and then refined like crude oil. The tar sands developed so far have yielded high-sulfur fuel; for example, Athabascan bitumen contains 5.5% sulfur. This means that the oil must be purified during the refining process to remove pollution-causing sulfates.

Tar sands are quite abundant throughout the world, but only the shallow deposits have proved economical to develop so far. It has been estimated that tar sands contain the energy equivalent of four times the world's known oil reserves. But there are many problems associated with large-scale production of oil from tar sands because of the abrasiveness and stickiness of the sands. Extracting machinery tends to clog frequently and wear out prematurely. In addition, rubber in tires, conveyor belts, and machinery parts break down quickly because of the solvent qualities of bitumen.

Another major obstacle to tar sand development is the serious environmental pollution associated with strip mining. Scarcity of water and potential water pollution also present obstacles to the development of tar sands. Finally, large-scale production of tar sands is a costly and energy-intensive process. Small projects have been undertaken in Trinidad, Albania, Romania, and the USSR. In Alberta, Canada, a major project was undertaken in 1967. But for the most part, large-scale development of tar sands has not been economically justifiable.

3.5.3 Coal Gasification and Liquefaction

Coal is an abundant fuel, but it is basically very dirty. Coal burning produces air pollutants such as sulfur dioxide and other toxic gases, radioactive particles, and other particulate matter. Burning of high-sulfur coal is restricted by the National Environmental Policy Act (NEPA) and the Clean Air Act.

Converting coal to gas and oil may solve not only the problem of air pollution but also other difficulties and inconveniences associated with the use of a solid fuel such as coal. Problems with handling and distribution, combustion control, and solid-waste disposal are expected to be alleviated by coal gasification and liquefaction.

Although there are only a few current commercial efforts under way in the United States to convert coal to gas and oil, coal conversion is not a new technology. Coal gas was manufactured in the United States commercially beginning in 1820 and was used for lighting and cooking. But with the large-scale development of natural gas in the United States after World War II, the coal gas industry collapsed in the United States.

Other countries are somewhat ahead of the United States in coal conversion. World War II saw a large-scale coal-conversion operation in Germany. In South Africa, the Sasol synfuel plant has been operational for almost 25 years, producing both gas and oil from coal in large quantities. Every day, the

Figure 3.5.4 The Synthane Pilot Plant, located near the Pittsburgh Energy Research Center, can process 72 tons of coal per day into 1.2 million cubic feet of gas. (*Source:* U.S. Department of Energy.)

Sasol plant converts 3500 tons of coal to 300 million cubic feet of gas, a substantial part of which is converted to oil (10,000 bbl/day).

In the United States, a model plant is being constructed at Newman, Kentucky, to convert coal to 20,000 barrels of oil a day. Another $2 billion project has been partially completed near Beulah, North Dakota. When this plant goes into operation (scheduled for 1985), it is expected to convert 14,000 tons of coal to 125 million cubic feet of gas per day. Another plant is being built at Morgantown, West Virginia, and other pilot projects are planned in most coal-producing regions.

Principles of coal conversion. Coal, oil, and gas are basically composed of carbon and hydrogen in different forms and proportions. For example, the ratio of hydrogen to carbon is 0.8:1 in coal, 1.75:1 in oil, and nearly 4:1 in natural gas. Thus, in order to convert coal to liquid or gaseous form, it is necessary to add hydrogen. This is achieved through the use of steam as a source of hydrogen. The hydrogen molecules must be disassociated from the water molecules and attached to the coal molecules. This process requires energy in the form of heat, which must be obtained from the coal being treated if the process is to be economically justifiable.

Four basic methods have been developed for coal conversion (see Fig. 3.5.5). They are:

1. *Carbonization.* When coal is heated to a temperature of 450 to 500 °C in the absence of air, the coal decomposes into gas and tar. Coke is also pro-

duced as a by-product. In other variations of the process, the coal is made to react with hot air, resulting in producer gas, which has a lower heat value because it incorporates nitrogen from the air.

2. *Hydrogenation.* Hot coal slurry is made to react with pressurized hydrogen derived from steam and char solids (by-products of the conversion process). A tungsten and molybdenum catalyst assists the process of conversion, which occurs under pressure at 450 to 500 °C. The result is a liquid which can be distilled and refined to produce synthetic oil and gas. This method, called the *Bergius process,* was invented and used commercially in Germany before and during World War II to supplement Germany's inadequate supplies of petroleum.
3. *Extraction.* Here a solvent containing hydrogen is used to dissolve the coal. The hydrogenation process produces synthetic oil along with the solvent, which is recycled. The synthetic oil can be refined to produce the desired products. Undissolved coal is gasified to produce hydrogen for the process.
4. *Fischer-Tropsch synthesis (FTS) method.* Coal is burned in an atmosphere of steam and oxygen to yield *water gas,* which is composed primarily of carbon monoxide and hydrogen. Then the gases are purified and refined, by means of a catalyst, into methane gas and a variety of other useful hydrocarbons. This process is used in South Africa to produce 10,000 barrels of synthetic oil per day.

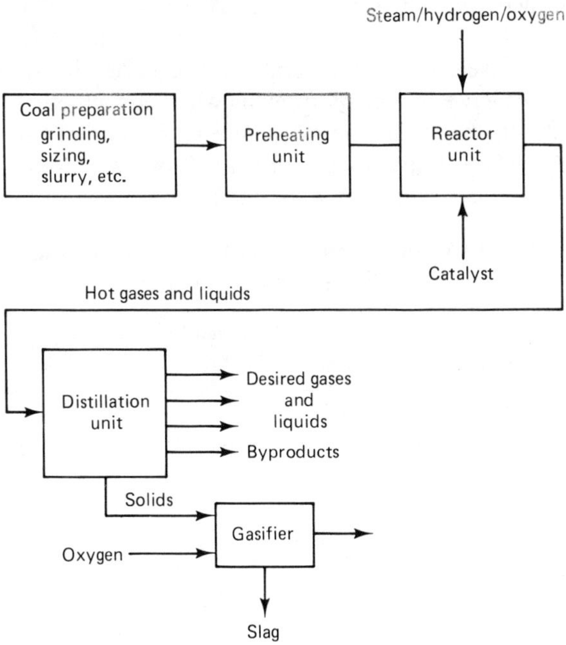

Figure 3.5.5 Block diagram of coal-conversion technology.

Prospects for coal conversion. Although coal gasification and liquefaction are clearly workable technologies, they face significant environmental problems which must be addressed before large-scale development can take place. One of the main justifications for coal conversion is the fact that coal burning creates serious pollution problems. But not enough data are currently available to determine whether coal gasification and liquefaction will increase or decrease pollution levels. The *products* of coal conversion are clean burning, but the *process* of conversion can release significant quantities of pollutants to the environment. This has the effect of localizing the environmental effects to the area where the conversion is carried out rather than causing environmental pollution in the cities where the products are used.

Using the principle "out of sight, out of mind," one can claim that the pollution problem has been addressed. But a more serious attempt at emissions control is necessary if coal conversion is to solve the environmental problems associated with coal. And even if the emissions problem is solved, there will be serious environmental and safety hazards associated with the mining and transport of the coal that is used in the conversion process.

Economic factors also stand in the way of full-scale development of coal-conversion technology. Conversion plants are very expensive to build (an estimated $2 billion for a plant to produce 125 million cubic feet of gas per day). The price of oil and gas must go up substantially before coal conversion can be economically feasible. One step taken by the U.S. government is the creation of the Synthetic Fuels Corporation of America to provide financial support and subsidies for these plants. But even with these incentives, coal conversion (and synfuels development in general) has proceeded slowly and irregularly. At the moment, we have no way of knowing whether synthetic fuels can contribute significantly to the world energy picture by the end of the twentieth century. However, the U.S. Department of Energy is optimistic about the prospects for synthetic fuels. See Table 3.5.1 for their projections to the year 2000.

TABLE 3.5.1 DOE Projections of U.S. Synfuels Production (1000 bbl/day equivalent)

Resource	Projected Production[a]		
	1985	1990	2000
Shale oil	30	225	1100
Tar sands	—	Negligible	Negligible
Coal liquids	20	150	1100
Coal gases			
High Btu	—	75	340
Medium to low Btu	30	110	340
All fuels	80	560	2880

[a]These estimates will vary depending on the price of oil.

3.6 NUCLEAR ENERGY: FISSION

Nuclear energy is certainly the most controversial of all energy resources. This is true for a variety of reasons. Probably the most important is that nuclear energy, besides being used for the production of useful electricity, is also used for the production of weapons of mass destruction. But the controversy over nuclear power centers around a number of other issues as well, including questions of safety, waste disposal, and economics. The debate over nuclear power is far from over, and this textbook can hardly present more than a general introduction to the controversy. Our primary purpose in this section will be to describe the history and general principles of nuclear power production. If readers wish to pursue the subject in greater depth, a wealth of resources is available on both sides of the issue.

3.6.1 History of Nuclear Energy

Nuclear energy was never understood even as a theoretical possibility until the early twentieth century, and its potential as a major source of power was first perceived during World War II. In the late 1930s, scientists discovered that it might be possible to produce nuclear weapons. Concerned that German scientists were following the same lines of research, the scientists urged President Franklin Roosevelt to develop such weapons ahead of the Axis powers. A highly secret program known as the Manhattan Project was initiated, drawing on scientists and engineers from every part of the country. After years of secret research and development, these researchers succeeded in producing a sustained nuclear chain reaction in 1942.

The successful production of a chain reaction proved the possibility of nuclear power production. However, the immediate focus of research was on

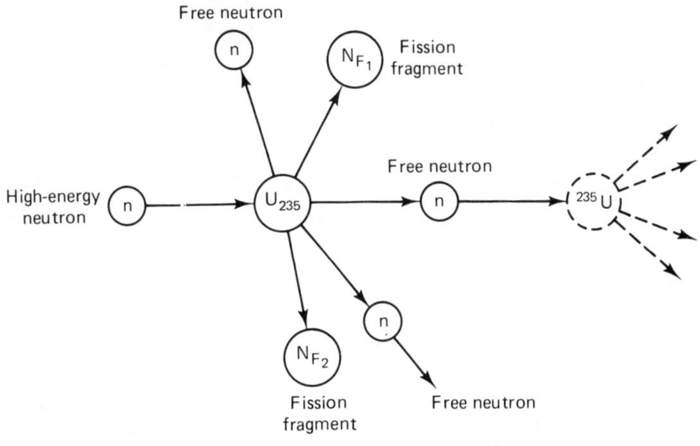

Figure 3.6.1 Uranium-235 chain reaction.

the production of nuclear weapons. It was not until 1953 that the United States initiated a program to use nuclear energy for peaceful purposes. Power reactors were soon developed, modeled on the highly successful reactors used to power submarines, and the first commercial power-producing reactor was commissioned in Shippingsport, Pennsylvania, in 1957.

Following the successful commissioning of the Shippingsport reactor, the nuclear power industry expanded rapidly for nearly 20 years. Projections of nuclear power capacity expanded rapidly as the number of operating nuclear power plants increased from 40 in 1973 to 71 in 1978. There were projections of 1600 to 1800 reactors by the turn of the century.

However, 1975 was the last year of the great expansion in nuclear power. During the early 1970s, opposition to nuclear power had been building, and the economic advantages of nuclear power did not appear as overwhelming as during the late 1960s. In 1978, orders in the United States stopped completely, and dozens of existing orders were canceled. The initial boom years appeared to be over as utilities concentrated on finishing the construction projects already under way.

As the 1970s continued, nuclear power experienced increased difficulties. Critics of nuclear power attacked the economics and safety record of nuclear power production. One government-sponsored safety study was suppressed and then released by antinuclear groups, and a second safety study (Wash-1400, the so-called Rasmussen Report) was withdrawn by the Nuclear Regulatory Commission in early 1979. Then, in late March 1979, a serious loss-of-coolant accident occurred at the Three Mile Island nuclear plant in Pennsylvania. There was no catastrophic meltdown, and the effects on the general public were certainly not as severe as the worst possible consequences of a nuclear accident. But public fears were increased greatly by the accident. At the same time, the owners of the Three Mile Island nuclear plant suffered a very serious economic loss as a result of the accident. This contributed to the already growing perception of nuclear power as a risky investment. The pace of reactor cancellations increased.

In the early 1980s, some of the reactors that had been under construction were completed, and most of these were put into operation. [(By July 1983, 81 nuclear power plants were operational in the United States (see Table 3.6.1).] But further shocks were in store for the nuclear power industry. The Diablo Canyon nuclear power plant, which had been fiercely opposed by environmental groups, finally received an operating license, but it was found that some of the plant's support structures had been constructed incorrectly, with the result that the plant could not withstand a severe earthquake with assured safety. In Cincinnati, Ohio, the Zimmer nuclear plant was 97% complete when the Nuclear Regulatory Commission ordered a halt to construction because of irregularities in safety and quality control by the contractors. Current plans call for the plant to be converted to coal burning. In January, 1984, the Nuclear Regulatory Commission denied an operating license to a plant in Illinois—the first such denial in the industry's history. In early 1984, the total number of domestic nuclear power plants in all stages of planning, construction, or operation was 140, with an aggregate design capacity of 134 GW.

TABLE 3.6.1 Status of Nuclear Power Industry, January 1, 1984

Licensed commercial reactors, 86
Construction permits granted, 52
Units on order, 2
Units announced, 0
Total power capacity installed, 69.8 GW
Total power capacity designed, 130 GW

Source: Atomic Industrial forum, Inc.

The future of nuclear power is uncertain. Critics feel that the rising prices of nuclear plants, together with the unresolved safety factors, will lead to the abandonment of nuclear power as a major energy resource. Supporters of the nuclear industry claim that development of nuclear power is the only ecologically sound energy resource capable of producing energy supplies in sufficient quantities for modern technological societies. It is difficult to make projections for the long term, but for the next two decades, it seems highly improbable that more than 150 nuclear power plants will be operating in the United States (Table 3.6.2). After that, energy growth trends should be more firmly established, and we will all have a better idea whether nuclear energy is a necessity, or whether it has been made obsolete by falling demand or other energy resources.

TABLE 3.6.2 Projections for Nuclear Capacity to the Year 2000

	1985	1990	1995	2000
Number of reactors	96	128	145	175
Electricity produced				
Trillion kWh	0.51	0.7	0.8	1.0
Quads	1.7	2.4	2.75	3.35
Capacity factor	0.61	0.63	0.63	0.64

Source: U.S. Department of Energy.

3.6.2 Principles of Nuclear Energy

Nuclear fission is a process of splitting the nucleus of an atom into two or more parts. Very heavy atoms are used because their large size makes them unstable. Uranium-235 (^{235}U), a rare isotope of uranium, is the most widely used fuel for nuclear reactions, although ^{233}U and plutonium-239 (^{239}Pu) can also be used. Of these three, ^{235}U is the only one that occurs naturally, and it occurs only in minute quantities.

A nuclear chain reaction begins when a neutron is absorbed by the nucleus of an atom. This produces a very unstable nucleus. For example, if the nucleus of a ^{235}U atom absorbs a neutron, it becomes a highly unstable ^{236}U nucleus. This unstable nucleus immediately breaks into two fragments and releases several neutrons, together with other types of radiation (see Section 7.5.4). The neutrons can react with other nuclei, and the resulting reactions produce still more nuclei. Thus the reaction is self-sustaining.

Although ^{235}U is the only naturally occurring *fissionable* isotope, ^{238}U and thorium-232 (^{232}Th) are called *fertile materials* because they can be converted into fissionable isotopes. ^{238}U absorbs a neutron and emits two electrons, forming ^{239}Pu (an element that does not occur in nature). Similarly, ^{232}Th can be transmuted into ^{233}U, which is a fissionable isotope. These reactions are not currently used for power production on a large scale; at present, the ^{235}U reaction is the basis of nearly all of the world's nuclear-produced electricity.

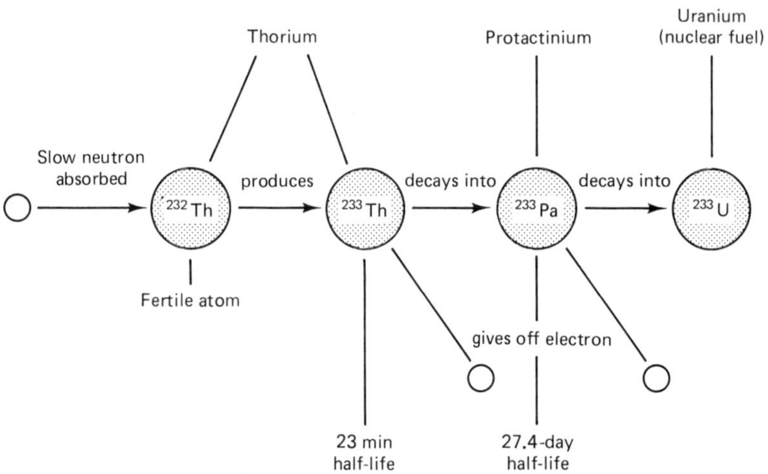

Figure 3.6.2 Nuclear reaction: fertile thorium-232 to fissionable uranium-233. (*Source:* U.S. Geological Survey.)

3.6.3 Supplies of Nuclear Fuels

In the United States, uranium is found primarily in the Southwest. New Mexico produces 45% of the nation's uranium, Wyoming provides 30%, and Colorado, Texas, Utah, and Washington account for the remaining 25%. Total U.S. uranium oxide ore reserves are estimated at 680,000 tons.

In the noncommunist world, uranium is also mined in Canada, South Africa, Australia, France, Gabon, Nigeria, and Zaire. Thorium occurs naturally as *thoria,* and it is three to four times more plentiful than uranium. In the United States, reserves of thoria are estimated at 100,000 tons. World reserves of thoria are approximately 1,000,000 tons. Thorium deposits are plentiful in India, Brazil, Canada, Australia, and Malaysia.

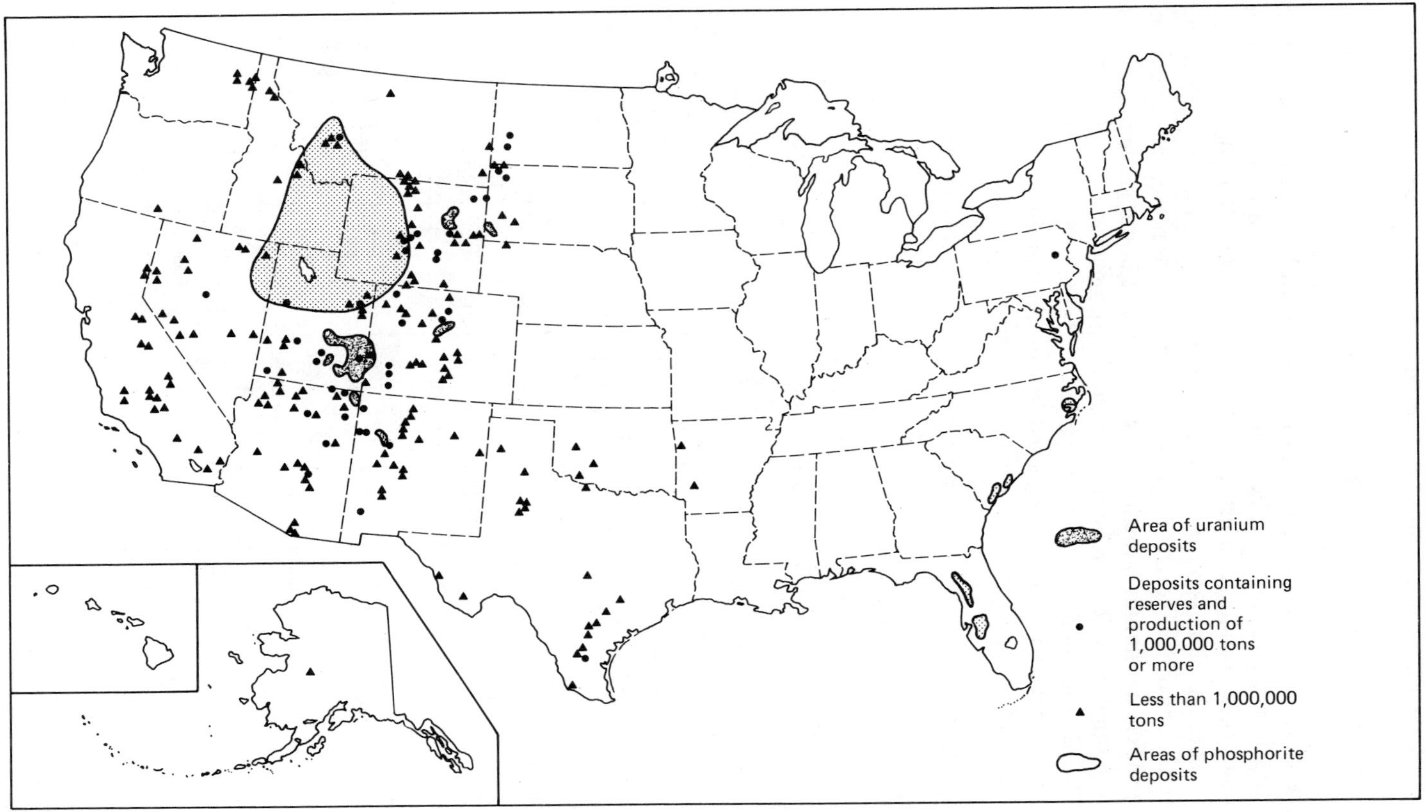

Figure 3.6.3 Nuclear materials, 1967. (*Source*: U.S. Geological Survey.)

3.6.4 Fuel Enrichment and Fabrication

Naturally occurring uranium contains only 0.7% ^{235}U. It also contains 0.005% ^{234}U and 99.3% ^{238}U. For most reactors, a concentration of 3% ^{235}U is necessary, so the fuel must be *enriched*. This can be done in several ways. In the United States, uranium hexafluoride gas is passed repeatedly through a series of tiny openings in a process called *gaseous diffusion*. Because ^{235}U is lighter than ^{238}U, it passes through the openings more readily. Eventually, the concentrations of ^{235}U reach adequate levels for nuclear power reactions. Some of the enriched uranium is further enriched to make bomb-grade materials; in fact, the same facilities can be used for both purposes.

Another method of enriching uranium is through the use of a gas centrifuge. This process is more widely used in Europe than in the United States. Yet a third method, laser enrichment, is still in the developmental stages, but

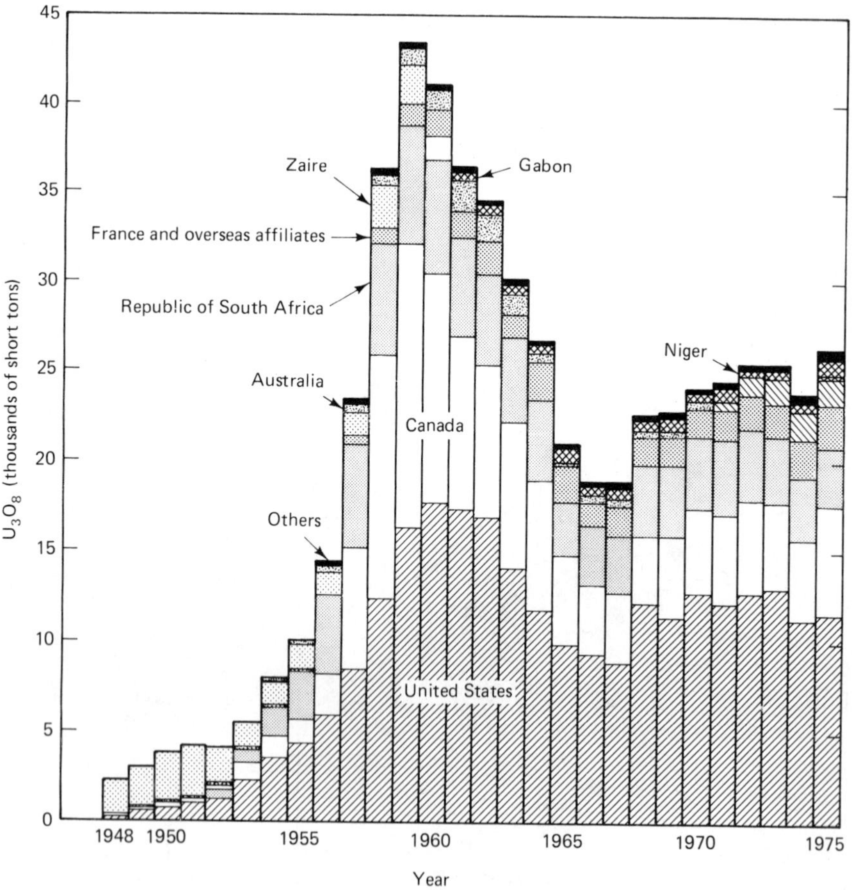

Figure 3.6.4 Uranium production in the noncommunist countries, 1948–1975. (*Source*: U.S. Geological Survey.)

it promises to be more efficient than either of the other methods.

Once the nuclear fuel has been enriched to the desired concentration of ^{235}U, it is converted to uranium oxide and shaped into cylindrical pellets. The pellets are then enclosed in cylinders made of a zirconium alloy and combined into *fuel assemblies.* Several dozens of these assemblies are placed in a very precise geometrical configuration to form the fuel core.

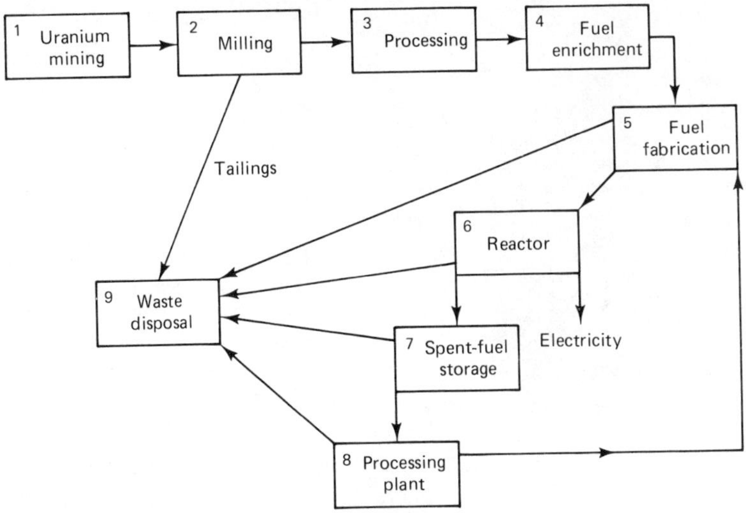

Figure 3.6.5 Nuclear fuel cycle: 1, uranium ore is mined; 2, ore is processed to uranium oxide (yellowcake); 3, uranium oxide is converted to uranium hexafluoride gas; 4, fuel is enriched to increase the concentration of uranium-235; 5, fuel is assembled into small pellets and pellets are combined into fuel assemblies; 6, fuel is used in reactor to produce electricity; 7, spent fuel is stored underwater at the plant site; 8, fuel is reprocessed to separate plutonium and uranium for future use; 9, unrecoverable wastes are placed in permanent repositories. (Steps 8 and 9 of the fuel cycle have not been implemented in the United States.)

3.6.5 Features of Nuclear Power Plants

Nuclear reactors come in a variety of designs: gas-cooled, boiling-water, pressurized-water, heavy-water, and liquid metal. Each design has its promoters. But all nuclear reactor designs share a few common features:

1. *Concentrated core of reactor fuel.* For a nuclear reaction to be self-sustaining, a *critical mass* of fuel must be brought together. If the mass is insufficient, not enough of the neutrons released by a decaying nucleus will interact with other nuclei, and the chain reaction will not be sustained.

2. *Neutron moderator.* When neutrons are released from a decaying nucleus, they move very rapidly—too rapidly to interact with other nuclei. They need to be slowed down by passing through a *moderating medium.* Graphite was used as the moderator in the first sustained nuclear reactor, and it is still

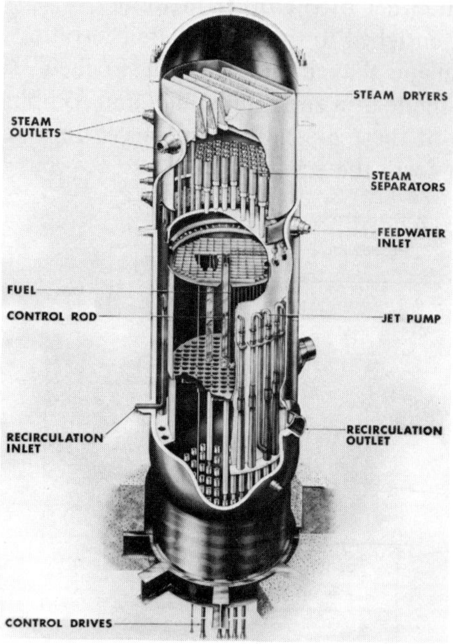

Figure 3.6.6 Cutaway section of a nuclear reactor. Fuel rods, control rods, and monitoring rods are arranged together, with the control drives at the bottom. The upper part of the reactor is devoted to steam separation and drying. (*Source:* General Electric Company of the United States.)

used in some British power reactors. Water is also a good moderator and is preferred by American reactor designers. In some breeder reactors, liquid metals are used because they allow the neutrons to move at a somewhat higher velocity, allowing the reactor to produce more fuel from infertile isotopes at the same time that it is using up its own stock of fuel.

3. *Controls on the reaction.* If a nuclear core were simply assembled and allowed to react uncontrollably, the reaction might go out of control and cause serious problems of overheating. To prevent this from happening, all reactors have a system of controls to reduce the rate of the fission reaction. Typically, metal rods are inserted into the fuel core. These rods absorb neutrons, thus effectively dividing the fuel core into a number of subcritical masses. When the rods are partially inserted, the reaction slows down, and when they are withdrawn, the reaction speeds up. When the rods are inserted fully, enough neutrons are absorbed to stop the reaction altogether.

4. *Coolant.* Much of the energy generated by nuclear fission is quickly converted into heat. This heat needs to be drawn away from the core for two purposes: to prevent overheating and to provide energy to drive the turbines. The coolant can be a gas, a liquid such as water, or a liquefied metal such as liquid sodium.

5. *Safety devices.* Although a nuclear power reactor cannot explode like an atomic bomb, it can cause very serious environmental damage in the event of a serious accident. For these reasons, all power reactors being built today have a number of safety features incorporated into them. In many cases, these safety systems are "redundant." That is, if the cooling system, for instance, should happen to fail, a secondary cooling system will be brought into play. These redundant features have sometimes helped to prevent nuclear accidents from going out of control. Here is a list of safety features found on most nuclear power reactors:

a. *Secondary containment.* Although the reactor vessel itself is very strong—it is typically made of 8-in.-thick steel—it may fail to contain radioactive by-products, especially gases, in the event of an accident. For this reason, all American nuclear power reactors are now built with secondary containment structures. Early secondary containment structures were made of steel, but newer power plants have very thick containments built of reinforced concrete. During the nuclear accident at Three Mile Island in 1979, large quantities of highly radioactive steam and gases escaped from the reactor vessel into the secondary containment structure. Without the secondary containment, the accident would have been much more serious than it was.
b. *Emergency core cooling.* In the event that all cooling systems fail, newer reactors have emergency core cooling systems. These systems inject extra water into the fuel core in the event of a loss-of-coolant accident such as the one at Three Mile Island. In that accident, the emergency core cooling system worked properly and would probably have prevented a significant accident, but it was turned off by a plant operator.

Figure 3.6.7 Nuclear power plant. In this plant, the domed containment structure houses the reactor under water. The large rectangular building contains the turbines and generators capable of producing 630 MW of electric power. (*Source:* Atomic Industrial Forum, Inc.)

3.6.6 Types of Reactors

Light-water reactors. The most common types of power-producing reactors are *light-water* reactors. Nearly all commercial reactors in the United States are of this type. They are called light-water reactors to distinguish them from *heavy-water* reactors, which use deuterium oxide as their coolant. (Deuterium is a rare isotope of hydrogen.) Light-water reactors fall into two types: *boiling water* and *pressurized water.* These terms refer to the ways that the heat energy is taken from the reactor core.

Boiling-water reactors do exactly what their name suggests: they boil water. Liquid water is pumped into the reactor core, where the heat from the reactor turns it to steam. The steam then powers a turbine connected to an electric generator. After its power has been used, the steam is condensed back to liquid water and returned to the reactor to be boiled again.

A 1000-MW boiling-water reactor is more than 6 m (20 ft) in diameter and 20 m (70 ft) high. The reactor requires more than 160,000 kg (350,000 lb) of enriched uranium oxide fuel. It is shut down for refueling. The steam pressure at the turbines is 67 to 70 kg/cm^2 [950 to 1000 pounds per square inch (psi)].

Boiling-water reactors are simpler that other types of water-cooled reactors because they have only one *coolant loop.* The same water moderates the neutrons in the core, cools the reactor, and drives the turbines. This simplicity is a great advantage of boiling-water reactors and gives them a slight efficiency advantage over pressurized-water designs (34% for a boiling-water power plant compared to 33% for a pressurized-water plant).

One drawback of boiling-water reactors is that radioactive materials from the core soon contaminate the entire power-generating system, making turbine repairs expensive and potentially dangerous. The systems have also come under criticism because of fear that some of the welds in the pressure system could

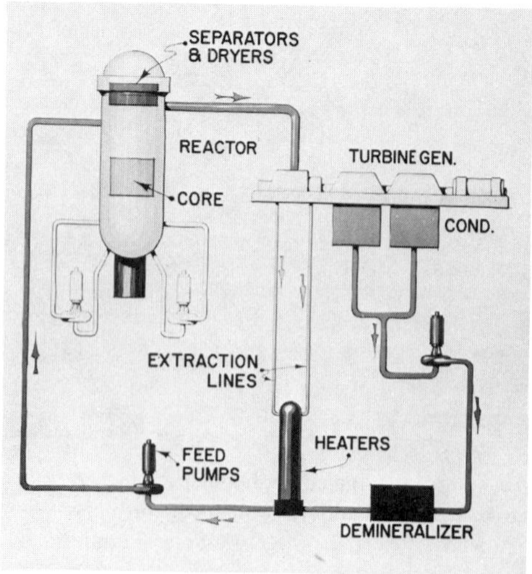

Figure 3.6.8 Direct-cycle boiling-water reactor. The enriched nuclear fuel undergoes a chain reaction in the reactor core and produces a large amount of heat, which is carried away by the water in the form of steam. The steam turbine drives a generator. (*Source:* General Electric Company of the United States.)

Figure 3.6.9 Nuclear fuel loading. An automatic fuel-handling machine positions each fuel bundle very carefully in the core of a nuclear reactor. The power plant must be shut down for weeks for such an operation. Safety checks, repairs, and other maintenance are also performed during this period. The radiation levels must be closely monitored during the entire process. (*Source:* Combustion Engineering, Inc.)

fail, releasing radioactive coolant. In pressurized-water designs, this particular type of accident is less likely to cause serious releases of radiation.

Pressurized-water reactors were originally designed for nuclear-powered submarines. They are somewhat more complex than boiling-water reactors because they have two coolant loops. In one loop, pressurized water circulates to moderate the neutrons and cool the reactor. The water pressure in this loop is maintained at 160 kg/cm^2 (2250 psi). The water in the primary coolant loop does not boil at all, although it reaches a temperature of 320 to 500 °C. The heat from this primary coolant is transferred through heat exchangers to the *secondary coolant,* another closed-water loop system. In the secondary coolant loop, water is boiled, producing steam at temperatures of 280 to 300 °C and pressures of 70 kg/cm^2 (1000 psi). The steam is used to drive the turbines and is then recondensed.

An 1100-MW reactor will be approximately 4.5 m (15 ft) in diameter and 13.5 m (45 ft) tall, with walls 20 to 28 cm (8 to 11 in.) thick. The reactor uses about 9000 kg (200,000 lb) of uranium oxide and must be shut down for refueling, which occurs every 10 to 12 months.

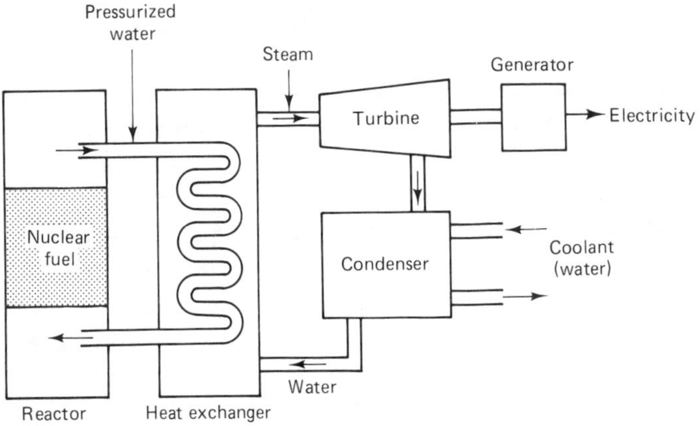

Figure 3.6.10 Pressurized-water reactor.

Liquid metal reactors. These reactors use a liquefied metal, usually sodium, as the neutron moderator and primary coolant. This is done so that the neutrons will be able to maintain a greater kinetic energy, an essential feature of the *fuel breeding* process (see Box 3.6.1). Sodium has a boiling point of 880 °C, so there is no need for the primary coolant loop to be pressurized. However, sodium is a very chemically active metal and reacts violently on contact with water. To reduce the danger from sodium reactions, liquid metal reactors use two water coolant loops in addition to the sodium loop. The secondary, or intermediate, coolant loop contains pressurized water, and the final coolant loop functions very much like the secondary loop of a pressurized-water reactor.

A full-scale commercial liquid metal reactor would be 20 m in diameter and would contain over 7 million kilograms (3300 tons) of liquid sodium at a temperature of 525 to 550 °C. The reactor core itself is relatively small: only about 10 m^3.

BOX 3.6.1 BREEDER REACTORS: MAKING THEIR OWN FUEL

It is commonly said that breeder reactors make more fuel than they consume. This is true in a very specialized sense. Of course, reactors cannot create matter, or at least not in significant quantities. What they do, however, is convert *fertile* materials to fissionable materials.

To accomplish this conversion, breeder reactors use plutonium, rather than uranium, as their primary fuel. Plutonium is a human-made element which undergoes fission readily; in fact, it is a common fuel in nuclear weapons. Plutonium reactors do not require the neutrons to be moderated to the same extent as do uranium reactors, so the neutrons can travel at a higher rate. This is what is meant by the term *fast* breeder reactor.

Surrounding the reactor core, breeder reactors contain a layer of nonfissile ^{238}U, the most common isotope of uranium. ^{238}U is not useful as a nuclear fuel, but it can be converted to plutonium by capturing a fast-moving neutron and emitting two electrons from the nucleus. This plutonium can then be used as fuel in a fast-neutron reactor. The amount of plutonium produced in a fast breeder reactor is somewhat greater than the amount used up during the same period of time. This is why this type of reactor is called a *breeder;* it breeds its own fuel by converting another material.

Heavy-water reactors. Nearly all U.S.-built power plants use enriched uranium as a fuel and normal water as a coolant. But other types of fuel and coolants can be used. One type of Canadian reactor, known as CANDU, uses *heavy water* as a coolant and neutron moderator. Heavy water is similar to regular water except that it contains *deuterium,* a heavy isotope of hydrogen. This special coolant permits the use of natural, unenriched uranium oxide. The coolant (heavy water) flows through horizontally placed, pressurized tubes containing uranium oxide fuel. These tubes are made of a zirconium alloy.

Figure 3.6.11 Fueling floor of a 250-MWe Phénix prototype fast breeder reactor operating since 1973 in France. (*Source:* Atomic Industrial Forum, Inc.)

Figure 3.6.12 Artist's concept of the proposed Clinch River Breeder Reactor at Oak Ridge, Tennessee. Funding for this project was terminated in 1983. (*Source:* U.S. Department of Energy.)

The temperature of the heavy water as it leaves the reactor is approximately 300 °C, and the pressure inside the tubes is 100 kg/m² (1450 psi). This is similar to the outlet temperature and pressure in a pressurized-water reactor. And like a pressurized-water reactor, a CANDU reactor has a secondary coolant loop containing ordinary "light" water. This loop provides the steam to drive the turbines.

Efficiencies in a CANDU reactor are only 30%—somewhat lower than in light-water reactors. But the cost of fuel is much lower because there is no need for enrichment. Also, the reactor is designed to work at full power even during refueling. All light-water reactors must shut down for refueling.

The reactor vessel in a CANDU reactor measures approximately 8.5 m (28 ft) in diameter and 6 m (20 ft) in length. Approximately 300,000 lb of fuel is required for a 750-MW reactor.

Gas-cooled reactors. Like the first nuclear "pile," modern gas-cooled reactors use graphite rather than water as a neutron moderator. Today, three types of gas-cooled reactors are in use. They have somewhat different characteristics and operate at different temperatures (Table 3.6.3). However, all three types share the same basic operating principle. The gas coolant is passed through the reactor fuel inside the moderator chamber, where it is heated to a high temperature. The hot gas is then used to heat water and produce steam to drive the turbines.

TABLE 3.6.3 Characteristics of Gas-Cooled Reactors

Type	Coolant	Temperature (°C)	Pressure (kg/cm³)	Efficiency (%)	Size
Magnox	CO_2	400	21	31	14 m diameter, 8 m high
Advanced	CO_2	650	42	42	9 m diameter, 8.5 m high
High-temperature	Helium	1366	50	42	8.5 m diameter, 6 m high

Source: Robert L. Loftness, *Energy Handbook* (New York: Van Nostrand Reinhold, 1978).

Gas-cooled reactors are considered somewhat safer than liquid-cooled reactors for two basic reasons:

1. The core of nuclear fuel is more dispersed than in other reactor types.
2. Graphite expands at higher temperatures. This means that if an accident causes a loss of cooling capacity, the core will expand. This will reduce the density of the nuclear fuel and moderate temperature increases in the reactor core.

Gas-cooled reactors are much larger than liquid-cooled reactors because gases have a much lower *heat capacity* than water or other liquids (see Box 4.1.3). So in order to transfer the same amount of heat from the reactor core, a much greater volume of gas is required. Gas-cooled reactors, like light-water reactors, must be shut down for refueling.

3.6.7 Nuclear Fuel Reprocessing

Unlike a regular fossil-fuel plant, a nuclear plant does not "use up" its fuel. Rather, it poisons itself with its own by-products. The fission products slow down the reaction, and when the concentration of ^{235}U reaches approximately 1%, the reactor needs to be refueled.

Spent nuclear fuel contains approximately 450 synthetic radioactive nuclides. Some of these materials can be used as nuclear fuel. For this reason, fuel reprocessing has been proposed to recover these materials. The spent fuel elements are chopped up and the useful materials are chemically separated out; then the wastes are disposed of at special waste sites.

Reprocessing makes economic sense because it permits the recycling of costly and valuable fuel. It also reduces the need to mine additional uranium. But it has serious drawbacks. Perhaps the most serious is the threat that recycling of plutonium would lead to proliferation of nuclear weapons. Only 25 kg of plutonium is needed to make a good-sized bomb, and that much plutonium could disappear from a reprocessing center without attracting much attention. Because of this concern, the United States does not permit reprocessing of fuel which it exports, although some other nuclear-exporting countries, including France, do not insist on such restrictions.

Reprocessing has never been commercially successful in the United States. Plants have been proposed or constructed at West Valley, New York; Barnwell, South Carolina; and Portsmouth, Ohio. But none of those facilities has operated successfully. Serious environmental problems have arisen, and the companies engaged in reprocessing has been unable to make a profit. For the near future, at least, reprocessing is at a standstill in the United States.

3.6.8 Disposal of Nuclear Wastes

Perhaps the most troubling aspect of nuclear power is the production of highly radioactive wastes. These wastes must be safely contained for many years, even centuries. So far, none of the proposed methods has been found entirely satisfactory.

The most promising technique for nuclear waste disposal is *vitrefaction* and burial. High-level waste products are mixed with other materials to form a glassy substance. These vitrified wastes can be formed into small capsules for easy handling. Because the radioactive materials are incorporated directly into the vitreous materials, they will not be able to migrate freely as liquefied wastes have done in the past.

Although the technology of vitrefaction has been successfully demonstrated, there is significant disagreement over the best way to dispose of the vitrified wastes. Some proposals call for placing the wastes in natural salt formations, which are believed to be quite stable geologically. But pilot projects in salt formations were abandoned when it was found that water migrated through the formations. Other plans call for burial in granite or in specially designed concrete burial vaults. Although many of these concepts seem promising, none has yet been demonstrated well enough to be approved for the permanent disposal of high-level wastes. Final requirements for disposal of high-level wastes in geologic repositories were set by the Nuclear Regulatory Commission on June 13, 1983.

Besides high-level wastes, the nuclear industry generates large quantities of low-level nuclear wastes. These include mine tailings, tools and clothing used by nuclear plant workers, and even the nuclear plants themselves after they have reached the end of their useful lifetimes. Land disposal is the principal method for disposing of most low-level wastes, although other plans have been suggested. Very few nuclear plants have been decommissioned, so little is known about the best methods for dealing with old plants. Perhaps equally important, little is known about the costs of safe decommissioning of nuclear plants.

3.6.9 Prospects for Nuclear Power

As recently as 1975, the prospects for nuclear power seemed to be very good. But since then, economic factors, public concern, nuclear accidents, and declining electrical demand have put a complete halt to new orders for nuclear plants and caused the cancellation of over a hundred proposed plants. Even those under construction are not immune from cancellation, although a few plants have come on line since the accident at Three Mile Island.

Even if electrical demand begins to increase, nuclear power may not share in the market for new generating capacity. The financial consequences of the Three Mile Island accident have been extremely serious for the utility that operated the plant, and the WPPSS debacle (see Section 6.5) has made utilities even more reluctant to invest in nuclear generating facilities. It will require significant changes in the economic climate of the United States before nuclear power enjoys the kind of popularity it experienced in the early 1970s.

3.7 NUCLEAR ENERGY: FUSION

Whereas nuclear fission produces energy by splitting the nuclei of large atoms such as uranium, fusion power is produced by joining two light atoms to form a heavier atom. The fusion process is much harder to achieve than the fission process because at ordinary temperatures and pressures, the atoms will not join together in a nuclear reaction.

To produce fusion, the two nuclei must be brought extremely close to each other—close enough to overcome the repulsive force between the two positively charged nuclei. This requires the nuclei to be moving extremely fast (i.e., to have very high kinetic energy). Only at extremely high temperatures or concentrations will these conditions be met.

So far, only uncontrolled fusion reactions have been achieved, in the form of hydrogen bombs. In a hydrogen bomb, a fission reaction is used to provide sufficient heat and compressive force to bring the nuclei together. Clearly, this is not an acceptable method of producing electrical power because the destructive forces unleashed in a nuclear bomb would destroy the power plant. More controlled fusion reactions are still in the experimental stages. No fusion reactor has yet produced as much power as it consumes, and researchers do not expect to produce significant amounts of power until the twenty-first century.

BOX 3.7.1 CONDITIONS FOR SUSTAINED FUSION REACTIONS

To produce an amount of fusion energy that is at least as great as the amount of energy required to maintain the reaction, certain critical conditions must be met. Temperature, density, and confinement time must reach or exceed a certain threshold. This threshold value is called the *Lawson criterion* (named after a British scientist, J. D. Lawson). At 100 million °C, the value of the criterion is equal to the following:

$$N \times T > \frac{10^{14} \text{ atoms/cubic centimeter}}{\text{second}}$$

where N = number of atoms per cubic centimeter

T = confinement time, s

As this criterion makes clear, a fusion reaction can be achieved either by concentrating the fuel at a very high density for a short period of time or by reaching a somewhat lower concentration for a longer time. A density of 10^{15} particles per cubic centimeter would have to be maintained for 10^{-1} s (one-tenth of a second), while a density of 10^{17} particles would only require a confinement time of 10^{-3} s, or 1 millisecond.

Laboratory-produced results have come within one order of magnitude: that is, values of $N \times T = 10^{13}$ have been observed. It seems likely that the break-even point will be reached within the next decade; however, successful production of commercial power will not be feasible for at least a few decades.

3.7.1 Types of Fusion Reactions

A number of possible fusion reactions have been identified. Here are four possible reactions:

$$
\begin{aligned}
{}_1D^2 + {}_1D^2 &= {}_2He^3 + \text{neutron} + 3.27 \text{ MeV} \\
{}_1D^2 + {}_1D^2 &= {}_1T^3 + {}_1H^1 + 4.03 \text{ MeV} \\
{}_1D^2 + {}_1T^3 &= {}_2He^4 + \text{neutron} + 17.6 \text{ MeV} \\
{}_3Li^6 + {}_1D^2 &= 2\ {}_2He^4 + 22.4 \text{ MeV}
\end{aligned}
$$

where

$$
\begin{aligned}
1 \text{ MeV} &= 1 \text{ million electron volts} \\
&= 1.6 \times 10^{-13} \text{ J} \\
&= 5.9 \times 10^{-8} \text{ Btu}
\end{aligned}
$$

The third of these reactions, where deuterium and tritium interact to form a helium nucleus plus one neutron, is the reaction that is being given the most study for possible commercial production. Deuterium occurs naturally in the world's water, and tritium can be manufactured from lithium, a readily available material. There is certainly no shortage of fuel. In fact, it can almost be said that nuclear fusion is a procedure for burning ocean water, since that is the expected source of the deuterium for fusion reactors. Ocean water contains enough deuterium to produce 10^{13} quads of energy, more than we could use in a billion years.

3.7.2 Creating the Conditions for a Fusion Reaction

Because deuterium is a naturally occurring material, it was feared that the first hydrogen bombs might set off reactions that would turn the world's oceans into enormous bombs and turn the earth into a star. But these fears were not justified. Under normal conditions, atoms will not react in the ways described above because of the negatively charged electrons in the orbits, which prevent a sufficiently close interaction between the nuclei. The only way to overcome these electrical forces is to maintain an extremely high temperature for a sufficient period of time and with a sufficient concentration of nuclei (see Box 3.7.1). Temperatures on the order of 100 million °C must be achieved and maintained for a few thousandths of a second. This presents several technical difficulties.

1. The particles must not be allowed to touch any cooler surfaces because this would make them lose too much heat and might melt the walls of the container.
2. The materials containing the reaction must be capable of sustaining extremely high temperatures.
3. The fuel must not become contaminated with other materials.

3.7.3 Current Status of Fusion Research

To solve the problems outlined above, researchers have proposed two methods of confining the fuel sufficiently to produce the needed reaction. These two methods are broadly categorized as *magnetic confinement* and *inertial confinement*. Within each general approach, there are several different design approaches. We will outline these two approaches and discuss the designs that are being explored most seriously.

Magnetic confinement. Magnetic confinement research is being actively pursued in both the United States and the USSR. The essence of this approach is to concentrate the high-temperature fuel by employing an enormous magnetic field. At very high temperatures, the electrons dissociate from the nuclei of the atoms, creating a special material state called *plasma*. Plasma has special properties unlike those of gases under ordinary conditions. In magnetic confinement schemes, the electrons can be drawn off by a magnetic field, leaving only the nuclei in the center of the confinement. This permits the fuel to be further confined; and in principle, a fusion reaction can take place.

One promising design is called a *TOKAMAK*. This word is made up of the first letters of a Russian phrase describing a *toroidal,* or doughnut-shaped magnetic chamber. As the name suggests, this design was first developed in the USSR. In the United States, a large experimental TOKAMAK is in operation at Princeton University, and an even larger one is being proposed (see Figure 3.7.1). The advantage of this type of confinement is that the fuel material can flow freely in a circular path around the torus, and a uniform magnetic field can be applied throughout the fuel chamber. Large magnetic coils, located at regular intervals around the outside of the torus, provide the desired magnetic field, which can be intensified until the fuel reaches the necessary temperature and confinement. Once the fusion reaction takes place, the energy released can be used to make steam and turn turbines, as in any other thermal plant. It seems likely that in the near future, a TOKAMAK-type reactor will sustain a nuclear fusion reaction. However, no one is certain whether the design will prove economically feasible in the long run.

Another magnetic confinement type of reactor under study is the "theta pinch confinement" reactor. This type involves so-called "magnetic mirrors." In this system, two intense magnetic fields are applied at opposite ends of the fusion chamber, and the plasma is suspended in space between them. It appears that such a system can achieve higher particle densities than a TOKAMAK but with a much lower confinement time. Work is being done on this design at Los Alamos, New Mexico, and at the Massachusetts Institute of Technology.

Inertial confinement. Some researchers believe that a fusion reaction can be produced by abruptly compressing small particles of fuel, producing very high densities and high temperatures. This is the principle that is used in hydrogen bombs, so we know that it can produce a fusion reaction. The problem is how to create a controlled fusion reaction.

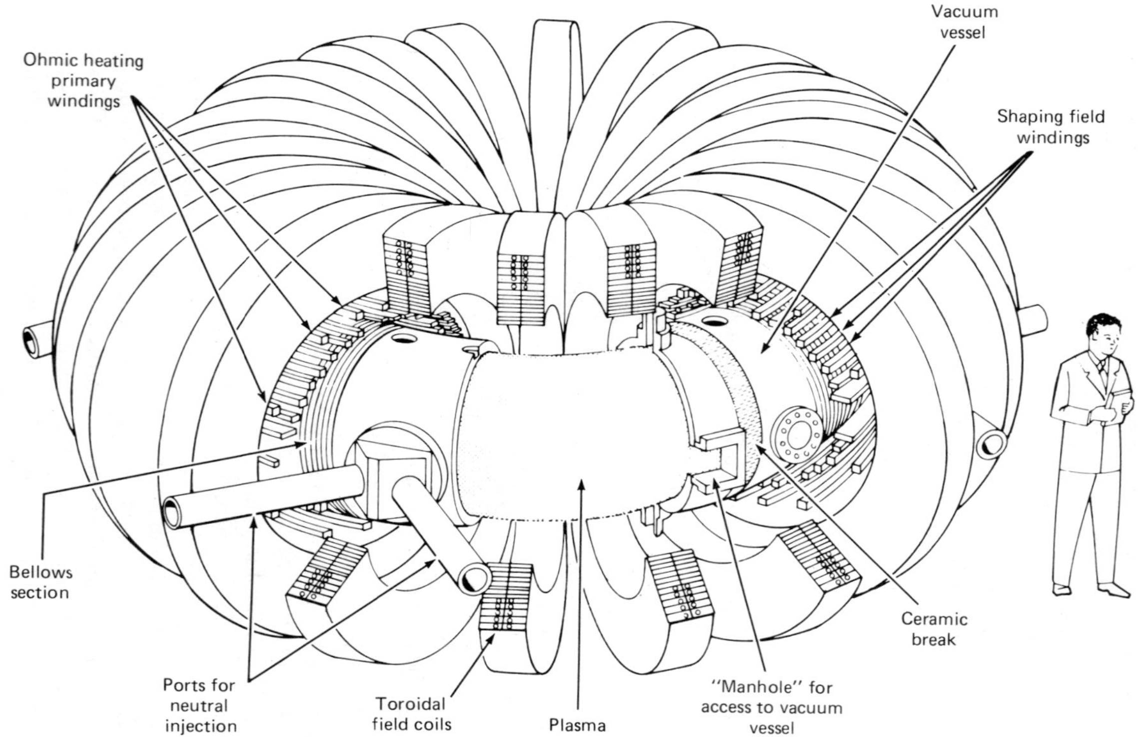

Figure 3.7.1 This fusion research device, known as the Princeton Large Torus (PLT), uses hydrogen as its fusion fuel; later generations will be expected to use deuterium and tritium as fuels. The fuel radius is 45 cm. Large fusion reactors will have a fuel radius of 170 cm. (*Source:* Princeton Plasma Physics Lab.)

Two basic approaches have been proposed for accomplishing this. In each case, a tiny particle of material is suspended in the center of a fusion chamber and bombarded from all sides by powerful bursts of energy. The energy can be in the form of lasers or particle beams.

Laser Confinement. Lasers produce high-energy, very precisely organized beams of light. When a large number of powerful beams are concentrated on a 1-mm-diameter fuel pellet, energy densities on the order of 1 MJ per gram per nanosecond (1 nanosecond = one-billionth of a second = 10^{-9} s) could compress the fuel pellet enough to achieve the Lawson criterion, and a fusion reaction could take place. The energy from this reaction could be used to generate electricity.

Progress on development of a laser confinement device has been very slow. One problem is that the energy of the laser beams has been transmitted to the particle with very low rates of efficiency: on the order of 5%. Also, the optical devices required to focus the laser beam's energy must be able to withstand very high temperatures and large energy flows. This, too, has been an obstacle to development of a workable laser fusion experiment.

Particle Beam Confinement. Particle beam technology has produced some rather promising results. At Sandia National Labs, an enormous generator is under development which will be capable of delivering a 6-MV pulse of energy in a millionth of a second. This device is scheduled for operation in 1986. However, the production of an actual fusion reaction has not yet been accomplished by particle beam confinement.

Figure 3.7.2 This Marx generator will be used in the Particle Beam Fusion Accelerator-II at Sandia National Laboratories, scheduled to begin operation in 1986. Each generator will be capable of delivering a 6-million volt pulse of energy in a millionth of a second. (*Source:* Sandia National Laboratories.)

3.7.4 Safety of Fusion Power

Nuclear fusion is believed to be somewhat safer than nuclear fission. For one thing, if the confinement in a fusion reactor is damaged, the nuclear reaction stops immediately. This is not true in a fission reactor, where spontaneous decay of reactor products can continue for many years. However, some critics of nuclear fusion point out that the inside of the reactor will still become radioactive from being bombarded with extremely high levels of energy. Also, most proposed fusion reactors call for the use of tritium, a radioactive form of hydrogen. Tritium is very difficult to confine; like hydrogen, it will migrate through most materials because its atoms are very tiny. And tritium can easily be combined with oxygen to form water, which is a highly biologically active material. Thus nuclear fusion may present some danger of radioactive exposure for the general public.

3.7.5 Prospects for Fusion Power

It can be safely assumed that a practical fusion reactor will involve tremendous engineering difficulties. Despite decades of research, scientists have so far been unable to achieve any net gain from a fusion reaction, even on the laboratory scale. And even if this is achieved, transferring laboratory methods to commercial-scale power generation will not be easy. According to the most optimistic projections, nuclear fusion may produce a small amount of electrical power by the year 2025. It has been estimated that by the end of the twenty-first century, fusion could produce up to 18% of U.S. energy needs. But until then, we will have to find other ways to cope with the energy situation.

3.8 GEOTHERMAL ENERGY

The earth is a huge reservoir of thermal energy. The interior of the planet is like the interior of a huge blast furnace. Hot molten metals and bubbling gases (called *magma*) make up the core of our seemingly temperate planet. In fact, the earth has cooled down considerably. A few billion years ago, it was a ball of fire.

If we reach sufficient depths—about 5 km under the ocean floor and 30 to 40 km beneath the land—the temperature is approximately 1000 °C. Even at relatively shallow depths, like mine shafts, the temperature gets warm enough to be uncomfortable. This interior heat is slowly and constantly being released to the surface—at an average rate of about 0.062 W/m^2, for a worldwide total of 960 quads per year. But in some locations, the heat from the interior reaches the surface more rapidly. Bubbling steam, hot springs, and volcanoes are all vents through which the earth's interior heat reaches the surface. The recent eruptions of Mt. Saint Helens in the state of Washington, which blew off 1500 ft of the mountaintop, were an indication of how much energy lies in the in-

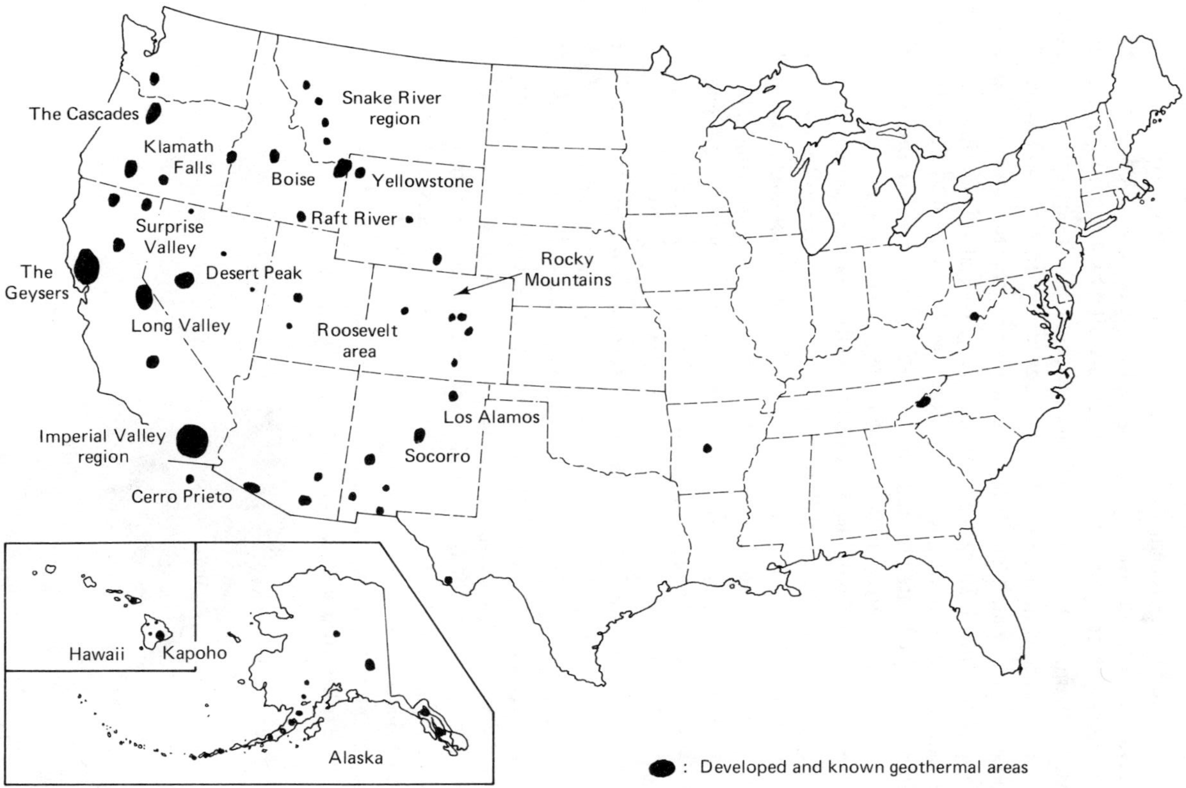

Figure 3.8.1 Geothermal energy in the United States. (*Source:* U.S. Geological Survey, 1979.)

terior of the earth—and that eruption released only a tiny fraction of the energy that is available.

No one knows how long humankind has used hot springs and pools for bathing and recreational purposes. But utilization of geothermal heat for other purposes is a relatively recent phenomenon. In 1812, steam from geothermal fields in Italy was used in the manufacture of boric acid. In 1904, at Larderello, Italy, a steam-operated generator produced electricity successfully using natural geothermal steam. In 1913, a power plant with a capacity of 250 kW was built at the same site, and now this field produces approximately 400 MW of electrical power—nearly half as much as a large new thermal power plant.

Despite its short history, geothermal energy is relatively common throughout the world. For a number of years, Reykjavik, the capital of Iceland, has been almost entirely heated by geothermal energy. In New Zealand, two power plants were built in 1960 and 1963 to produce 170 MW of electric power. Geothermal energy is also being used, both for heating and electric generation, in the USSR, Japan, Mexico, and Kenya. At least 80 countries of the world have favorable geological formations for potential geothermal sites. Out of these, at least 25 countries have already started developing geothermal plants.

In the United States, the first attempt to harness geothermal energy was a technical success but an economic failure. But in 1960, at the same site—the "Geysers" north of San Francisco—a 12.5-MW electric power plant was built and successfully operated. Since its initial development, the site has been expanded until today the 15-unit power plant has an output power of 908 MW, consuming more than 2.75 million kilograms of dry steam every hour. This is the world's largest geothermal power plant, with a power output as great as that of a new coal or nuclear plant. Other geothermal installations in the United States include a system for residential space heating in Klamath Falls, Oregon, and a successful 5-MW plant at Raft River, Idaho. Many research and demonstration projects are currently under way in southern California, Oregon, New Mexico, and Idaho, where 2.3 million acres of federal land have already been leased for exploration and development.

Figure 3.8.2 Wet cooling towers at the "Geysers" geothermal area in Ukiah, California District. The Bureau of Land Management has leased public land to the utility for steam development. (*Source:* Bureau of Land Management, U.S. Department of the Interior.)

3.8.1 Types of Geothermal Fields

Geothermal fields are classified according to the form in which the geothermal energy can be found. The four basic types of geothermal fields are: hot water fields, dry steam fields, hot rock fields, and geopressurized zones. Methods of extracting the geothermal energy vary depending on the type of field, and so do the uses for the geothermal energy once it has been developed.

The most common type of geothermal field, and the first to be developed,

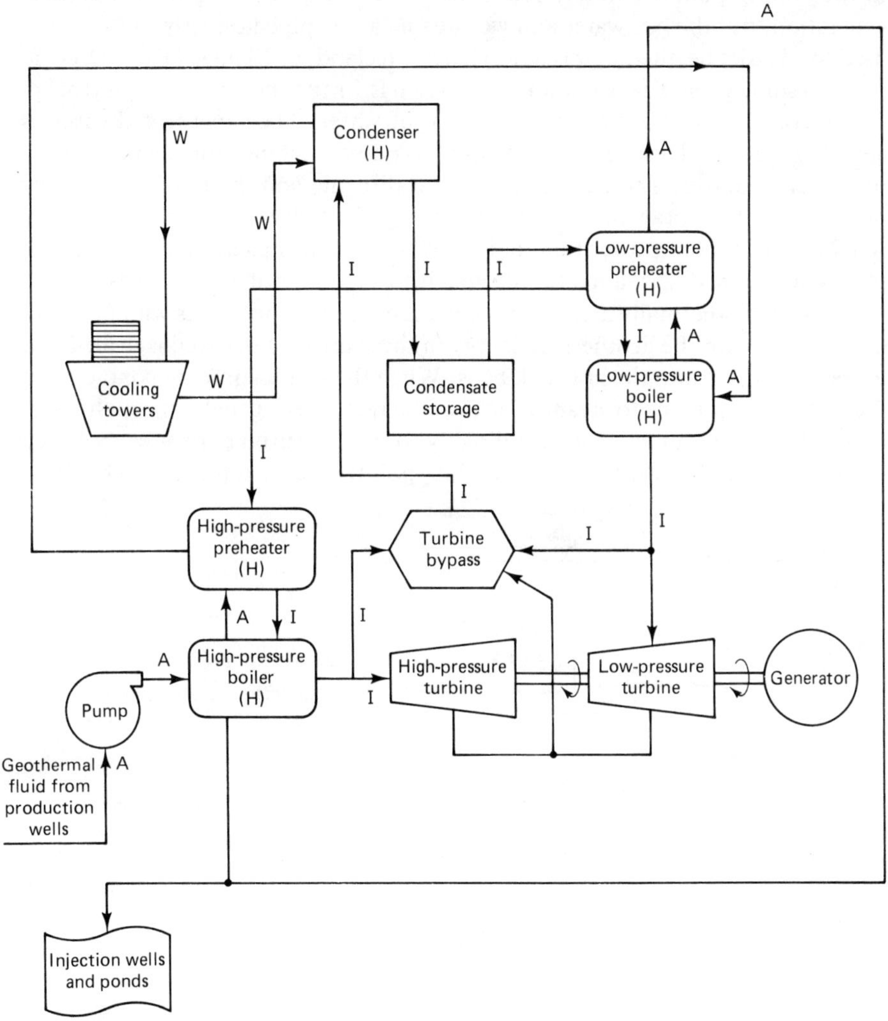

Figure 3.8.3 Schematic of a geothermal power plant. A, geothermal fluid; I, isobutane or similar low-boiling-point fluid; W, cooling water; H, heat exchangers. (*Source:* Raft River 5-MWe Pilot Geothermal Power Plant; C. J. Bliem; E.G. & G., Idaho; and U.S. Department of Energy.)

is the *hot water* or *wet steam* field. Naturally occurring underground water reservoirs come in contact with geological "hot spots" where heat from the interior has penetrated almost to the surface. The water is heated to a high temperature, which forms steam and raises the pressure in the underground reservoir. If there are cracks or weak spots in the earth's crust, the mixture of hot water and wet steam escapes to the surface, producing a hot spring or a geyser. A drilled well can also provide a passageway for this hot water and steam to reach the surface. The temperature of this water and steam mixture will be very close to the boiling point of water: 100 °C or 212 °F. This is too cold for electrical generation, but the hot water and wet steam can be piped directly to a city and used for district heating, as it is in Reykjavik, Iceland, or Klamath Falls, Oregon.

In some geological situations, the steam is further heated so that it reaches a temperature far above the boiling point of water. The superheated vapor is called *dry steam*. This steam can be used to power turbines for the production of electrical power. The Geysers plant in California and the Larderello power plant in Italy both draw their geothermal energy from dry steam wells. Dry steam wells produce more power than hot water wells and thus are more profitable to develop, but dry steam sites are much rarer than hot water sites.

Most geothermally active areas consist entirely of hot rocks with no water or steam. To tap the geothermal energy in these areas, a dry well is drilled and water is pumped into the hot rock beds. When the water comes in contact with the hot rocks, it turns to steam which can be recovered, usually through a second well. This steam can be used directly to drive a turbine, or it can be used to heat another fluid with a lower boiling point, such as ammonia. This fluid

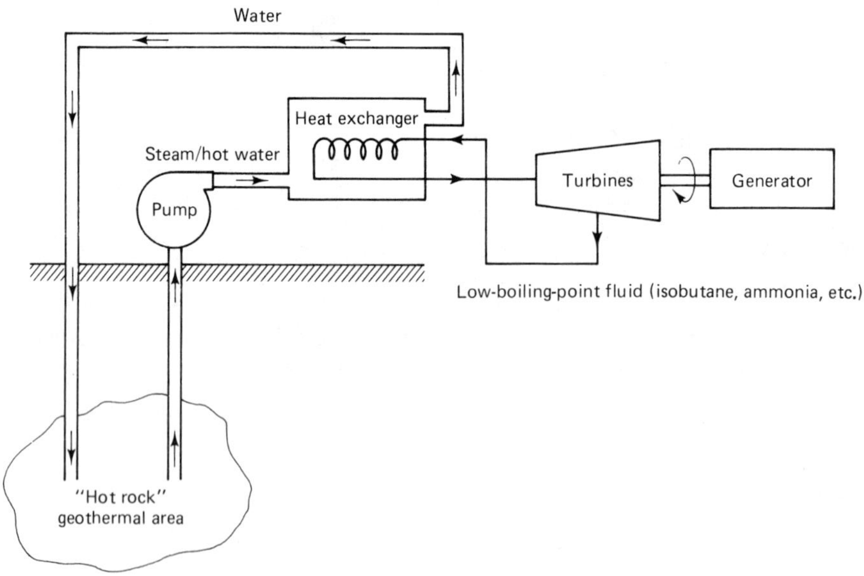

Figure 3.8.4 Schematic of a hot-rock geothermal power plant.

can be used to drive a turbine while the water is circulated back to the thermal area through the system of wells.

A fourth type of geothermal field is called a *geopressurized zone.* Geopressurized zones are found on the eastern seaboard along the Gulf of Mexico. Little is known about these geothermal fields. Apparently, these hot rock deposits contain large quantities of hot water under extreme pressure, with natural gas dissolved in the water. More research must be carried out before these resources can be developed safely and effectively.

3.8.2 Environmental Concerns

Geothermal energy resources are not pollution-free. Care must be taken to avoid water, air, thermal, and noise pollution. The steam and water from geothermal fields contains many toxic chemicals, including the following:

Hydrogen sulfide gas; the source of the "rotten egg" smell characteristic of many geothermal regions

Sulfur dioxide, also called "brimstone"; a major air pollutant which has been linked to the production of acid precipitation

Mercury compounds

Radioactive gases and particles, especially radium and radon

Disposal of spent steam and water into surface streams can cause serious thermal pollution, altering the ecology of the streams. The loud hissing noise created by escaping steam can also be a source of annoyance if the geothermal site is located in a heavily populated area.

In general, these pollution problems are less serious than the problems associated with fossil fuels, and solutions are not hard to find. Pumping the spent water or steam back into the underground formations minimizes the water and air pollution problems associated with geothermal energy projects. Another solution would be to build chemical treatment plants on the site to extract some of the more valuable minerals from the hot water and steam. This would create a useful by-product: fresh water, which is a scarce commodity in some geothermal regions.

3.8.3 The Future of Geothermal Energy

In the strictest sense of the word, geothermal energy is not *renewable:* that is, the energy that is taken out of the ground will not be replaced in the future. Following this line of argument, it could be said that geothermal energy is even less renewable than fossil fuels. However, if we consider the *scale* of geothermal energy use compared to the world's total resources of geothermal energy, we can see that there is very little danger of exhausting our geothermal energy resources. So in terms of energy supply, geothermal energy is a nearly inexhaustible energy resource, unlike fossil fuels, which can be exhausted in a few centuries or even decades. Production of geothermal energy is still in its experimental stages, but there appear to be good prospects for future use of this vast resource. Multipurpose applications, including *cogeneration* (see Chapter 11), could add to the effectiveness of geothermal energy resources. Particularly in the western United States and Gulf coast regions, development of geothermal resources could make a substantial long-term contribution. Also, the economics of geothermal energy development is more favorable than for more exotic technologies such as wave power or ocean thermal conversion. Currently, the Department of Energy projects that 3.5 million kilowatts (GW) of geothermal electrical generating power will be available for electrical production by 1990, and 6.0 GW of production by the year 2000. This means a total electrical production of about 0.1 quad in 1990 and 0.2 by 2000. The total energy (heat and electricity) provided by geothermal electric power is expected to reach 0.3 quad by 1990 and 0.6 quad by the year 2000.

STUDY QUESTIONS

General Questions

1. Which three energy resources are most widely used in our economy? What qualities make these resources so important? Can you think of adequate substitutes for these resources?
2. Which nonrenewable energy resources are stored forms of solar energy? Which are not? Why are these resources used instead of direct solar energy?
3. What are fossil fuels? How are they related to other kinds of fossils?
4. Is there a limit to our use of nonrenewable resources? How far are we from reaching that limit? What should we be doing now to plan for the time when we can no longer depend on nonrenewable resources?

Coal

5. Investigate the history and early use of coal. How has the use of coal changed over the years?
6. Compare the different methods of mining coal. Which are the most efficient? Which are the safest? Which cause the most environmental damage?
7. Compare the different methods of transporting coal. Which methods are most widely used today? Which are the most efficient? Do you expect any changes in the transport of coal during the next 20 years?
8. The primary use of coal today is for generation of electricity. Why is this so? What are some other uses of coal? How do they compare with electrical generation in terms of efficiency and environmental effects?

Oil

9. Oil provides us with more energy than any other resource, if we discount the global heating effects of solar radiation. Why is oil such an important resource? What makes it more widely used than more abundant fuels such as coal?
10. Distinguish between proven reserves and potential reserves of petroleum. Show how this can give a significantly different picture of resource availability, depending on the intention of the person giving the report.
11. What methods are used to explore for petroleum? How many of these methods were available 50 years ago? Why are the newer methods used today?
12. Many older wells are being brought back into production using enhanced recovery techniques. What has changed to make these methods practical today?
13. What are the products of a typical oil refinery? How are they used? Why does a refinery choose a particular "mix" of products?
14. What percentage of petroleum is used to produce plastics and other petroleum-based chemicals? What does this suggest about the energy impact of using plastic products?
15. Trace the history of U.S. oil imports. When did the United States become a net importer of oil? What factors led the United States, a major oil producer, to depend so heavily on imports?
16. What steps could be taken to eliminate oil imports by the year 2000? What is the likelihood that these steps will be taken?

Natural Gas

17. Why has natural gas replaced fuel oil as the nation's primary fuel for space and water heating?
18. Why is natural gas "flared" in the Middle East rather than being refined and sold?
19. How is natural gas transported over long distances?
20. Why does the United States import large quantities of Middle Eastern oil, but negligible amounts of Middle Eastern natural gas? Is this situation likely to change in the near future?
21. How would you compare the environmental impacts of natural gas and coal? Can we use natural gas as a replacement for coal? Give your reasons.

Synthetic Fuels

22. Is it a good idea to develop alternative fuels to replace oil and natural gas? Give your reasons.
23. What is kerogen? Where is it found? How is it extracted?
24. Shale oil has somewhat greater environmental impacts than coal during production, but the environmental impacts of coal are greater during use. Overall, which resource presents the fewest environmental problems?
25. Early predictions for oil shale development in the United States were highly optimistic, but during the early 1980s, no new shale oil projects were started and some

of the existing projects were canceled. What are the causes for this slowdown? What are the prospects for renewed interest in shale oil development?

26. What are the main regions where tar sands are found? What geographical characteristics do they have in common? How does this affect the prospects for production of tar sands?
27. Coal gasification technology has existed for over 100 years. Why has there not been widespread interest in coal gasification until recent years?
28. Compare the environmental impacts of coal conversion and direct burning of coal.
29. Why did the government create a Synthetic Fuels Corporation? Why was synthetic fuels development not left to private companies?

Nuclear Fission

30. There are several types of nuclear fission reactions. What elements are common to each of them?
31. Why do nuclear engineers speak of a nuclear fuel "cycle?" What makes the process a cycle rather than a once-through process?
32. What is the current status of nuclear waste disposal plans? What is done with the spent fuel from nuclear power plants? How is this likely to change in the next 20 years?
33. Why is it necessary to have so many safety features in a nuclear plant compared to a coal plant?
34. No public fatalities or serious injuries have been directly attributed to the operation of nuclear power plants. Why, then, is there so much concern over the safety of nuclear plants?
35. What were the causes of the accident at Three Mile Island? Did this accident show that nuclear power is unsafe, as critics claim, or that emergency safety systems work, as supporters of nuclear power claim?
36. Why have there been no orders for new nuclear plants since 1978? What would have to change in order for the nuclear industry to regain its former level of activity? What are the chances that this will happen?
37. Some people have called for the immediate shutdown of all nuclear plants. In your view, how realistic is this demand? How would electricity be supplied to areas like Massachusetts and Chicago, which get a large percentage of their electricity from nuclear power plants?

Nuclear Fusion

38. What is fusion? What are its advantages over nuclear fission?
39. What is the current status of nuclear fusion research? When do researchers expect to produce commercial fusion power?
40. What are the main sources of fuel for nuclear fusion reactors? Why has fusion not been developed before if the fuel is so plentiful?
41. What obstacles stand in the way of commercial development of nuclear fusion? Do you expect these obstacles to be overcome by the year 2000? 2050?

Geothermal Energy

42. Of the different types of geothermal systems, which type do you believe to be the most economically feasible? Is this the type where the greatest development efforts are being concentrated?
43. What are the geographical characteristics of a good geothermal site? Are there any developed geothermal sites near your home? Can you think of any potential sites nearby?
44. Are there any environmental problems associated with the production of geothermal energy? If so, how can those problems be dealt with in an operating geothermal plant?
45. District heating (using one heating system to heat several buildings) is an especially popular application of geothermal energy. Can you see any advantages of district heating as compared to geothermal systems for individual houses?

PROBLEMS

Coal

1. In 1981, total U.S. coal production was 823.75 million metric tons. At that rate of production, how long would U.S. coal reserves last? Economically recoverable reserves are estimated at 255 billion metric tons.
2. One kilogram of coal has a heat content of 8000 kcal. How much coal will be needed to operate a coal stove for 4 hours if the stove has an output rating of 5000 kcal/h? (Assume that the stove operates at an efficiency of 50%.)
3. If 81% of 1981 U.S. coal production was used to produce electricity at an average efficiency of 32%, how many kilowatthours of electricity were produced? (One metric ton = 1000 kg.)

Oil

4. U.S. proven reserves of oil are 25 billion barrels. If average daily use is 15 million barrels, and if the United States imports 40% of its oil supplies, how long will the U.S. oil reserves last?
5. The world's proven reserves of petroleum are close to 660 billion barrels. World oil consumption equals 52 MBD. How long can this rate of consumption be continued assuming that no new reserves are discovered?
6. Using the figures from Problem 5, and assuming that each year new reserves are discovered which add 5% to the world's total reserves at the time of discovery, how long will the world's oil reserves last?
7. A new American automobile gets 26 mpg on the average and is driven 9500 miles per year.
   - **(a)** What is the annual gasoline consumption in barrels?
   - **(b)** If the car lasts 7.6 years, what is its gasoline consumption over its lifetime?
   - **(c)** What is the total cost of fuel over the car's lifetime if gasoline averages $1.50 per gallon?

8. One gallon of home heating fuel has a heat value of 140,000 Btu. If a house loses 10,000 Btu per degree-day in a 7000-degree-day climate, how much fuel oil does it use per year? The furnace delivers heat to the house at 70% efficiency.

Natural Gas

9. U.S. total reserves of natural gas are estimated at 200 TCF. One cubic foot of gas has an energy content of 1000 Btu.
   - **(a)** How many quads of energy are contained in the U.S. reserves?
   - **(b)** If the United States tried to meet all its 1982 primary energy requirements (71 quads) using natural gas, what percentage of the national reserves would be used up?
10. The United States currently meets approximately 28% of its energy needs with natural gas. If energy use averages 70q per year, how long will domestic gas reserves last?
11. If additional unconventional resources and improved technology bring domestic reserves up to 8000 TCF, how long can we keep on using natural gas at 70q per year?
12. **(a)** If the owner of the house in Problem 8 decides to switch to a natural gas furnace with an efficiency of 85%, how many cubic feet of natural gas will be used per year?
   - **(b)** Using prices obtained from a local fuel oil dealer and a natural gas supplier, estimate the annual savings from switching to natural gas.

Synthetic Fuels

13. If the fuel content of shale oil is equal to 25 gallons per ton, how much shale would have to be mined to replace U.S. oil imports (currently 5 MBD)?
14. If oil shale has a specific gravity of 1.7, how many cubic feet of shale would have to be mined in a year to replace all U.S. oil imports?

Nuclear Fission

15. The electrical rating of a nuclear plant is 1000 MW. The plant operates at an efficiency of 30%. For cooling purposes, the plant uses river water. Assuming that the outgoing water is 4 °C warmer than the incoming water, how much water is required per hour for cooling?
16. A radioactive material has a half-life of 25 years. How long will it be before the material decays to 1% of its original radioactivity?

Nuclear Fusion

17. A plasma composed of tritium and deuterium fuel is confined at a density of 3.4×10^{17} atoms per cubic centimeter. How long must it remain confined to meet the Lawson criterion?
18. The plasma in Problem 17 has been successfully confined for one-tenth of a millisecond. How much will this performance have to be improved in order to meet the Lawson criterion?

Geothermal

19. Assume that in the year 2000, total U.S. primary energy consumption is 100 quads. According to the latest government projections, geothermal energy is expected to produce 3×10^{14} Btu of energy per year. What fraction of U.S. energy needs would this provide?

20. A geothermal area produces hot water at a temperature of 200 °F. The water is used in a district heating system to keep buildings at a temperature of 65 °F. If the total seasonal district heating needs are 10 billion Btu, how many gallons of hot water per year must be used?

FURTHER READINGS

COOK, EARL. *Man, Energy, Society.* San Francisco: W. H. Freeman, 1976.

Energy. A special report. Washington, DC: National Geographic Society, February 1981.

KRUGER, PAUL, and CAREL OTTE, eds. *Geothermal Energy: Resources Production, Simulation.* Stanford, CA: Stanford University Press, 1973.

MCMULLAN, J. T., R. MORGAN, and R. B. MURRAY. *Energy Resources.* New York: John Wiley, 1977.

OSTERHOUT, MARILYN M. *Decontamination and Decommissioning of Nuclear Facilities.* New York: Plenum Press, 1980.

SCHURR, SAM H., ET AL. *Energy in America's Future.* Baltimore: Johns Hopkins University Press for Resources for the Future, 1980.

SINGER, S. FRED, ed. *Energy: Readings from Scientific American.* San Francisco: W. H. Freeman, 1979.

STOBAUGH, ROBERT, and DANIEL YERGIN, eds. *Energy Future.* New York: Random House, 1979.

U.S. DEPARTMENT OF ENERGY. *Monthly Energy Review.* Washington, DC. Published monthly.

4

Renewable Energy Resources

Sunlight provides the great majority of the energy that is used on the earth. In fact, with the exception of meteors (which provide little useful energy) and tidal energy, sunlight is the only significant source of outside energy reaching the earth. For this reason, only energy sources based on the use of sunlight are genuinely *renewable*. Besides direct use of sunlight, renewable energy resources include wind, ocean thermal, hydro, and biomass energy. These other renewable energy sources will be discussed in a later section. Some people refer to geothermal energy sources as renewable, but strictly speaking they are not. The use of geothermal heat causes a slight cooling of the earth's core.

4.1 DIRECT SOLAR: WATER AND SPACE HEATING

Most of our uses of solar energy are simply taken for granted. For example, consider the use of sunlight for lighting. Windows, skylights, and reflectors can all be used to direct sunlight to places where it is needed. During the day, sunlight can provide most of the light that is needed in most buildings. Although we take this light for granted, it provides the equivalent of several kilowatthours of electric light per day per building. This adds up to a substantial contribution when we consider all the buildings that use this form of lighting during the day.

Most people also take for granted the enormous contribution of heat that we receive from the sun. One way to understand the size of this contribution is to consider what would happen if no sunlight reached the earth. Since heat energy would escape from the earth into outer space, the earth's temperature would drop close to absolute zero— –273 °C or –456 °F. Most of us are concerned with much smaller temperature differences, such as the difference between outdoor winter temperatures and the temperatures we would like to maintain in our houses and buildings. But we should also be aware of the enormous amount of solar energy which the earth receives continuously from the sun.

Each year, 15 *quintillion* (1.5×10^{18}) kilowatthours of solar energy reaches the earth's atmosphere. Slightly more than one-third of this energy is reflected back into space, another one-sixth is absorbed by the earth's atmosphere, and a bit less than one-half reaches the earth's surface. This is many thousands of times more energy than people use for all purposes. Capturing and converting this sunlight to usable forms is one of the most difficult challenges of solar technology. In some areas, especially large-scale industrial applications, progress to date has been disappointing. But in other areas, particularly domestic water and space heating, solar energy has already begun to make a significant contribution.

4.1.1 Solar Water Heating

In the United States, solar water heaters were popular during the early and mid-twentieth century, but they were rapidly replaced by simpler, less expensive natural gas water heaters as natural gas was introduced. However, that was in the days of cheap fossil fuels. Today, energy prices have increased so that solar water heating systems are economically competitive with electric water heaters in most parts of the country and may become significantly more popular as fuel costs increase.

Most commercially available solar water heaters use roof-mounted collectors and freeze-resistant heat-exchange fluids, although owner-built systems use other principles as well.

BOX 4.1.1 FREEZE PROTECTION

One of the most difficult problems in designing solar water heaters concerns the problem of freezing. In cold weather, especially on cold winter nights, the temperature inside the collector may drop far below freezing. If plain water is used as the heat-exchange fluid, freezing temperatures can cause serious damage to the collector and to the entire heating system. To prevent this problem, many manufacturers use freeze-resistant fluids, such as automotive antifreeze, as the heat-exchange fluid. This requires the use of specially designed heat exchangers to prevent mixing of the heat-exchange fluid with the water in the tank. Another common solution to this problem is to allow the water to drain out of the collector when the collectors fall below a certain temperature. This approach requires a greater amount of energy to pump the water back up into the collector when the collector temperature increases, but it prevents the danger of water contamination.

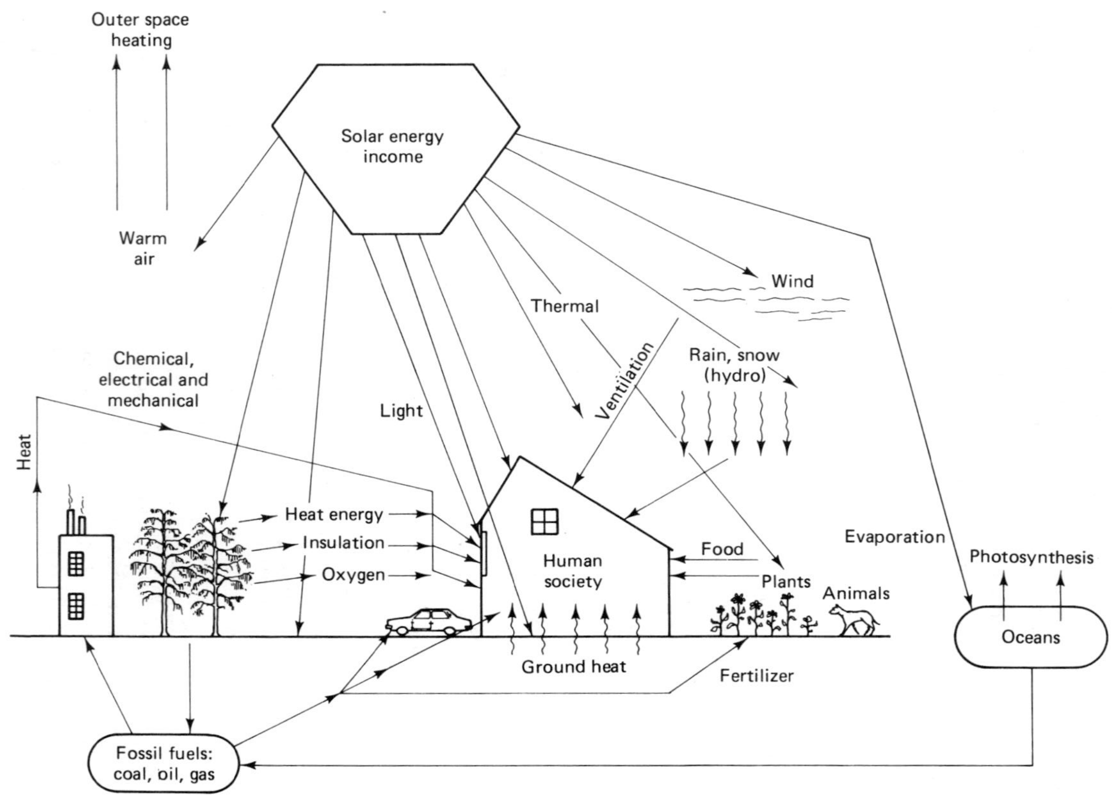

Figure 4.1.1 Solar energy: the lifeline.

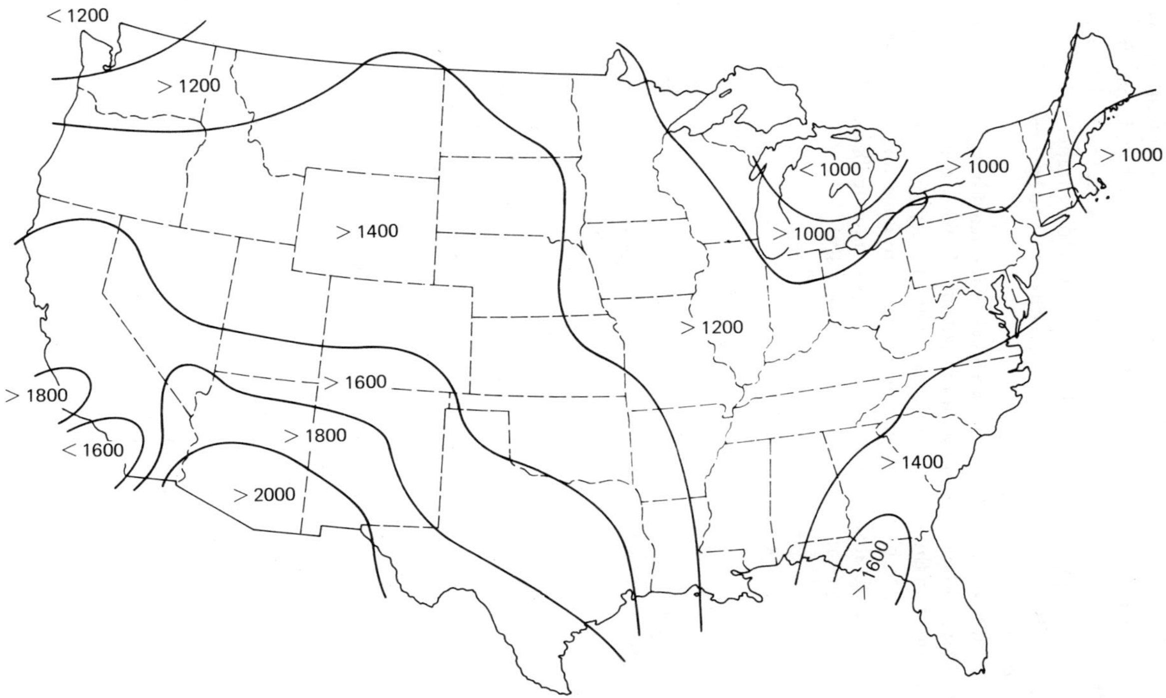

Figure 4.1.2 Average solar energy availability (Btu/ft^2 per day). (*Source:* U.S. Department of Energy.)

BOX 4.1.2 COLLECTOR DESIGN

Most solar collectors depend on the *greenhouse effect* to capture heat. Many materials, including glass and many plastics, will allow visible light to pass through readily. But when that visible light is converted to heat energy, the heat cannot be reradiated through the glass. This can cause heat to be trapped. If you have ever left a car with its windows rolled up on a sunny day, you have certainly noticed the greenhouse effect in action.

A typical solar collector is simply a shallow insulated box with glass on top and a dark-painted *absorber plate* on the bottom. Fluid (either air or a liquid) is circulated through the box or behind the plate to carry away some of the heat that has been trapped by the greenhouse effect. In air collectors, the air usually flows under the dark absorbing surface, while most liquid collectors have channels embedded in the absorber plate (see Figure 4.1.4).

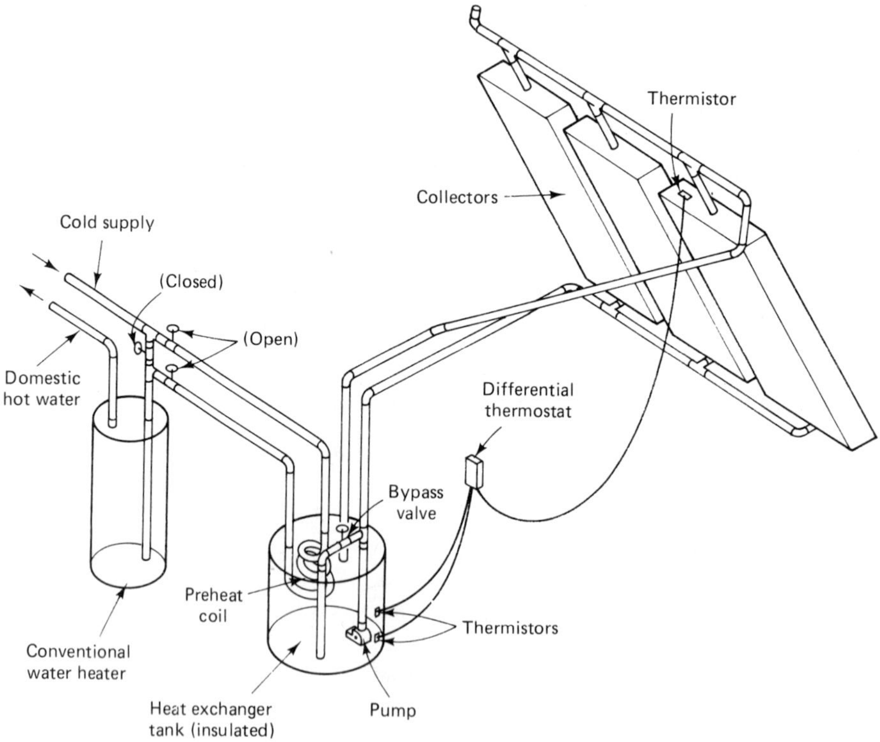

Figure 4.1.3 Solar hot water system. [Reproduced with permission from *The Solar Survey* (Butte, MT: National Center for Appropriate Technology, 1979).]

When the sun shines on the solar collector, the heat-exchange fluid in the collector is heated to a high temperature—as high as 150 to 180 °F (see Box 4.1.2). A control called a *differential thermostat* compares the temperatures in the collector and in the water tank. When the fluid in the collector is hotter than the water in the tank, the control turns on a pump which drives the heated fluid from the collector down into the water tank, where it gives off some of its heat. Meanwhile, the cooler fluid which had been in the tank moves up into the collector to be heated (see Figure 4.1.3). At night and on cloudy days, the differential thermostat keeps the pump shut off, preventing the system from pumping heated fluid from the water tank into the cool collector.

Heat storage and backup heat. All solar heating technologies depend on storage, because the sun shines less than half the time. Most of the time, the sun is on the other side of the earth or blocked by clouds. So solar heating systems need to store heat for use during those times when sunlight is not directly available. In a solar water heater, the water itself provides the necessary storage. Because of its high *heat capacity* (see Box 4.1.3), the water can store large quan-

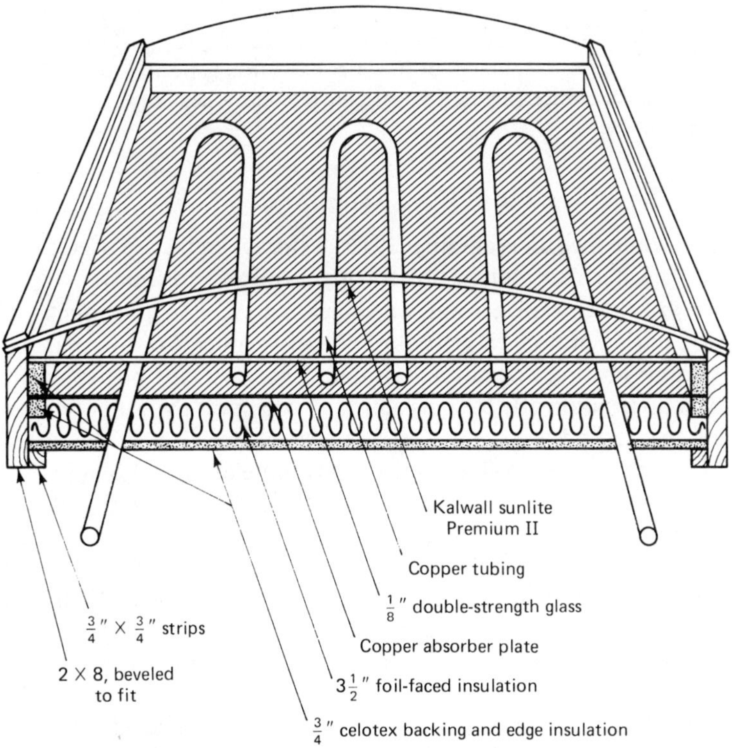

Figure 4.1.4 Design of a solar collector. [Reproduced with permission from *The Solar Survey* (Butte, MT: National Center for Appropriate Technology, 1979).]

BOX 4.1.3 SPECIFIC HEAT AND HEAT CAPACITY

As everyone knows, when materials absorb heat energy, they increase in temperature. The relationship between a material's mass, the amount of energy it absorbs, and its increase in temperature, is called the *specific heat* of the material. For example, water has a specific heat of 1. This means that 1 kilocalorie of energy will raise the temperature of 1 kilogram of water by 1 degree Celsius. In the British system, water also has a specific heat of 1 because 1 Btu of heat energy will raise the temperature of 1 pound of water by 1 degree Fahrenheit.

The specific heat of water is fairly high compared to that of most other materials. For example, only 0.20 kcal (Btu) is required to raise the temperature of a pound of concrete by 1 degree Celsius (Fahrenheit). So concrete has a specific heat of only 0.20.

In many instances, volume rather than weight is the most important consideration. For example, the volume of a concrete wall or floor slab can be computed much more easily than its weight. *Heat capacity* refers to the relative amount of heat that can be absorbed in a given volume of a material. For example, 1 cubic foot of water requires 62.4 Btu to raise its temperature by 1 degree Fahrenheit, whereas 1 cubic foot of concrete requires only 22 Btu to produce the same temperature change. So a cubic foot of water has a *heat capacity* of 62.4 Btu per degree Fahrenheit, while 1 cubic foot of concrete has a heat capacity of 22 Btu per degree Fahrenheit.

tities of heat while the sun is shining. If large storage tanks are used (80 or 100 gal capacity), water can remain hot for a period of 1 to 2 days.

Even with large tanks, solar water-heating systems lose much of their effectiveness during prolonged periods of cloudy weather. For this reason, some kind of backup heating system is desirable. The most obvious solution is to provide backup heat directly to the water tank whenever the water temperature falls below the desired level. But tankless heaters, which heat the water as it passes through the hot water lines, are actually more efficient because they provide the backup heat only for the water that is actually being used.

Water heating currently consumes 2.13 quads of nonrenewable energy per year in the United States. If solar water heaters become more widely used, this figure can be reduced by at least two-thirds. But because of high initial costs, solar water-heating systems have not caught on rapidly in the United States. Even generous tax credits have failed to kindle significant interest among the American public. In other countries, though, interest has been much higher. In Japan, several million solar water heaters have been installed. In Israel, only solar water-heating systems may be sold. As running hot water becomes more common in less developed countries—which are clustered in the tropical regions where solar water heating is most effective—solar water heating is likely to be the preferred method.

4.1.2 Solar Space Heating

Besides heating water, solar energy is widely used to provide space heating for houses and other buildings. In fact, this is the main use of solar energy in the United States.

Two basic approaches are used in solar space heating. *Active* solar heating systems work on the same principles as the water-heating system described in the preceding section. *Passive* systems have fewer moving parts and rely on architectural design to collect, store, and distribute solar energy. Technically, a passive solar heating system requires no input of outside energy in order to function, whereas an active heating system requires outside energy. A third type of solar heating system, a *hybrid* system, does not require outside energy in order to function, but it uses fans or pumps to enhance its performance; thus it is a kind of hybrid between active and passive systems.

Both types of space-heating systems are widely used in all parts of Europe and North America. Passive systems are generally regarded as the best choice for new construction because of their greater efficiency and lower cost, while active systems are most appropriate for incorporation into existing structures.

Active space heating. Although active systems work on the same principle as water-heating systems, they are somewhat more complex because of the addition of a system to control the distribution of heat to the building being heated. Collectors are usually mounted on the roof with a differential thermostat to control the movement of the heat-exchange fluid, which can be either air or a liquid.

In air systems, the heated air can either be blown directly into the building or routed through a storage bin filled with fist-size rocks. If the building needs

BOX 4.1.4 EXTERNAL POWER REQUIREMENTS

All active solar systems require external power to move the heat-exchange fluid from the collector to the storage, and vice versa. Water is much heavier than air, so more energy is required to move a cubic foot of water than to move a cubic foot of air. However, the heat capacity of 1 ft^3 of water is 62.4 Btu per degree Fahrenheit, while the heat capacity of 1 ft^3 of water is only 0.018 Btu per degree Fahrenheit. Over 3000 ft^3 of air must be moved to carry the amount of heat that could be transported by 1 ft^3 of water. Thus an air system requires more external energy than a water system.

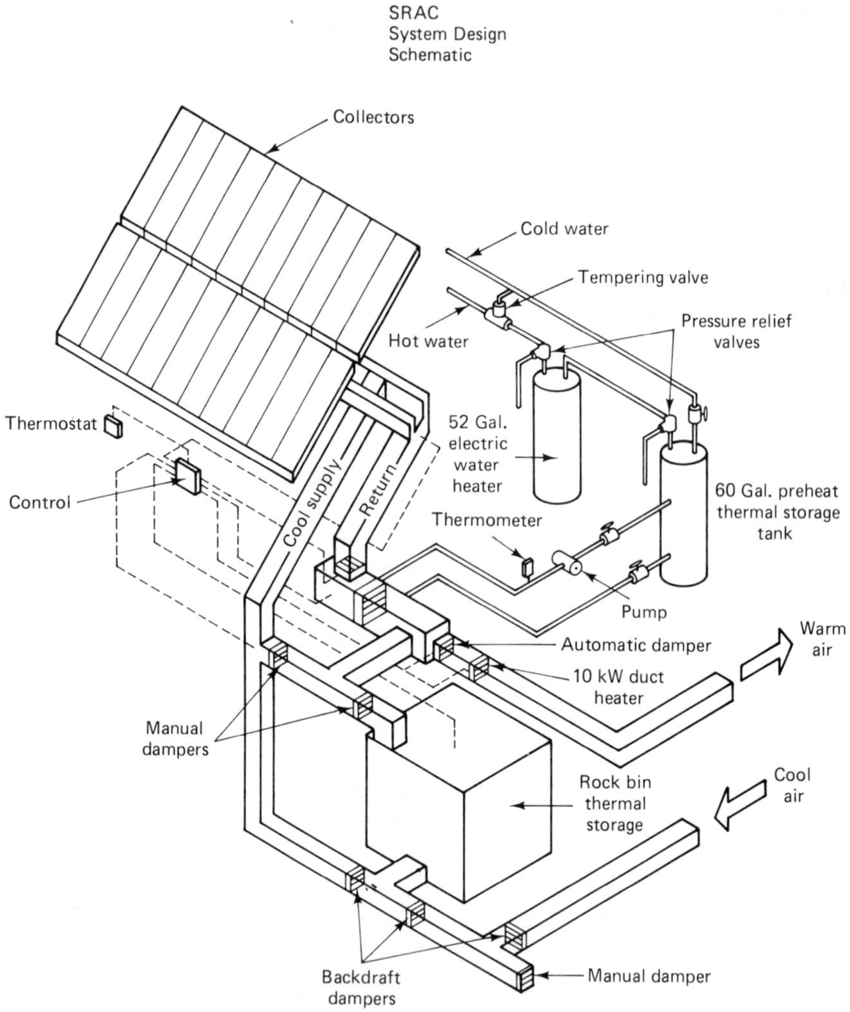

Figure 4.1.5 Active solar heating system. [Reproduced with permission from *The Solar Survey* (Butte, MT: National Center for Appropriate Technology, 1979).]

heat at a time when the collectors are cold, heat can be drawn out of the rock storage. Air systems do not require freeze protection and can easily be integrated with forced-air heating systems. They are also easier to install than liquid systems. A minor leak in an air system will cause only a small reduction in the system's efficiency, whereas a leak in a liquid system can create a major problem.

Liquid systems are generally used in buildings that have hot water or steam heating systems already in place, although they can also be used in conjunction with forced-air heating systems. Like water-heating systems, liquid-based space-heating systems require freeze protection and skilled installation. Also, because of the high operating temperatures of steam and hot water heating systems, liquid systems can be inefficient. However, liquid systems require little external energy to circulate the heat-exchange fluid, whereas forced-air systems require more powerful fans because of the low specific heat of the air.

Passive systems. As new buildings are constructed, they can easily be designed to take advantage of the sun for heating. By simply placing most windows on the south side, a home builder can provide a reasonable amount of solar heat to the dwelling. By including thermal mass and systems for distributing the collected solar heat, the builder can easily reduce the house's heating requirements by more than half.

BOX 4.1.5 COLLECTOR TILT ANGLES

As Figure 4.1.9 indicates, solar collectors function differently at different tilt angles. For example, a horizontal flat collector absorbs much more solar energy in summer than in winter. Thus it is not a good choice for space heating or water heating. However, a horizontal collector could be used effectively to power an absorption-type air conditioner.

A slanted collector collects a large amount of sunlight year-round. It is a good choice for water heating, where heat must be provided in all seasons of the year. But for space heating, vertical surfaces are most appropriate because they collect the most heat in the winter, the time of greatest need. In cold climates, vertical collectors (such as south-facing windows) can also take advantage of sunlight reflected from the winter snow cover—a source of sunlight unavailable to slanted or horizontal collectors. This extra reflected heat can amount to as much as one-fourth of the sunlight that falls directly on the vertical surface.

A typical house has both slanted (roof) and vertical (walls) surfaces. If the pitch of the roof faces south, it can provide an ideal surface for a solar water-heating system. The south-facing wall of a house, on the other hand, is a perfect location for a passive solar heating system such as a large group of windows or an attached greenhouse.

Direct-Gain Systems. The simplest solar heating systems are simply south-facing windows. In fact, a south-facing window is the most efficient type of solar collector for space heating. But a conventional house with large south-facing windows will experience severe temperature fluctuations, with some overheating on sunny fall and spring days. For this reason, *thermal mass* is usually incorporated in the design of a passive solar house. A concrete floor, a large brick fireplace, or barrels of water can absorb solar energy and help to prevent

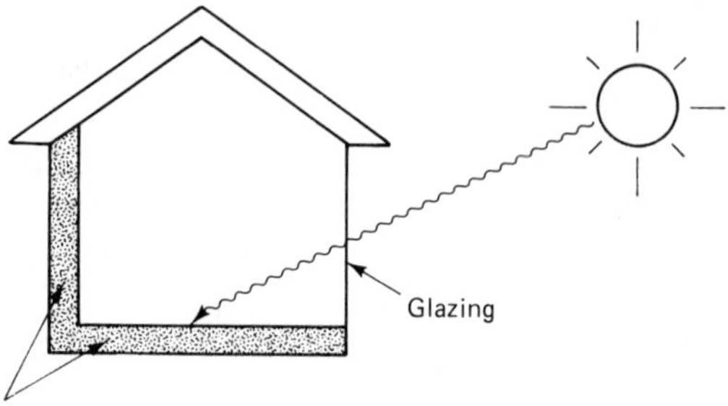

Figure 4.1.6 Direct-gain solar heating system.

the dwelling space from overheating. Other difficulties with south-facing windows are somewhat more difficult to resolve: on sunny days, the glare from large south-facing windows can be considerable. Fabrics can fade and weaken when exposed to large amounts of sunlight, and large areas of glass sometimes cause a loss of privacy.

Indirect-Gain Systems. To prevent some of the problems associated with large south-facing windows, many passive solar designs have large stone, brick, or concrete walls, or walls built of water-filled containers, directly inside the large south-facing glass surfaces. This reduces temperature fluctuations and helps to prevent overheating. Sometimes, vents are located in the top and bottom portions of the wall to permit circulation of the heated air which collects between the wall and the glass surface. A wall with these vents is called a *Trombe wall,* named after a French solar designer, Felix Trombe.

Attached Greenhouses. Attached solar greenhouses can also provide heat to, or at least reduce heat losses from, all kinds of buildings. Anyone who has worked in a greenhouse has surely noticed that on a sunny day, large amounts of excess heat can build up. This heat is usually vented to the outside. But in an attached solar greenhouse, this excess heat is used in two ways. Some of the heat is absorbed by the thermal mass which is placed inside the greenhouse. Planting beds, concrete floors and walls, and water barrels can all provide thermal mass. This stored heat will keep the greenhouse warm at night. The remaining heat is transferred to the attached building, except on very warm days, when it can be vented to the outside.

Combined Systems. Passive solar heating systems can be combined in a number of ways. For example, a house may have a Trombe wall and south-facing windows, perhaps with an attached greenhouse along one section of the south wall. These combined systems, if they are well designed, can provide steady, efficient solar heating in most climates.

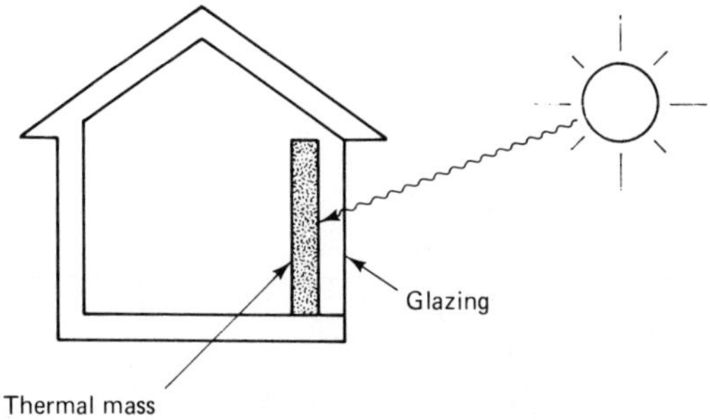

Figure 4.1.7 Indirect-gain solar heating system.

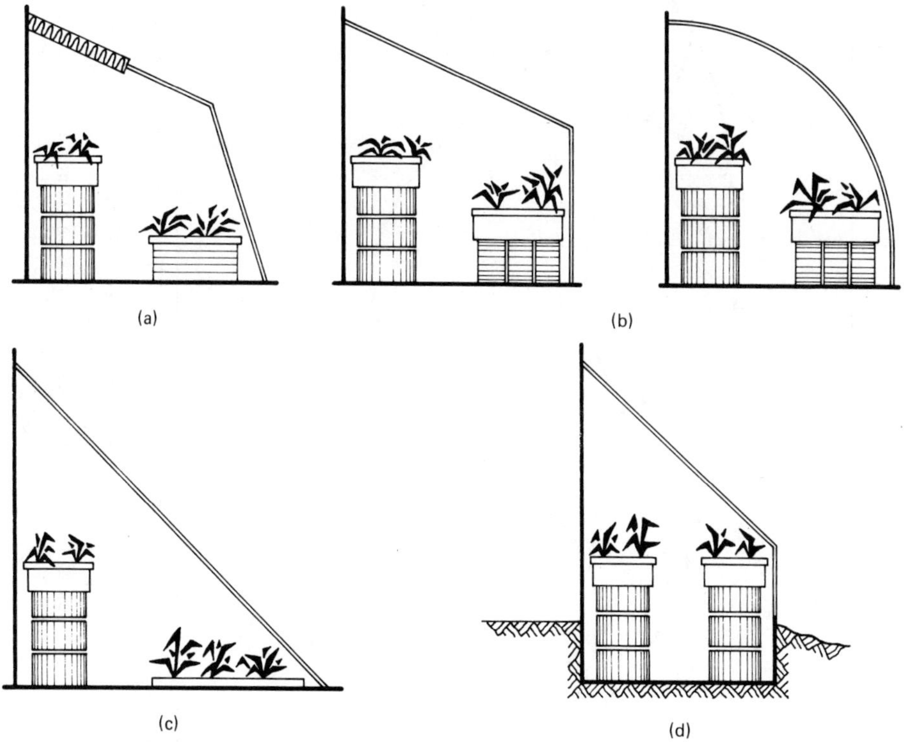

Figure 4.1.8 Attached greenhouse systems. (a) Good for sunny winters and for areas with hot summers. (b) Good for areas with cloudy winters and areas with winter snow cover south of the greenhouse. May need some type of shading in areas with hot summers. (c) Good for areas with cloudy winters and little snow. May need some type of shading in areas with hot summers. (d) Good for south-sloping sites or flat sites to cover part of second story of house. Large, steeply angled upper glazing can be good for heat collection. May need some type of shading in areas with hot summers. [Reproduced with permission from *Low Cost Passive Solar Greenhouses* (Butte, MT: National Center for Appropriate Technology, 1980).]

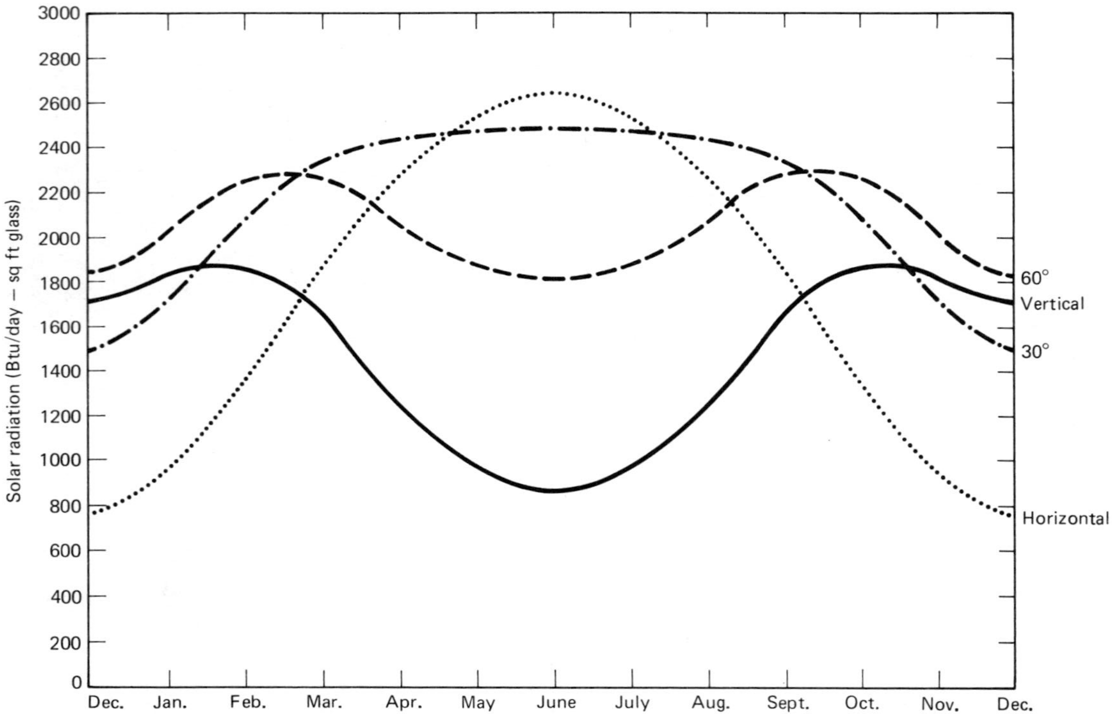

Figure 4.1.9 Collector tilt angles and solar gain. This graph represents clear day solar radiation values, on the surfaces indicated, for 40° north latitude. (Reprinted from *The Passive Solar Energy Book* © 1979 by Edward Mazria. Permission granted by Rodale Press, Inc., Emmaus, PA 18049.)

BOX 4.1.6 MOVABLE INSULATION

The importance of a well-insulated house is becoming more apparent every day. In the case of a passive solar house, excellent insulation is an essential feature of the heating system. Glass transmits solar energy very efficiently, but it is a poor insulator. Therefore, to prevent excessive heat loss at night, most designers of passive solar buildings provide for some kind of movable insulation to cover the glass areas at night. Heavy, quilted window shades, polystyrene panels, and even polystyrene beads blown between layers of double windows, all can provide substantial amounts of insulation at night. And in any house, whether heated by solar energy or by conventional means, good insulation in the walls, ceiling, and floors or basement is well worth the money and effort required to install it.

Overview of Passive Solar Heating. Passive solar heating is quite inexpensive and is technically less complex than active solar water or space heating. For these reasons, it is rapidly gaining popularity in the United States and Europe. For several years, the city of Davis, California, has required all new buildings to incorporate passive solar heating features. Many architecture schools offer courses in passive solar design. Solar greenhouse kits and prefabricated solar house kits are widely available. Of all the many solar space-heating technologies, passive solar space heating is the most solidly established.

BOX 4.1.7 BASIC SOLAR DESIGN PRINCIPLES

1. Reduce unnecessary heat loss.
   a. Put heat barriers and insulation in walls, ceilings, and floors.
   b. Use movable insulation on all large glass surfaces.
2. Optimize solar energy input.
   a. If possible, orient the building to provide a large south-facing surface.
   b. Use tilted surfaces for water-heating collectors and vertical surfaces for space-heating collectors.
3. Use thermal mass.
   a. Prevent overheating during the day.
   b. Store excess heat for use at night.
4. Circulate the heat evenly throughout the dwelling space; use natural circulation whenever possible.
5. Use landscaping.
   a. Locate trees to provide shade in summer without blocking the winter sun.
   b. Use evergreens and shrubs to block north winds.

4.2 DIRECT SOLAR: PHOTOVOLTAIC CELLS

Photovoltaic cells seem almost too good to be true. They produce electricity with no moving parts, without consuming any nonrenewable fuel, and without emitting any pollutants. They can be located anywhere the sun shines, including very remote installations and spacecraft. And they can be made of silicon—a very abundant element found in common sand. With all these advantages, it may seem surprising that solar cells are not already in use everywhere.

There are, however, two major barriers to widespread use of solar cells. First, the technology required to manufacture the cells is very complex. Very high temperatures must be maintained within a narrow range for long periods of time. As a result, the cost of photovoltaic cells is quite high: over twice as high as the most expensive nuclear power plants.

The economics of solar cells is not as bad as it was 10 years ago, when photovoltaics cost over 100 times as much as nuclear power plants for the same amount of power. Since then, prices have dropped enough that photovoltaics are competitive with diesel electric generation in many parts of the Third World. Proponents of photovoltaic cells report significant progress in simplifying the manufacturing process and in reducing costs. It seems quite possible that photovoltaically produced electricity will be competitive with conventionally generated electricity within a few years.

4.2.1 Some Theory

Electrons and electron shells. To understand how a photovoltaic cell functions, it is necessary to understand a bit about the structure of the atom. Basically, an atom can be compared to a circular amphitheater, with rings of seats. Each ring of seats corresponds to an electron "shell," and each seat is an electron "hole"—a space that can hold exactly one electron. The inner shell of an atom has only two seats or electron "holes." The second and third shells have eight spaces each. After that, the situation becomes more complex, although the principle is the same. Electrons will fill the inner shell first, then fill the second shell, and so forth. Only the electrons and electron holes in the outermost partially filled shell are available for bonding with electrons or electron shells from other atoms.

Since a hydrogen atom has only one electron, that electron occupies one of the two electron holes in the first shell. The other electron hole is empty. Thus hydrogen has one available electron and one available electron hole (see Figure 4.2.1). Silicon, on the other hand, has a total of 14 electrons. This means that the first and second electron shells are completely filled (two electrons in the first shell and eight in the second), and the third shell contains four electrons and four electron "holes." Thus a silicon atom has four electrons and four electron holes available for combining with other atoms.

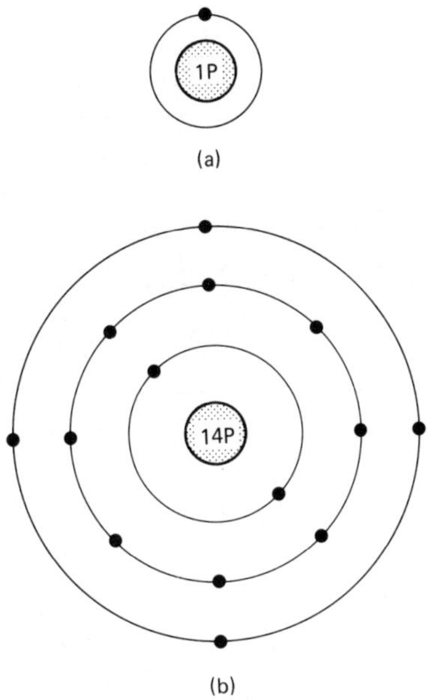

Figure 4.2.1 Atomic structure of hydrogen and silicon: (a) hydrogen atom: nucleus has 1 proton, first shell has 1 electron; (b) silicon atom: nucleus has 14 protons and 14 neutrons, first shell has 2 electrons, second shell has 8 electrons, third shell has 4 electrons.

Electrons can migrate from atom to atom depending on the availability of electron holes and the electrical conductivity of the material. In a similar way, electron holes can be transferred from atom to atom through a somewhat more complex process. (In fact, the electron holes do not move, but the effect is very much the same.) The movement of electrons and of electron holes is the basis of electricity in general, and of the photovoltaic process in particular. Readers should refer to a text in modern physics or electronics for a more detailed understanding of this electron motion.

Impurities. In a pure silicon crystal, the number of electrons exactly matches the number of electron holes, and there are no free electrons or electron holes. But when an impurity is introduced, this balance can be upset. For example, arsenic has a valence of 5: it has five electrons available for bonding and only three electron holes. When an arsenic atom replaces a silicon atom in a pure silicon crystal, the number of available electrons exceeds the number of available electron holes. The extra electrons can move through the crystal structure, while all the electron holes are locked in place. This material can be used in the lower layer (the *N* layer) of a cell. The opposite takes place when the crystal is contaminated by a material with a valence of 3, such as boron (three available electrons and five available holes). In this case, the electrons are locked in place by the crystalline structure and the electron holes are free to move. This material will be used in the upper layer (*P* layer) of the cell.

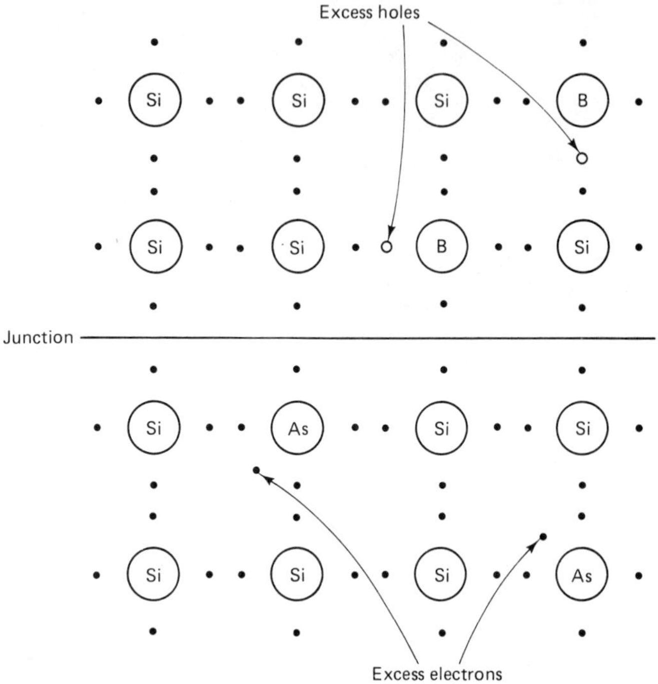

Figure 4.2.2 Chemistry of a photovoltaic cell: Si, silicon; As, arsenic; B, boron.

4.2.2 Structure of a Cell

Several designs exist for photovoltaic cells. One very common type of photovoltaic cell is a crystalline "sandwich" composed of two layers. The bottom layer is made of a material with movable electrons and immovable electron holes, while the top layer has movable electron holes and fixed electrons (see Box 4.2.1). The cell also has pieces of metal called *electrodes* attached at the top and at the bottom.

BOX 4.2.1 MANUFACTURE OF A CELL

At present, the manufacture of a photovoltaic cell is an expensive and inefficient process. Several different processes can be used; the one described here is one of the most common. First, the silicon must be melted and chemically purified, since impure crystals will not function properly. Next, the silicon is "doped" with a small amount of arsenic (about 1 part per million). A tiny "seed" crystal of silicon is introduced into the molten silicon–arsenic mixture, and a cylindrical crystal is "grown." This crystal must be cut into thin cross-sectional slices. A diamond saw is used for this process, which wastes approximately half of the silicon. Then each slice is placed into a furnace filled with boron trichloride vapor. The boron combines with the outer portion of the wafer, producing a boron-contaminated layer around the entire outside of the wafer.

Next, the boron-contaminated layer is removed from the bottom and sides of the wafer, leaving a layer on top of the cell. (This wastes a bit more silicon.) Finally, metal electrodes are attached at the top and bottom of the cell (see Figure 4.2.3).

This process is quite wasteful, and new processes are being developed to make the process of manufacturing solar cells more economical. One promising technique is the production of a large, thin "ribbon-shaped" crystal rather than a cylindrical crystal. Other processes focus on the idea of producing lower-efficiency but cheaper cells based on noncrystalline silicon. Still other techniques call for the deposition of millions of tiny silicon crystals on a metal substrate. None of these new techniques has been conclusively demonstrated, but the prospects for improved manufacturing techniques are promising.

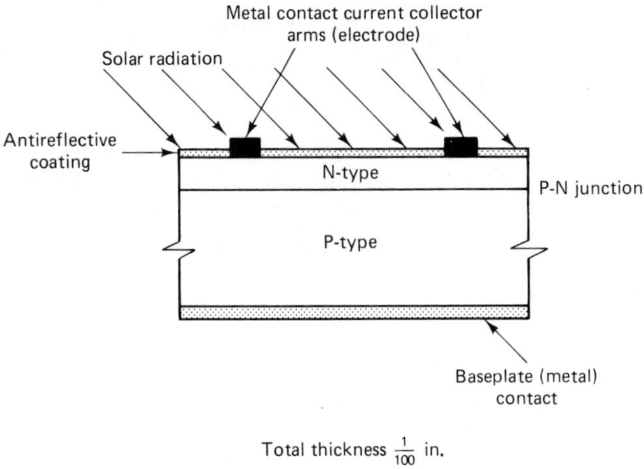

Figure 4.2.3 Solar cell cross section.

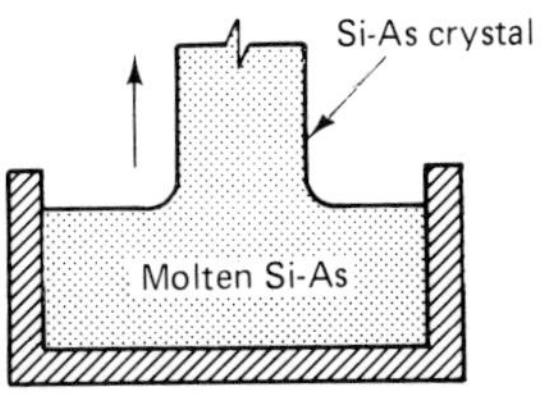

Figure 4.2.4 "Growing" silicon crystal.

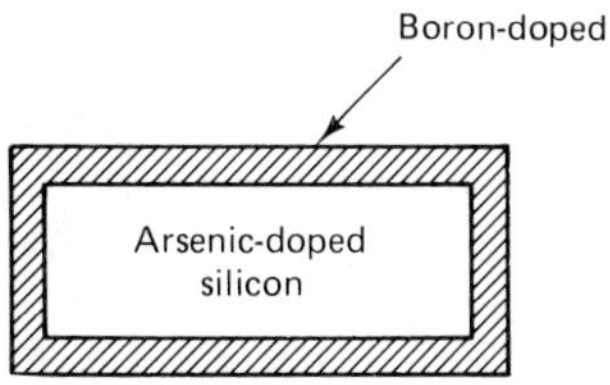

Figure 4.2.5 Cross section of a doped silicon wafer.

4.2.3 Formation of the Barrier Layer

As the free electrons from the lower layer migrate and diffuse through the cell, they occasionally cross the boundary into the upper layer, where they immediately combine with the available electron holes in the upper layer. This leaves a negatively charged particle (ion) in the upper layer and a positively charged ion in the lower layer. These ions are attracted to each other, just as the opposite poles of a magnet are attracted to each other, so they form *bonded pairs* across the boundary. As a large number of bonded pairs form across the boundary, they present a barrier to the free movement of electrons across the boundary. The actual physics involved is fairly complex. Electrons can cross the boundary, but they require extra energy to get past the barrier. This barrier, as we shall see, is essential to the proper functioning of a photovoltaic cell.
to the boundary and formed a barrier layer of bonded pairs, the concentration of electrons and electron holes is no longer uniform throughout the cell. In the top layer, there is an excess of electron holes. In the newly formed barrier layer, electrons and electron holes are equally concentrated. In the bottom layer, there is an excess of electrons (see Figure 4.2.6).

4.2.4 Production of Electricity

When a particle of light (a photon) interacts with the material in either layer of the cell, it produces a movable electron and a movable electron hole. If this happens in the top layer, the electron hole is repelled by the excess of

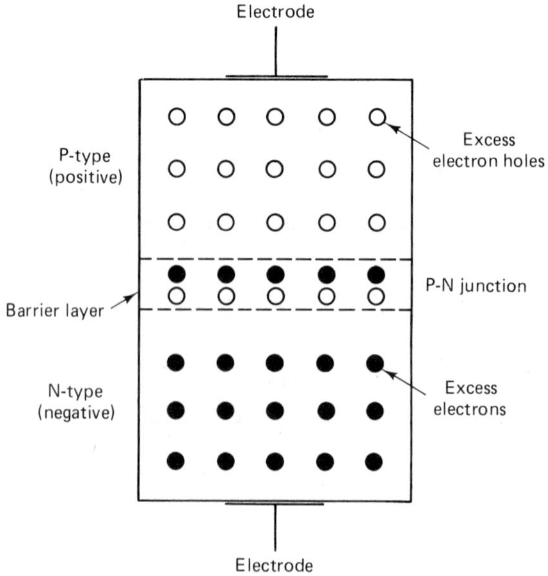

Figure 4.2.6 P-N junction.

electron holes in the vicinity, just as two similar poles of a magnet repel each other. Since the electron hole cannot cross the barrier layer to reach the bottom layer, it migrates to the upper electrode in order to escape the excess of positive charges in the top layer. If the interaction occurs in the bottom layer, the electron migrates to the lower electrode. After several interactions, a positive charge builds up on the top electrode and a negative charge builds up on the bottom electrode. This difference in charge produces an electric potential difference (voltage), which can be used to produce electric current.

BOX 4.2.2 EFFICIENCY AND CONCENTRATION OF LIGHT

Photovoltaic cells use less than one-fourth of the sunlight that strikes them. Combined with the fact that the sun does not shine most of the time, this means that solar cells do not operate at a particularly high efficiency. However, photovoltaic cells can produce electricity from *concentrated* sunlight—in fact, their efficiency tends to improve somewhat as the intensity of the light increases, reaching its highest point at a concentration of 20 to 100 times the normal intensity of sunlight. Since reflective materials are much simpler and less expensive than photovoltaic cells, it makes sense to use reflectors to concentrate sunlight from a large area onto a smaller area of photovoltaic cells. Experiments with aluminized Mylar and other reflective surfaces have produced promising results, although no reflective arrays have yet reached the point of commercial marketability.

4.2.5 Linkage of Photovoltaic Cells

Each photovoltaic cell can provide only a tiny amount of electricity. But cells can be linked together to provide an increased amount of power. If they are connected in series (each top layer of one cell connected to the bottom layer of the next cell), the voltage produced will be increased. If the cells are connected in parallel (each top layer connected to another top layer, each bottom layer connected to another bottom layer), the voltage will not increase, but the amount of electric current supplied will increase. By connecting several cells into a single *array,* and then by connecting several arrays together into a larger group of cells, it is possible to produce electric potential and current suitable for most domestic and commercial uses.

For most domestic uses, the space on a roof of a house provides sufficient area to power the house with photovoltaic cells. But for industrial applications, larger arrays are required. Suggested approaches include huge arrays located in desert areas, banks of collectors located in specially designated collection areas, and orbiting satellites carrying large arrays of photovoltaic cells and beaming their electricity back to Earth in the form of microwaves. Most theorists believe that the modular nature of photovoltaic cells favors a decentralized distribution of cells rather than large-scale, centralized plans like the solar satellite (see Section 4.3). At present, however, photovoltaic cells are still in the developmental stage, so it is difficult to predict how they will be used in the future.

BOX 4.2.3 THE PROBLEM OF STORAGE

As with most solar technologies, a major technical problem with photovoltaics is the need for storage. Since electricity can be produced only while the sun is shining, some form of storage is needed to provide electric power at other times. In some applications, such as pumping water or oil, storage is not needed. But if photovoltaics are to contribute substantially to the world's electrical supply, reliable means of storage must be developed. Battery storage, pumped hydro storage, and chemical storage are all being considered.

One novel approach, which appears especially promising when used in conjunction with photovoltaic cells, is the electrolytic production of hydrogen from water. Hydrogen can be used as a fuel for highly efficient fuel cells first developed for the space program, and the waste heat from fuel cells can be used for other purposes such as water heating.

The problem of storage has not been completely solved for any solar technologies. But in spite of these problems, photovoltaic cells clearly have important uses in today's energy picture.

4.2.6 Uses of Photovoltaics

Because of the high cost of photovoltaic cells, they are used primarily in remote areas without centralized electrical systems. They have been used to power isolated lighthouses, pumps, and other mechanical devices. These applications are feasible only when the cost of batteries or diesel generating equipment exceeds the cost of photovoltaic cells. But as the cost of photovoltaic cells declines, their range of uses will grow. The next probable area of application will be to provide electricity for telecommunications equipment in remote locations, and to provide electric power for isolated desert and tropical villages.

Because of their decentralized nature, photovoltaic cells are also appropriate for incorporation into utility grids, but costs will have to drop significantly before photovoltaic cells can compete with electricity generated at centralized power stations.

4.2.7 Prospects for Photovoltaics

The U.S. Department of Energy has set goals for the development of cost-effective photovoltaic cells. The revised goal for 1986 is a cost of \$4 per peak watt, up from an earlier goal of 70 cents. (A peak watt is the amount of equipment needed to generate 1 W of electricity at optimum collection on a sunny day.) Eventually, costs are expected to drop below $1 per peak watt, which is substantially lower than the capital cost of coal or nuclear power plants and would make photovoltaic cells quite attractive economically. If these goals are met, photovoltaic cells will probably be an important source of electricity by the end of the century.

4.3 DIRECT SOLAR: LARGE-SCALE SYSTEMS

Some energy analysts believe that solar energy is best suited for small-scale, decentralized purposes such as space and water heating, and low-power production of electricity for domestic use. But several schemes have also been proposed for large-scale generation of electric power. At present these are not as well developed as the applications described in earlier sections. But significant quantities of research funding are being applied to some of these technologies, and they may yield significant results in the future.

Although direct conversion of sunlight to electricity is fairly inefficient (around 10%), conversion to heat can be more efficient—as high as 95% in extremely high-technology designs. This heat can be converted to electricity with an efficiency near 40%, resulting in a sunlight-to-electricity efficiency of as much as 35%. This is much higher than direct conversion and comparable to thermal power plant efficiency. Technologies to perform this conversion have been proposed, but at present they are quite complex and extremely expensive.

4.3.1 Technologies for Central Power Generation

Energy farm. One plan for central solar electric generation, proposed by the Meinels of the University of Arizona, calls for thermal collectors to be distributed over a large area. For a 250-MW plant (one-fourth the size of a typical thermal plant), approximately 2 square miles of desert land would be required. Liquid sodium is pumped through evacuated collector tubes and transported back to the central power plant, where its heat is used to produce steam for a conventional steam turbine.

In the energy farm concept, temperatures near 550 °C (1000 °F) must be achieved in order for the system to generate electricity with a reasonable degree of efficiency. But such temperatures are impossible with conventional flat-plate collectors such as those used for domestic space and water heating. To overcome this problem, the collectors proposed for this system use curved reflectors to focus sunlight on the collecting fluid.

To keep the heat from escaping back to the environment, two methods are used. First, the collecting medium is contained in a cylinder with a *selective absorption surface.* This surface absorbs solar energy readily but radiates only a tiny portion of its heat energy back to the environment. As a second measure to prevent the escape of heat, the collecting fluid is contained in an evacuated

Figure 4.3.1 Solar furnace located near Odeillo in the Pyrenees Mountains of southern France. In the foreground, an array of 63 mirrors (heliostats), reflects sunlight onto the curved mirror surface of the office building in the background. This in turn focuses the sunlight on an aperture in the center tower, where temperatures of 7000 °F can be produced. (*Source:* U.S. Department of Energy.)

cylinder, working on the same principle as a Thermos bottle to prevent the escape of heat. When these two methods are combined, very high operating temperatures can be achieved. Temperatures of 550 °F are attainable in principle.

Power tower. In a second proposal for central power generating, it is the sunlight, rather than a heat-transfer fluid, which is transmitted to the power generating station. This "power tower" plan calls for arrays of mirrors in the desert, oriented so that the sunlight from a very large area is focused on a central collecting site, where it is used to heat a boiler, producing steam to drive a turbine generator. The large amounts of concentrated sunlight can produce a high enough temperature to produce electric power with a reasonable degree of efficiency.

Solar satellite. A third proposed approach to central electric power generating calls for huge arrays of photovoltaic cells to be mounted in a satellite in stationary orbit around the earth. In one visionary scheme, the cells are to

Figure 4.3.2 The solar power tower shown here has a field of 222 heliostat arrays (foreground), each containing 25 A-ft^2 mirror facets, which concentrate the sun's rays on solar receivers (boilers) mounted on the 200-ft-high tower. (*Source:* Sandia National Laboratories.)

cover an area as large as Manhattan Island. The intensity of sunlight is much greater outside the atmosphere, solar energy is available virtually all of the time, and overheating is not a problem. Photovoltaics have been used successfully to power spacecraft, so the electrical conversion technology presents few difficulties.

The main design problem for the solar satellite concept is the difficulty of transmitting the electricity back to earth. Most plans call for the electricity to be converted to microwaves and sent to receivers located in remote desert regions. Then the microwaves can be converted back to electricity.

Critics of the solar satellite plan claim that the system is overpriced, inefficient, and even dangerous. Microwaves are believed to cause health problems even at relatively low levels, and the solar satellite plan calls for the transmission of extremely intense beams of microwaves. If the microwave beam were misdirected because of mechanical failure or a collision between the sending station and a meteor, populated areas could be inundated by strong beams of microwave energy.

During the early 1980s, the U.S. Department of Energy downgraded the research status of solar satellite plans. At the moment, no designs are being seriously pursued. So for the immediate future, the solar satellite plan appears to be little more than a technological curiosity.

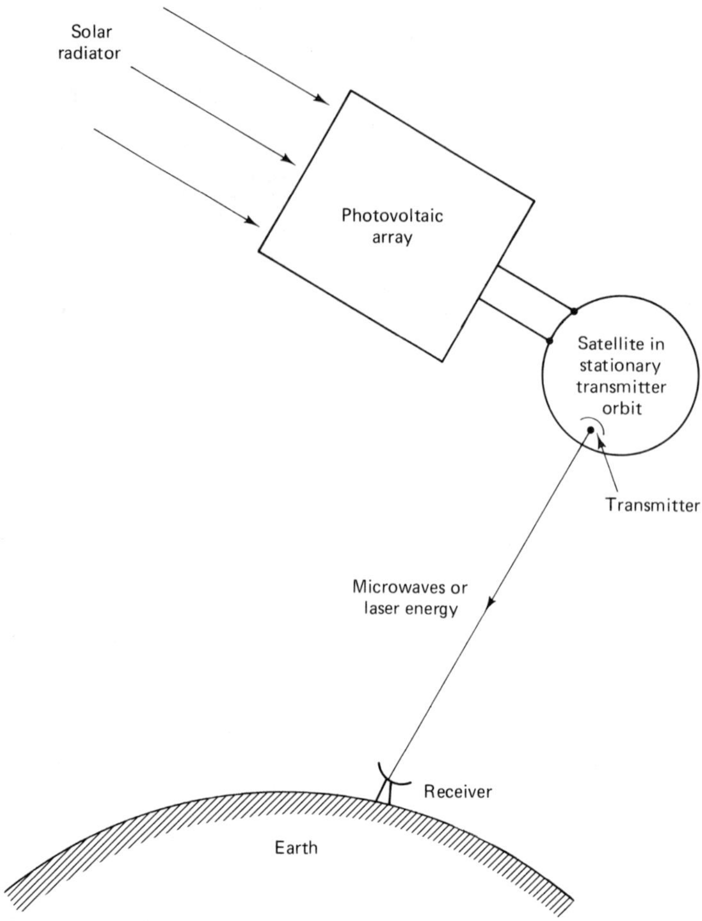

Figure 4.3.3 Solar satellite scheme.

4.3.2 Are Central Solar Technologies Workable?

In all three of these plans, the main difficulty is encountered in transporting solar energy to the central power station. In the energy farm concept, dozens of miles of evacuated silicon tubing must be produced and maintained. The heat-transfer fluid is a highly corrosive material which burns violently on contact with water, so any spills could have fairly serious consequences. In the power tower concept, hundreds of reflectors must be kept in the proper orientation to focus all the sunlight on the central collecting point. This requires sophisticated tracking controls which can keep hundreds of reflectors at the proper orientation to the sun.

Opponents of these central solar power technologies fall into two categories: those who claim that solar energy in general is an unworkable technology, and those who claim that simple, decentralized solar energy systems are better than large, complex central facilities. Opponents of solar energy claim that the systems are inefficient because solar energy is not available at night or during periods of cloudy weather. But supporters of the plans counter that the systems produce the greatest amounts of power when demand for electricity is highest: on hot, sunny days when demand for air conditioning pushes up electric use. Those who favor simpler systems argue that the complexities of the systems make them unworkable, even dangerous because of the use of liquid sodium and intense microwaves. But supporters of the plans claim that some kind of central electric generation is necessary, and that large-scale solar energy systems offer a reasonable alternative to nuclear power plants. They claim that accidents, even those involving escaped sodium, present much less hazard to the public than nuclear accidents.

4.3.3 Prospects for Large-Scale Solar Energy Systems

At present, these technologies are highly speculative, so no reliable cost estimates are available. In general, capital costs can be expected to be high, but operating costs should be extremely low once initial startup problems are solved. The success of these various systems will not be known for a number of years, but these prototype designs demonstrate that solar central power stations are at least technically feasible.

Perhaps the greatest challenge to these plans for central solar electric generation comes not from opponents of solar energy, but from those who believe that central generation of electricity is an inappropriate use of solar energy. Some argue that the decreasing costs of photovoltaic cells will soon make all central electric power stations obsolete, whereas others argue that conservation and a shift away from electric heating will make them unnecessary. Considering the trends in electrical consumption that have developed over the past few years, it must be said that prospects for central solar electric generation are not as good as they appeared 10 years ago. However, a renewed growth in demand for electricity could still bring about a revival of interest in large-scale solar electric systems.

4.4 INDIRECT SOLAR: HYDRO ENERGY

Hydro energy has been used by humanity for thousands of years. Waterways have been used for navigation since prehistoric times. But other uses of water energy also have a long history. Over 2000 years ago, the kinetic energy of moving water was harnessed to provide mechanical power. Since then, the energy of moving water has been continuously used to run sawmills, grease mills, textile mills, cereal mills, and other kinds of machinery designed to provide for basic human comforts and necessities. It is no coincidence that essentially all the world's great population centers are located along seacoasts or major waterways.

Today, hydro power is used primarily for electric power generation. But hydroelectric power is only a recent phenomenon. The world's first hydroelectric power plant was built at Godalming, England, in 1881—just over a century ago. In the United States, the first major hydroelectric installation was constructed at Niagara Falls in 1895 with a rated output of 3.75 MW, certainly an enormous power output at that time. Since then, hydroelectric plants have grown in size and output. There are 64 hydrosites in the world with a design capacity of at least 1000 MW each (Table 4.4.1). Today, the world's largest hydropower scheme is the James Bay project in northern Quebec. When completed, the installation is expected to produce 30,000 MW of electric power.

TABLE 4.4.1 Some Important Hydroelectric Projects

Location	Name of Project	River	Capacity (MW)
Washington	Grand Coulee	Columbia	6,500
Arizona	Glen Canyon	Colorado	1,021
Arizona/Nevada	Hoover	Colorado	1,344
New York	Niagara Falls	Niagara	2,000
Quebec	La Grande complex[a]	La Grande	10,300
Egypt	Aswan	Nile	2,100
Venezuela	Guri	Orinoco	10,060
Brazil	Itaiph complex	Parana	12,600

[a]Under construction.

Hydro energy provides about one-fourth of the world's electricity. In 1982, 3.245 quads (309.2 billion kWh) of hydroelectric energy was produced in the United States. This was nearly 15% of total U.S. electricity production, or slightly more than the share produced by nuclear power plants. In other countries, hydro's share is even larger: more than 80% in Brazil, 45% in India, 40% in Japan, and 15% in the USSR.

For the United States, the U.S. Department of Energy projects a gradual increase in production of electricity from hydro plants. DOE projections are as follows:

1985: 3.3 quads
1990: 3.4 quads
2000: 3.9 quads

It is interesting to note that the projection for 1985 is lower than the actual production in 1982. This suggests that the actual hydro capacity may be somewhat higher than the projections. It should also be observed, though, that annual precipitation makes a significant difference in total electrical production.

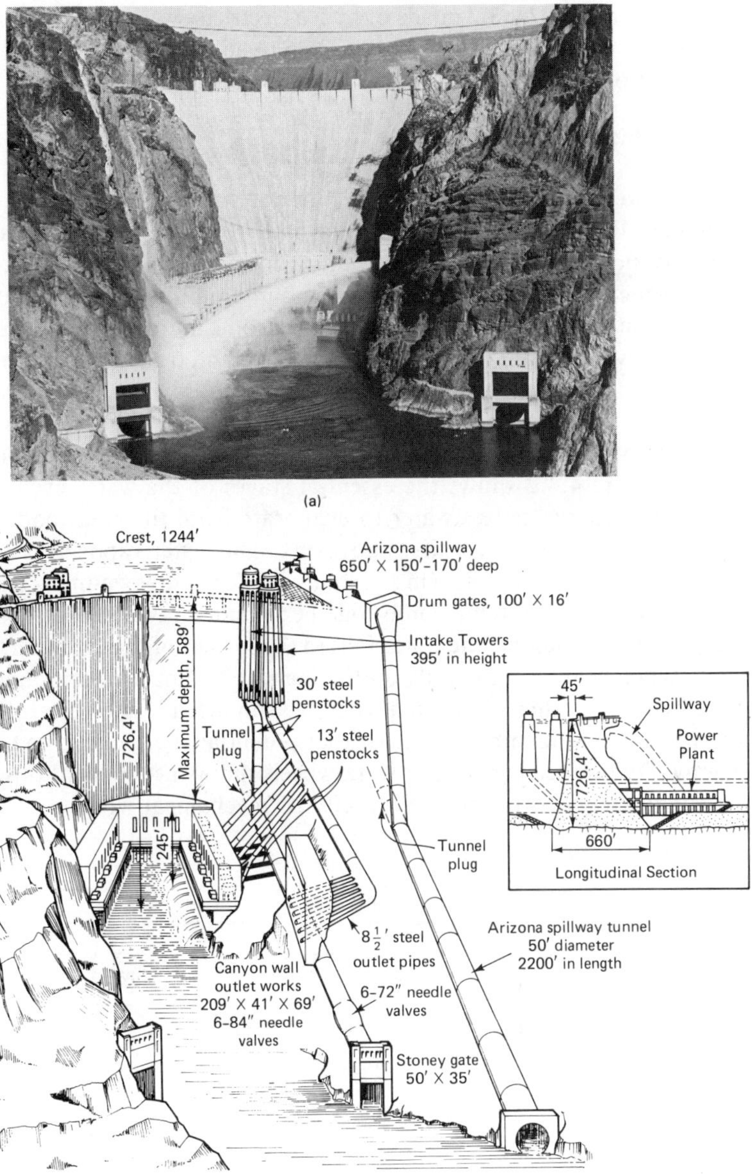

Figure 4.4.1 (a) Hoover Dam, Boulder Canyon Project, Arizona and Nevada. There are a total of 17 generators with a combined capacity of 1344.8 M-W. Hoover Dam is one of many hydroelectric facilities on the Colorado river. Others include Glen Canyon, Davis, and Parker Dams. (b) How Hoover Dam works. (*Source:* Bureau of Reclamation, U.S. Department of the Interior; photographer, E. E. Hertzog.)

For example, in 1974, 3.2 quads of hydro electricity were generated in the United States, but in 1977, the figure fell to 2.35 quads, primarily because of reduced precipitation levels.

4.4.1 The Water Cycle

As we mentioned in Chapter 2, there are two forms of energy, kinetic and potential. Both kinds of energy can be used in hydro power generation. When water is moving, in a flowing stream or in a waterfall, it has kinetic energy. When water is stored at a height, in a mountain pool or behind a dam, it contains potential energy. Both kinds of energy come ultimately from the sun, which lifts water vapor from the surface of the earth and transports it to higher elevations, from which the water runs back down to the ocean. The water cycle is extremely important, not only for its energy value, but also for the continued existence of life on this planet. Without it, life as we know it would not exist today. Figure 4.4.2 shows the essential stages of the water cycle.

Solar energy causes water to evaporate from the great bodies of water: oceans, seas, lakes, reservoirs, and rivers. The water vapor thus formed rises and is carried away by the wind. Clouds are formed when high- and low-pressure atmospheric systems come together. Under favorable atmospheric and temperature conditions, the water falls to the surface of the earth as rain or snow.

When rain or snow falls on the land, it retains a certain amount of the potential energy that it acquired when the sun lifted it into the atmosphere. As the water flows downhill, some of that potential energy is turned into kinetic energy. Hydroelectric plants or waterwheels simply use some of this kinetic and potential energy to produce electricity or mechanical work.

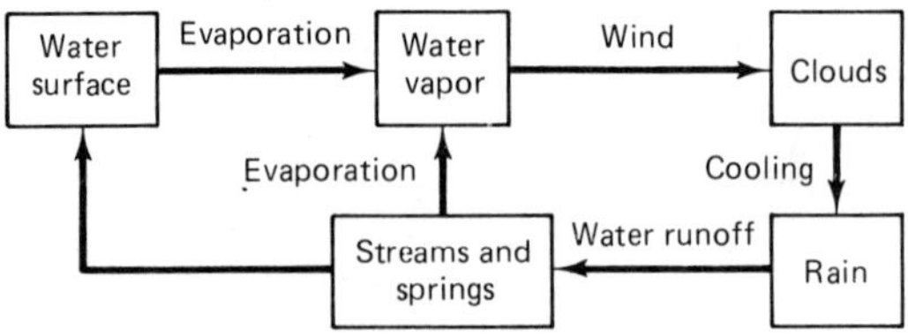

Figure 4.4.2 The water cycle.

4.4.2 Principles of Hydro Power Conversions

As we explained above, when water falls on the earth it contains potential energy which is turned into kinetic energy as it flows toward the sea. This kinetic and potential energy can be harnessed in a variety of ways. Moving water can be used directly to turn waterwheels, or water can be stored behind a dam to produce a reservoir of water to serve as a concentrated source of power. During the Middle Ages and the early parts of the Industrial Revolution, mills and factories were usually located near streams, where they could take advantage of hydro power. The most common conversion systems were mill wheels. Mill wheels are still used today for the direct production of mechanical power.

Mill wheels come in two basic types: overshot and undershot. In an overshot wheel, water from the stream is diverted onto a human-made *sluice,* a wooden trough that carries water from the stream to a point just past the center of the wheel (see Figure 4.4.3). The water falls from the end of the sluice onto the wheel, where it falls into compartments called *buckets.* The weight of the water in the buckets causes the wheel to rotate. As each bucket reaches the bottom of its revolution, the water spills out of it and back into the stream. Overshot wheels generally rotate very slowly, but they provide a large amount of torque for operating machinery. These are the wheels that are most commonly seen in restored mills. But in some regions of the world, overshot wheels still provide the necessary power for functions such as grinding grain.

Undershot wheels operate on a simpler principle. Here the stream is channeled and flows under the wheel, where it pushes against paddles attached to the wheel. This turns the wheel, which can be connected to machinery. Under-

Figure 4.4.3 Overshot wheel., *Source:* Reproduced with permission from *Microhydro: Reviewing an Old Concept* (Butte, MT: National Center for Appropriate Technology, 1979).

shot wheels require a greater stream flow than overshot wheels, and they do not use the flow of the stream as efficiently. But they work well in areas where there is not a sufficient drop in the stream level to construct a sluice for an overshot wheel.

4.4.3 Dams and Reservoirs

Waterwheels are effective devices for converting the motion of flowing water to steady mechanical power. But in many applications, particularly the generation of electricity, demand is not constant, and therefore a steady supply of constant power serves no good purpose. Also, stream flow is rarely constant, even in the most reliable stream. There can be an oversupply of hydro power during rainy periods and a shortage of power during dry seasons. To compensate for these problems, most large hydro installations include a dam and a reservoir. Dam storage is not absolutely essential for hydropower production, but it facilitates power production in three important ways:

1. It creates an artificially large *pressure head* (distance through which the water falls), thereby increasing the output of the system.

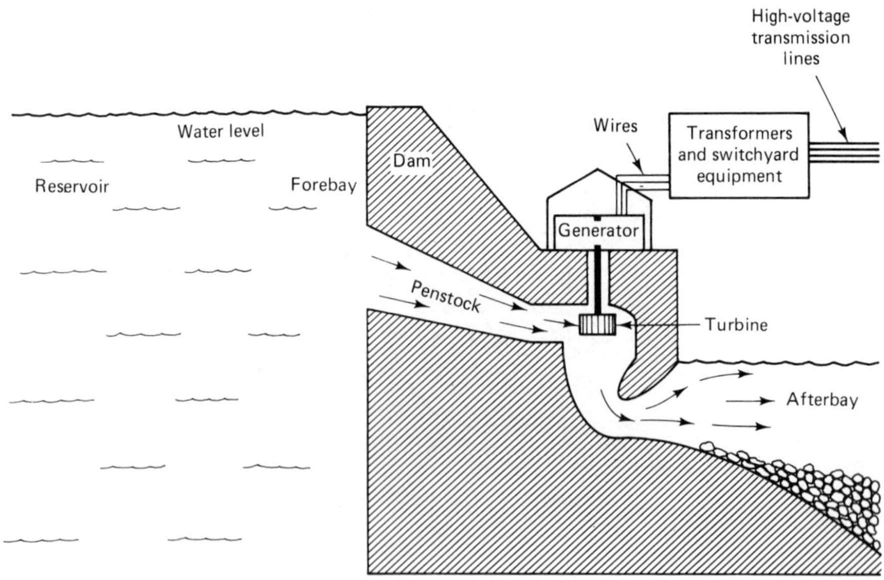

Figure 4.4.4 Hydroelectric site schematic. (*Source:* Adapted from *Hydro Power—Water at Work,* Bureau of Reclamation, U.S. Department of the Interior.)

2. It provides a storage system to maintain the output at an acceptable level during periods of high or low water flow.
3. It allows power supply to be adjusted as demand for power fluctuates.

To generate electricity, water from behind the dam is channeled through either a waterwheel or a turbine, a machine that converts the kinetic energy of the moving water into mechanical rotational energy. This rotary power is then used to turn an electric generator.

4.4.4 Site Selection

The development of hydropower depends on the availability of good sites. The quality of a site depends on the following characteristics:

1. The height of the waterfall, called the *head* (H).
2. The rate at which the water moves, known as Q, also called the *discharge rate.* (In general, a good site should have either a large flow or a large head.)
3. The area's suitability for constructing a dam and a storage reservoir.
4. The site's proximity to energy consumption centers.
5. The probability of flooding agricultural and residential areas when the reservoir is completed.

In general, these conditions are most likely to be met in mountainous or hilly country with large amounts of snow or rainfall.

It can be difficult to find a site where all these characteristics are favorable. Moreover, prime hydropower sites tend to be distributed unevenly through the world. In the Pacific coast area of the United States, good hydro sites are very common, and hydroelectric generation provides 30,000 MW of power—a large portion of the region's electric power. In the Southwest, the Colorado river system provides a large number of very high quality hydro sites, and the southeastern United States is blessed with a number of good hydro sites. But in other regions, such as the Great Plains or the deserts of northern Nevada, good sites are quite hard to find.

Although hydropower can theoretically be renewed forever, an individual hydro site has a limited lifetime. The water coming into the reservoir contains impurities, including substantial amounts of silt. This silt can fill up the reservoir in a few centuries, or, in severe cases, a few decades. This limits the useful lifetime of the site. A new site will then have to be located, or a method will have to be developed for removing the silt from the reservoir. (At present, no such method has been developed.) Thus a site has a maximum life span of about 100 to 400 years under normal operating conditions—and a much shorter life span under unusually bad operating conditions.

Figure 4.4.5 Conventional hydroelectric power in the United States. *Note:* The first number in the parentheses shows developed power capacity and the second number shows the full potential capability of the region identified; units are gigawatts = 1,000,000 kW. Example: The North Pacific region has an installed capacity of 22 GW or 22,000 MW, whereas the full potential is 62 GW or 62,000 MW. For the United States as a whole, the developed power capacity totals 65 GW, and the potential capability totals 180 GW. Average annual hydroelectric production totals 670 GWh. (*Source:* Bureau of Reclamation, U.S. Department of the Interior.)

Figure 4.4.6 Bureau of Reclamation power facilities in operation and under construction in the western United States. (*Source:* Bureau of Reclamation, U.S Department of the Interior.)

4.4.5 Turbine Design

Three mechanical designs are widely used today for converting hydro power to rotational motion. They are (1) the Kaplan turbine, (2) the Francis turbine, and (3) the Pelton wheel. The preferred design in a given situation depends on the head (H) and the flow (Q) at the particular site, because each design is suitable only for a certain range of incoming water pressures.

Both the Kaplan and Francis turbines can handle large flow rates (Q) effectively. For sites with low head (10 m or less), the Kaplan turbine is the preferred system. It is fitted with adjustable blades to maintain high efficiency over a wide range of flow conditions and power production requirements. These wheels use a complex valve system which must be custom designed for each installation. For sites with medium head (up to 100 m or so), the Francis turbine is most effective. In situations where the head is large (100 m or more) or where the flow rate Q is not large, the Pelton wheel design is preferred. The Pelton wheel is a highly efficient waterwheel with "no-splash" cups or buckets arranged around the circumference of the wheel. Splashing needs to be avoided when flow rates are low so that little kinetic energy will be lost in the transfer from the water to the wheel.

The principle of *conservation of momentum* tells us that to achieve the most efficient transfer of the energy from water to the wheel, the buckets or blades should travel at one-half the velocity of the water jet that is directed against the buckets. The wheel's diameter should be determined with this principle in mind.

In hydroelectric installations, the jet of water is usually focused through a nozzle for more efficient performance. The choice of nozzle type, like the design of the turbine or waterwheel, will vary depending on the output requirements and the characteristics of the site. Nozzles are available in a wide range of sizes and are usually interchangeable. At nearly all sites, a broad selection of nozzles is stocked to deal with varying flow rates or power production needs.

Figure 4.4.7 Bird's-eye view of four turbine-generator sets at the Parker Dam Power Plant on the Colorado River in Arizona. The plant has an installed capacity of 125,000 kW. (*Source:* Water and Power Resources Service, Bureau of Reclamation, U.S. Department of the Interior; photographer, E. E. Hertzog.)

BOX 4.4.1 OUTPUT POWER

The power obtained from a hydro system depends on a number of variables. Let us assume that the flow rate of the water directed at the buckets or blades is equal to W kilograms per second, and that the height through which the water falls is H meters. As we established in Chapter 2, the gravitational force acting on each kilogram of water equals 9.8 N. Then the rate of conversion from potential energy to kinetic energy is equal to the following:

$$\text{power} = (9.8 \times W \times H) \text{ newton-meters (joules) per second}$$

Variables used in calculating output power:

W = flow, kg/s

H = height of waterfall (head), m

Q = flow, m^3/s

D = density of water = 1000 kg/m^3

Because it is difficult to measure the flow directly in kilograms per second, the usual method is to measure the flow in cubic meters per second. This can be done by using special instruments or by computing the cross-sectional area of the stream and the speed of flow. Once the flow in cubic meters per second has been measured, it is easy to convert to kilograms per second. One cubic meter of water contains exactly 1000 liters of water, and each liter of water has a mass of 1 kg. Thus a flow rate of 1 m^3/s equals a flow rate of 1000 kg/s. Substituting in the earlier equation, we derive the following expression:

$$\text{available power} = 9.8 \times 1000 \times Q \times H \quad \text{J/s}$$

This is the total amount of available water power in the system.

Not all the water power is converted to electricity. Some of the power is lost in the pipes (*penstocks*) used to deliver water to the turbine because of friction. In general, these effects are taken into account by simply reducing the head (H) according to a standard formula.

Once the water reaches the generating system, more energy is lost in the turbine because no device can be 100% efficient. Turbine efficiency is typically about 80 to 90%. Therefore, the output of the turbine is equal to the efficiency (E) of the turbine system times the total amount of available water power. Stated mathematically:

$$\text{power output} = 9.8 \times E_{\text{turbine}} \times 1000 \times Q \times H \quad \text{J/s}$$

But we do not have electricity yet. The power of the turbine must be converted to electric power through a generator. This generator has its own operating efficiency, which will reduce the amount of electrical power even further. The final amount of electric power produced by the system equals the following:

$$\text{electrical output} = 9.8 \times E_{\text{generator}} \times E_{\text{turbine}} \times 1000 \times Q \times H$$

This output figure will be stated in joules per second, or watts.

BOX 4.4.2 ILLUSTRATIVE EXAMPLE

Suppose that a hydroelectric power plant is to be designed for a site where the available head is 100 m. The average discharge rate that can be expected for continuous operation is 3 m^3/s. A Francis turbine appears to be the best choice, and its efficiency is 0.88. The alternator (an electric generator producing alternating current) operates at an efficiency of 93%. Given this information, we can make the following substitutions:

efficiency of generator (E_g) = 0.93

efficiency of turbine (E_t) = 0.88

flow (Q) = 3 m³/s

head (H) = 100 m

$$\begin{aligned}\text{power output} &= (9.8 \times 0.93 \times 0.88 \times 3 \times 1000 \times 100)\ \text{W}\\ &= 2{,}406{,}096\ \text{W}\\ &= 2406\ \text{kW}\\ &= 2.4\ \text{MW}\end{aligned}$$

4.4.6 Pumped Hydro Storage Systems

Besides being used for direct production of electricity, hydro power can also be used for storage of electric energy which has been generated by other means. Because hydro power plants have a high efficiency of 80% or more, and because of their versatility and quick response, hydroelectric power plants can be used as energy storage devices. The power plant operates not only as a generating station but also as a pumping station, taking water from downstream and pumping it to a higher storage reservoir in times of low demand. This is possible because highly efficient, reversible pump-turbines are widely available.

A pumped storage system does not decrease the total amount of energy converted to electricity in the system as a whole. In fact, it is a net *user* of energy, requiring more electricity to be produced overall. But pumped storage is useful because of the fluctuating demand for electricity.

Thermal (fossil and nuclear) power plants are good for steady, 24-hour power production. In utility terminology, they are *base loaded.* It takes considerable time to adjust the output power of a thermal power plant, and there is a significant drop in efficiency if a thermal power plant is turned on and off too frequently. So these plants run all the time. But during periods of low demand, not all the electricity generated by these base-load plants is needed right away. The excess energy needs to be stored, or else it will be wasted. Because pumped storage can respond quickly and efficiently to changes in demand for electric power, it can be an effective way to store the excess energy generated by thermal plants during periods of low demand—energy that otherwise would have to be dumped—and release that energy for use when demand rises.

The efficiency of a pumped storage system is approximately 66%. This means that for every 3 kWh of electric energy used to pump water into the reservoir during off-peak hours, 2 kWh can be delivered when the demand rises.

Unfortunately, not every hydro site is suitable for pumped storage operation. The reservoir must be large enough to store a large amount of water during off-peak hours. Also, fluctuations in flow rates can cause environmental problems farther downstream.

4.4.7 Small-Scale Hydro

During the early part of the twentieth century, a number of towns in New England had their own hydroelectric generating plants. But as centralized electric utilities developed, small hydro sites were considered obsolete. Many dam sites were sold to utilities which removed the turbines. Other small hydro sites were simply abandoned. According to some estimates, if all the abandoned small hydro sites were restored to their former productivity, the amount of electric power produced would be greater than the amount currently produced by all the nuclear power plants in the United States.

Today, there is great interest in restoring these old hydro sites, and in developing new small-scale sites. In the town of Hyde Park, Vermont, a hydro project has been approved by the citizens, and the necessary state and local certificates have been acquired. When finished, the facility is expected to provide 23 MW of power to the region's electric utility. Five other municipal facilities have been licensed in Vermont, with nine applications pending. If all the hydro projects under consideration are completed, the state's installed hydro capacity will be expanded by 85 MW, almost doubling the current installed capacity of 91 MW.

To put these figures in context, in 1941, Vermont had an installed hydro capacity of 158 MW, which provided 90% of the state's electricity. But as fossil fuels became cheap after World War II, and the hydro plants aged, the sites were not refurbished. In fact, interest in local hydro projects has only been renewed during the past 5 to 10 years. It remains to be seen how many of the planned installations will actually go on line.

Individual landowners are also expressing renewed interest in restoring small-scale hydro sites or building new ones for electrical or mechanical power production. A number of manufacturers offer small turbines whose outputs range from a few hundred watts to many kilowatts for backyard hydro projects. Some federal tax incentives and guidelines have been approved to help bring about a hoped-for increase in small-scale hydro. For example, the Public Utilities Regulatory Policies Act of 1978 guarantees small-scale hydro producers that their power will be purchased by the utilities at a price comparable to what the utilities would have paid for power from other sources. This law is still being interpreted in the courts, but it appears to provide a very significant incentive for small power producers.

4.4.8 Constraints on Hydro Power Development

Hydro power, which has been an important energy resource since prehistoric times, will continue to be a major source of energy for the foreseeable future. Hydro energy is renewable, nonpolluting, and regionally abundant. Rising energy costs should make even small-scale hydro power an economically feasible option. But no energy system is entirely free of drawbacks. Hydro projects can have serious environmental impacts on land use, wilderness areas, and local climate.

In the Southwest, large-scale hydro power projects have created reservoirs in stretches of the Colorado River where the river once flowed freely through wild canyons. The most controversial of these projects was the Glen Canyon project. A little known wild canyon was selected as the site for a major hydro project, which changed the nature of the area for centuries. Environmentalists claim that the benefits of the project were exaggerated and that the costs were understated. Besides the irreversible loss of a wilderness area, they claim the project caused a reduction in water availability downstream, increased silt deposition upstream, and increased population pressures on fragile ecosystems.

Other proposed projects have been opposed bitterly by environmental groups. Perhaps the best known case is the hydro project that was proposed for the Grand Canyon. Largely as a result of pressures from environmental groups, plans for such a project are no longer under active consideration. But many other projects have been planned in wilderness areas where free-flowing rivers present the potential for significant hydro power development.

Not all potential hydro sites are located in such environmentally sensitive wilderness areas. In many cases, dams already exist but have been inactive for decades. In other cases, hydro power development can be carried out with little environmental damage, or even with some enhancement of the region's environmental qualities. But hydro power development, like development of other energy resources, must be carried out with a clear understanding of all its impacts, both beneficial and harmful.

4.5 TIDAL ENERGY

Tidal energy can be classified as a special kind of hydro energy. An ordinary hydro power plant captures some of the energy of water flowing from a higher to a lower level. As described in Section 4.4, this energy is used to drive waterwheels or turbines to obtain mechanical or electrical output (power).

Tidal power differs from regular hydro in three basic ways:

1. Tides occur in saltwater bays where land and ocean meet, whereas hydro sites are exclusively on freshwater inland streams.

2. Tidal energy is the result of gravitational forces exerted by the moon and to a lesser degree by the sun. Regular hydro power is a form of indirect solar energy.
3. Tidal flow is a two-way process. At high tide, water flows into the bays and estuaries, only to flow back again at low tide. This oscillation has a time period of 12.4 hours per cycle. In a regular hydro project, the flow is a one-way street: from a higher level to a lower level.

4.5.1 Tidal Resources

Not every bay and estuary can be made into a tidal power plant. Unless the site is large enough to support a large power plant, the construction costs cannot be economically justified. Only tidal sites with high differentials between high and low tides (generally 10 m or more) are considered to be adequate for economically feasible production of tidal power. Like any other energy resource base, tidal energy resources are unevenly distributed. Fewer than two dozen prime tidal power sites have been identified in the entire world. Thus it is obvious that tidal power can be an important resource only in a few areas of the world.

In the United States and Canada, potential tidal power sites can be found near Cook's Inlet in Alaska, Puget Sound in Washington State, Passamaquoddy Bay in Maine and Canada, and the Bay of Fundy in Nova Scotia, Canada. These are the areas where the mean tidal range is 10 m or more and the geography is suitable for construction of an effective tidal power station. The potential peak power capacity of the sites in the United States has been estimated at 100 GW, with energy production potential of about 350 billion kWh per year. But no power plant has yet been built in the United States.

4.5.2 Operating Tidal Power Stations

Two large tidal power stations are currently in operation. One is located in the USSR at Kislaya-Guba. The other is in France, at La Rance Estuary near St. Malo in the northwestern region. The tidal range at La Rance is 13.5 m (44 ft). The plant was commissioned in 1966 with an initial capacity of 240 MW, using 24 separate turbine units. The initial energy production was 550 million kilowatthours per year. Recently, the output capacity was increased to 320 MW. The plant recovers 25% of the total tidal energy present at the site.

To harness the tidal power, the La Rance tidal station incorporates a large damlike structure called a *barrage,* built across the mouth of a bay. The basin area enclosed by the barrage has an area of 22 km^2 (5400 acres or approximately 9 square miles). The plant uses bulb turbines with adjustable vanes to harness the flow of water in both directions. The cost of operation at a load factor of 32% is estimated at $740 per peak kilowatt.

Figure 4.5.1 Tidal power plant. The tide which rushes up the Rance estuary on the northern coast of Britanny is one of the highest in the world, reaching 13.5 m. The idea of harnessing tidal power on the Rance dates back to the twelfth century, when the incoming waters were used to turn paddle mills on the banks of the estuary. (*Source:* Michel Brigaud, the French Embassy.)

4.5.3 Basic Design of a Tidal Station

Although the design of a tidal power plant must be tailored to the specific site, all tidal power plant designs share a few basic characteristics. A system of dams and sluice gates or pumps is built to enclose a basin or system of basins. Turbines are installed in the sluice gates to harness the tidal power.

As the tide rises, the sluice gates are opened and the water is allowed to flow into the basin. In a two-way system, electricity can be generated by the incoming water. When the tide reaches its maximum height, the gates are closed and the water is held until low tide. Then the water is released through the turbines to generate electricity.

Improvements on this basic design have been proposed. In the two-basin scheme, water flows through a connected system of basins, beginning in a high water basin, which is filled at high tide. At periods of peak demand, water is released from the high to the low basin. The low basin is emptied at low tide.

A new tidal power plant design has been proposed recently which might make tidal power economically feasible by lowering the capital costs. This scheme, patented by Alexander Gorlov of Northeastern University, uses tidal energy to compress air, which in turn is used to drive compact high-speed gas turbines. The air compression is accomplished by allowing incoming tide to flow from the high side of a flexible dam into the bottom of a strong watertight chamber. As the tide rises, the level of water in the compartment rises, causing the air above it to be compressed. At periods of low tide, a similar technique is used to develop additional power.

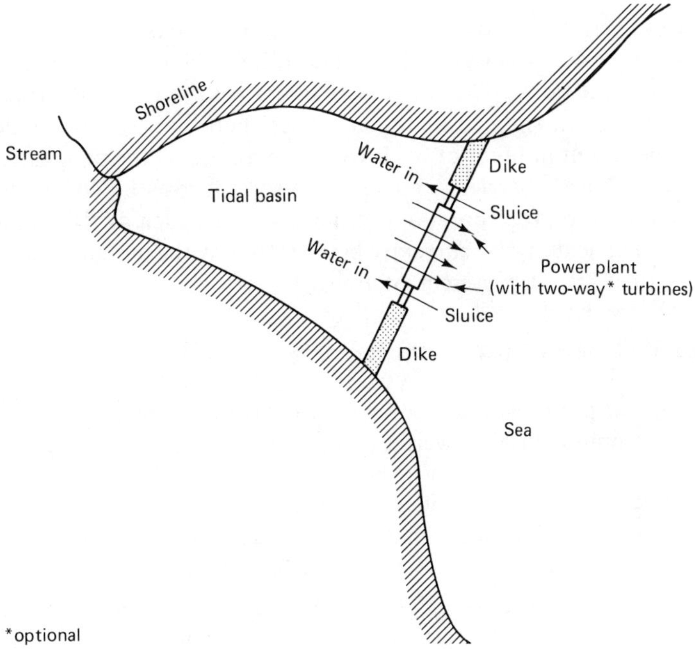

Figure 4.5.2 Basic tidal power plant site.

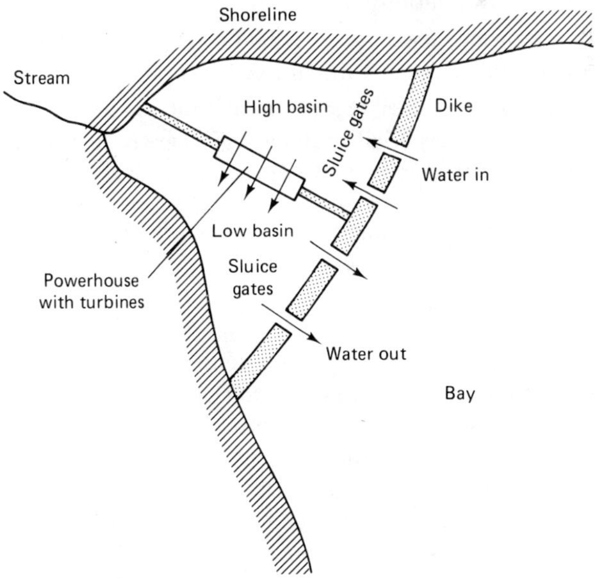

Figure 4.5.3 Two-basin tidal power plant site.

These experimental compressed-air generators can operate efficiently with tidal ranges as low as 1 m (about 3 ft). This allows the turbines to be operated during 60% of the tidal cycle, roughly 16 hours per day. Thus the total energy generated at a given site can be much higher than that with other designs.

Another point in favor of his low-head plan is that the dam can be made out of flexible plastic barriers, reducing the capital costs by 90%. The power plant can be built on the shore, and other expenses, such as fish ladders, can be eliminated. The design is currently being tested by the designer under a contract from the U.S. Department of Energy.

4.5.6 Problems with Tidal Power

Despite the promise of new designs, there are several major problems which must be overcome if tidal power is to become a significant energy resource.

1. Periods of high and low tides do not coincide with periods of high electrical demand. This means that the power generated by a tidal station may not be available during periods of peak demand.
2. All tides are not the same. There is a substantial variation between tides from one week to the next. There are also smaller seasonal variations. This causes variations in the electrical output.
3. Salt water is highly corrosive. Turbines and other electrical components are especially vulnerable to saltwater damage.
4. Severe storms occur along seacoasts. These can damage the machinery at tidal power installations.
5. Seacoasts are prime scenic areas. Tidal power stations change the appearance of a coastal region significantly.
6. Tidal power stations can create significant environmental problems. Coastal estuaries are especially rich biological areas, and tidal power plants can change the characteristics of the region. Fish migrations can also be impeded by the construction of barrages, although fish ladders can reduce the severity of this problem.

4.5.5 Prospects for Tidal Power

At the current stage of technological development, tidal power is not likely to provide large amounts of electric generating capacity. The high capital costs, the scarcity of good tidal sites, and the serious environmental problems make tidal power less attractive than other options. But as electrical costs increase, tidal power will become more attractive.

Even if tidal power does become economically feasible, it seems likely that it will only be a regionally significant energy resource. But in those areas that are blessed with high tides and a suitable structure of bays and estuaries, tidal power could provide an alternative to nonrenewable sources of electric power.

4.6 INDIRECT SOLAR: WOOD ENERGY

Wood is a renewable resource that provides more than just energy. It is a multipurpose resource like hydropower. Wood can be used for a variety of functions, including the following:

1. Construction material for buildings, furniture, and sporting goods
2. Thermal insulating material (cellulose fill)
3. Raw material for paper products
4. Packaging material

Wood also finds use in tool handles, railroad ties, artwork, and in many other areas of human endeavor. Trees give us food, fuel, recreation, and relaxation. Forests influence our climate, control soil erosion and flooding, provide a barrier to strong winds, and help reduce air and noise pollution. A forest also provides habitats for much of our wildlife.

It must be emphasized that wood is a renewable energy resource only if properly nurtured and cared for. Selective cutting, replanting, proper forestry management, conservation, and a little foresight are all essential if wood is to provide a significant quantity of continuously available energy.

4.6.1 Wood's Heat Energy Contribution

Wood has been used as a fuel for heating and cooking since prehistoric times. Up to the time of the Industrial Revolution, it was the world's main source of heat energy. However, during the Industrial Revolution, wood began to be replaced by coal. Early in the twentieth century, firewood use declined considerably because of its inconvenience as a fuel, compared to oil, gas, and electricity, But the Arab oil embargo of 1973, combined with the subsequent steady increase in energy prices, has reversed the trend.

The exact output of wood heat in this country is not easy to measure. Accurate statistics are hard to find for many reasons. The federal government has virtually ignored wood as a part of the national energy picture. The U.S. Department of Energy has not kept a record of the use of wood for fuel, as it has for other fuels, such as coal, natural gas, oil, and nuclear power. Meanwhile, many northern states have doubled their wood consumption for residential space heating in the past decade. For example, in Vermont, the number of homes using wood as their primary heating source has doubled since 1977 and now exceeds the number using oil heat. More than 40% of all households in Vermont are heated primarily by wood heat, according to a study by a state planning office. The New England Congressional Institute reports that wood displaced nearly 25 million barrels of oil in the six-state region in 1980; and they further predict that firewood consumption will more than double in the next four years (1981–1985). According to an office of the U.S. General Accounting Office, wood will provide up to 10 quads of energy per year by the year

2000 with proper forest management and federal encouragement. (By comparison, nuclear power now provides approximately 3 quads of energy per year.) According to the September 1983 *Monthly Energy Review*, published by the U.S. Department of Energy, wood has replaced liquefied propane (LP) gas as the fourth most widely used home heating fuel, providing 4.7 million households in the United States with primary heat. The total number of households using one-third of a cord or more during 1980 was 14.2 million, and total household consumption was 41.6 million cords.

4.6.2 How Wood Is Used

Most of the energy obtained from wood—about 1.4 to 1.5 quads—is used for industrial purposes, mostly for direct heat and process steam generation. The lumber industry in particular uses scrap wood in fairly large quantities for space heating. About 0.5 to 0.8 quad of wood fuel goes for residential purposes. Wood use in the residential sector is harder to measure because of the difficulty of estimating the amount of wood gathered from private woodlots.

On the average, industrial wood heat processes operate at an efficiency of 50%. In the residential sector, an airtight stove or furnace, depending on the design and construction, can have an efficiency of about 40 to 60%. A fireplace is not a good way to heat a house, because its efficiency can be as low as 10 to 20%. In fact, some studies show that if the outside temperature is lower than $-7\,^{\circ}C$ ($+19\,^{\circ}F$), fireplaces may cause a net *loss* of energy.

BOX 4.6.1 HOW WOOD IS MEASURED

For residential purposes, wood is sold by the cord. One cord is 128 ft^3 of wood carefully stacked. In other words, a cord is a stack of wood 4 ft wide, 4 ft high, and 8 ft long. Of this volume, 60 to 100 ft^3 is filled with wood, the rest with air. In the North, stories abound of wood being artfully stacked over stumps and boulders, with more air than wood in the piles. Besides cords, other units of measure exist, such as face cords, ricks (in the South), stove cords, pickup cords, fireplace cords, and truckloads. Consumer protection laws are almost nonexistent in the area of wood fuels, and it is up to the buyer to find out just how much wood is actually being purchased.

4.6.3 Categories of Firewood

Green wood is the term used for freshly cut, unseasoned, or wet wood. Green wood will shrink in volume by about 8% when dried. Thoroughly air dried or seasoned wood usually contains about 20% moisture: that is, the weight consists of 20% water and 80% wood. Green wood can be as much as 50% moisture, and when it is burned, a large amount of energy will be used up just to evaporate this moisture and send it up the stack. Thus less useful heat energy is available from green wood than from dry wood. It is also harder to start a fire with green wood—a fact that many campers have noticed. In a wood stove, green wood causes more air pollution than dry wood. For these reasons, burning green wood is not recommended. (It is interesting to note that in earlier times it was a husband's duty to provide dry wood for his family, and failure to do so was a sufficient reason for the wife to seek a divorce.)

BOX 4.6.2 CREOSOTE AND CHIMNEY FIRES

Creosote is a complex mixture of condensed wood gases. It forms when incompletely burned wood gases come into contact with cool surfaces such as stovepipes and chimneys. Most people are familiar with creosote as the black, sticky, smelly resinous material used to treat telephone poles and railroad ties. But what most people do not realize is that this substance, which can build up in chimneys, is highly combustible. A roaring fire can ignite creosote which has accumulated in a chimney, causing a fire in the chimney or stovepipe rather than in the wood stove or furnace. If the chimney or stovepipe is imperfect or passes too close to a combustible surface, a chimney fire can turn into a major fire that can destroy a house. For this reason, wood burners should clean their chimneys frequently.

Chimney cleaning is an old and honorable profession, but it is also a dangerous one. Chimney sweeps in London during the Industrial Revolution suffered from serious lung ailments similar to those that afflict coal miners. Professional chimney sweeps usually wear masks to keep small particles from lodging in their lungs. Amateurs should do the same. Soot particles can also irritate the eyes and skin, so precautions should be taken to minimize discomfort. Whether homeowners use a fireplace, an airtight stove, or a wood-burning furnace, they should have their chimneys cleaned and inspected at least twice a year, and more often if wood use is very heavy or the chimney is a metal pipe.

There are many varieties of wood. The densest firewood has the greatest heat content per unit volume, and the lightest wood has the least. A pound of maple gives the same amount of heat energy as a pound of birch, as long as both woods are equally dry. But as mentioned earlier, wood is sold by volume, not by weight. A cord of maple or beech will give off much more heat than a cord of aspen or pine. Also, light softwood tends to burn faster than hardwood and thus does not hold a steady fire for a long period of time. Table 4.6.1 gives relative fuel values for some common types of wood. It can easily be observed that the lighter the wood, the lower the Btu content of a cord.

TABLE 4.6.1 Common Types of Wood and Their Energy Contents[a]

Species	Specific Gravity (water = 1)	Million Btu per Air-Dried Cord	Cord Equivalent in Gallons of No. 2 Heating Oil
Hickory	0.70–0.74	32.3	230.6
Red oak	0.60–0.73	28.8	205.7
Beech	0.64–0.66	29.3	209.5
Hard maple	0.58–0.65	30.6	218.5
Yellow birch	0.55–0.64	27.6	197.4
Ash	0.57–0.61	27.0	192.9
Elm	0.50–0.59	25.8	184.6
Soft maple	0.47–0.54	25.3	180.8
Tamarack	0.49–0.53	25.3	180.8
Cherry	0.50–0.52	24.8	177.0
Spruce	0.41–0.44	19.1	136.4
Hemlock	0.40–0.42	18.8	134.8
Aspen (poplar)	0.37–0.39	18.7	133.4
Basswood	0.37–0.39	17.9	128.1
White pine	0.35–0.37	18.0	128.8

[a]These figures assume 95 ft^3 of solid wood in a cord. It is also assumed that the wood is air-dried to 20% moisture content.

Source: Vermont State Energy Office, 1978.

4.6.4 Principles of Wood Combustion

Wood is a form of stored solar energy. The solar energy is stored by means of *photosynthesis*, a chemical process that converts sunlight, water, and atmospheric carbon dioxide into carbohydrates.

Photosynthesis can be represented in simple form as follows:

$$CO_2 + \text{water} + \text{energy (sunlight)} \rightarrow \text{carbohydrates} + \text{oxygen}$$

This should not be understood to mean that wood is pure carbohydrate. There are also trace minerals in the composition of wood, such as potassium, sulfur, and sodium. However, wood is mostly carbohydrate, in the form of cellulose and lignin.

When wood is burned, the process of photosynthesis is reversed. The complex carbohydrate molecules are combined with atmospheric oxygen and broken down into simpler water and carbon dioxide molecules, releasing heat energy to the environment. Schematically, the process of oxidation can be represented as follows:

$$\text{Carbohydrates} + \text{oxygen} \rightarrow CO_2 + \text{water} + \text{energy (heat)}$$

The stored energy is converted into heat (and light, as anyone knows who has ever watched a campfire). The residue, or ash, is rich in minerals and can be used as fertilizer or insect repellent.

Unfortunately, wood combustion does not always occur under optimum conditions, and the process of oxidation is often incomplete. When this happens, carbon monoxide and a wide variety of hydrocarbons are released to the atmosphere. Slow combustion is especially likely to produce these by-products. In eastern U.S. forests, for example, slow oxidation of dead trees produces a blue haze in some mountainous regions. This haze is similar to the photochemical smog found in large cities, and it is the source of the name "Smoky Mountains."

4.6.5 Wood Stoves

Wood stoves are the most common type of residential wood heating device. A wood stove is usually placed directly in the space being heated. Various models of wood stoves are available on the market, and quality, price, and efficiency vary widely. A box of cast iron with an opening for wood and another for a chimney could function as a wood stove, and some wood stoves are little more than cast-iron boxes. But for efficient use of wood energy, more is required of a wood stove. For efficient operation, there must be complete combustion, which occurs at high temperatures, and there must be a relatively high transfer of heat from the firebox to the room being heated.

Higher efficiency means that less wood is consumed for the same amount of useful or delivered heat. Maximum efficiency can be achieved by making both the combustion and heat transfer as efficient as possible. Various designs have been used to achieve this goal. In general, the efficiency of a wood stove depends on the following factors:

1. Airtight stoves are generally more efficient than non-airtight models (although they tend to create more air pollution and creosote buildup than do non-airtight models).
2. Stoves that provide for air circulation to the room tend to be more efficient than stoves which have no provision for air circulation.
3. Stoves with long flame paths, baffles, and secondary air supply are generally most efficient.
4. Stoves made of heavy cast iron or thick plate steel usually have higher efficiencies than stoves made of thin sheet steel.

Other useful features often incorporated into wood stoves include automatic thermostatic controls for convenience and steady heat output, soapstone or similar heat storage surface material for longer and less variable heat storage, convection plates for better heat transfer and safer operation, and chimney dampers for heat control.

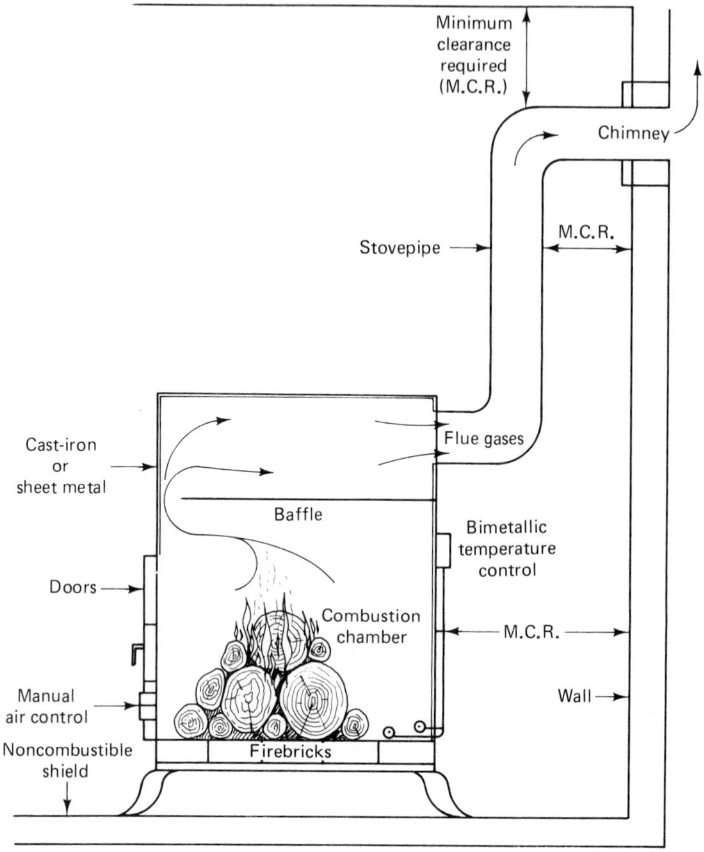

Figure 4.6.1 Wood stove system.

Ratings of stoves. Stoves can be rated in three ways:

1. Maximum useful heat output (to the inside of the house) in Btu per hour
2. Number of cubic feet of living space that can be heated by the stove
3. Number of average rooms that it can heat comfortably

The first of these is the preferred rating because it is more scientific and is most indicative of the output of the stove (or any other heating appliance). But the second and third are commonly used because often a buyer may not know what it really means to have a rating of 10,000 Btu/hr, especially when he or she may not be sure about the heating requirements of the dwelling.

It is obvious that heating requirements change from day to day, from region to region, and from house to house. The heat requirements of a house depend on many factors, including the following:

Outside temperature

Insulation and infiltration characteristics of the house

Desired temperature in living quarters

Height of rooms

Air circulation patterns in the living space.

A stove's output rating should match the heating requirements of a room or house as closely as possible because otherwise the temperature fluctuations indoors can be unbearable and efficiency can be down. For example, suppose that a large stove is being used to heat a small cabin. First, the owner builds a fire that heats up the entire stove. Then a large amount of heat is transferred to the dwelling space, making the room unbearably hot. In response, the owner opens the windows and lets the fire go out. Soon the room cools down and another fire needs to be built. If, instead, the owner had installed a smaller, properly sized stove, a fire could be maintained steadily and the cabin would be much more comfortable.

For continuous heating, as in northern climates, it is desirable to use a stove that can hold a number of large pieces of wood so that the stove will not have to be loaded (or stoked, as the process is often called) as frequently. A high-efficiency stove is definitely desirable when wood supply and costs are major considerations. A thick-walled stove is good for steady, continuous operation, whereas a thin-walled stove is good when a space must be warmed up quickly and for a relatively short period of time.

4.6.6 Wood Furnaces

Wood stoves are usually located in the living space, while wood furnaces are usually in basements and are equipped with forced-air or hot water distribution systems. Wood furnaces do not have to be good looking because they are not part of the living room furniture. They are usually bulky and have higher Btu ratings so that they can be used for central heating for an entire house, not just a room or two.

The same criteria mentioned for wood stoves apply to wood furnaces, as far as heating requirements and efficiency are concerned. Since wood furnaces are out of sight (and thus out of mind) for the most part, it is desirable to have a thermostatic control and a very large firebox. Also, since wood furnaces are not placed directly in the living space, overheating is less of a problem with wood furnaces.

Hot water heat distribution systems are preferrable to forced air systems in a wood-fired furnace, for the following reasons:

1. Water has a much higher specific heat than air. (See the discussion of liquid and air solar heating systems in Section 4.1.2.)

2. Water acts as a thermal mass, which reduces temperature fluctuations in the dwelling space.
3. If hot water is used as the heat distribution fluid, some of the heated water can be circulated through a water heater to produce domestic hot water.

Wood furnaces give even, predictable heat with a minimum of attention. With automatic wood-feeding systems using wood chips or pellets, a wood furnace can even be left unattended during cold weather for a period of several days. However, this convenience has its price. Wood furnaces require an outside source of power (usually electricity) to pump heated air or water through the distribution system, whereas wood stoves do not. No one system is best for everyone; the choice of systems depends on the preferences and life-style of the individual wood user.

4.6.7 Advantages and Drawbacks of Wood Heating

Wood heating is a labor-intensive method of heating and therefore does not offer the convenience of electric, gas, or oil heat. Wood requires a large amount of storage space, and that is hard to come by in urban areas. Ash removal can also be a problem. It is possible to bring termites or carpenter ants into the house along with a load of firewood. Some dwellings (especially multiunit apartment houses) are simply not suitable for the installation of wood heating devices.

Wood heating can be quite dangerous. Large numbers of house fires are caused every year by improperly installed or operated wood stoves. A hot stove can be a dangerous plaything for very young children. Chain saws and axes injure many homeowners every year. And fumes from wood stoves, especially airtight wood stoves, can cause a significant health hazard. The chemicals in wood smoke are similar to those found in tobacco smoke—and the health hazards of tobacco smoke are beyond dispute. Emission control remains a serious concern for thoughtful advocates of wood burning.

Wood heat is an important and irreplaceable source of energy for millions of people. It offers security and reliability in rural areas. Fire hazards can be minimized by proper installation of the wood stove or furnace and good maintenance of the chimney. An old proverb says that "cutting wood warms you twice": once when you cut the wood and once when you relax next to the cozy warmth of a stove. Cooking on a wood stove can be a challenge, but the results can be gratifying (it is said that the very best bread is cooked in a wood-fired oven), and space heating is a useful by-product.

Home heating with wood is definitely economical and will remain so for many generations, as long as proper care is taken in heating system operation and forestry management to ensure a continuous and renewable supply of wood for many uses.

4.6.8 Industrial Uses and Conversions

Wood has been used as an industrial fuel in various forms for different purposes. For example, charcoal is used extensively in underdeveloped regions of the world as an industrial fuel. The first steel produced in colonial America was manufactured in charcoal-fired furnaces. Charcoal is obtained by incomplete combustion of wood in an oxygen-poor environment. This process removes most of the moisture and volatile gases that cause smoke and creosote. Sawdust has also been used as a fuel, and in some cases as an insulating material in buildings, especially ice houses. Wood chips are being used increasingly as a fuel. The city-owned Burlington Electric Department in the state of Vermont is currently (1984) constructing a wood-fired electric power plant with a design capacity of 50 MW of electric power. A number of other utilities have been producing electricity from wood on a smaller scale for several years. Some critics believe that using wood to produce electricity is inefficient and wasteful. They point to the fact that thermal electric conversion is only about 30% efficient, whereas a good wood stove operates with an efficiency closer to 60%. However, the practice of burning wood to generate electricity appears to be gaining popularity in heavily forested regions. There have even been examples of wood-powered automobiles, and research is being conducted on the efficient use of wood gas, wood pellets, and wood chips as sources of energy for transportation.

We will close this section by pointing out that due to the nature of wood and its associated heating systems—particularly the need to restoke wood stoves and furnaces at frequent intervals—many people find it necessary to have a backup heating system. This is not a major problem in itself; in fact, all energy and power systems should have alternative backups. But woodburners tend to use electric baseboard systems for backup heating. This puts an extra load on electric utilities during periods of peak demand. This can result in a greater need for electric utility generating capacity. Thus, although wood burning can provide a substantial amount of energy, it must be used intelligently and with a clear understanding of how wood fits into the total energy picture.

4.7 INDIRECT SOLAR: BIOMASS AND BIOFUELS

The term "biomass" refers to a wide variety of energy resources based on the conversion of plant material to usable forms of energy. Wood burning can be regarded as a form of biomass conversion, but the term is usually reserved for more complex processes such as the production of gas or alcohol fuels from plant materials. Agricultural wastes, animal wastes, garbage, and forestry by-products are all classified as biomass. The idea is to get as much energy as possible out of these by-products before discarding them. Like most renewable energy resources, biomass is an indirect form of solar energy. But unlike certain other renewable technologies, biomass conversion does not require us to look in the oceans or distant regions of the earth. Biomass products are found in our cities, on our farms, and in our landfills.

Biomass can be used for energy needs through a variety of conversion processes, categorized under three general headings: (1) direct utilization, (2) thermochemical conversions, and (3) biochemical conversions.

4.7.1 Direct Utilization

Some biomass products, such as wood and forestry by-products, can be burned directly with very little treatment. Other waste products, such as paper and petrochemical wastes, can also be burned readily. Although the energy value changes with the chemical composition of the biomass resource, combustion of biomass generally produces heat energy of 4500 to 5000 kcal per kilogram of dry material.

When biomass is burned directly, temperature and air supply must be controlled to produce a clean combustion process. Because of the variability of biomass sources, waste-conversion plants do not always operate at optimum efficiency. Air polluting emissions can also be a problem if wastes are being burned inefficiently. Also, municipal wastes can contain hazardous materials. Few controls are possible because of the large number of sources for municipal wastes. This adds to the danger of air pollution from toxic materials.

Electrical generation from municipal wastes is a popular technology because it solves two problems at once. Solid wastes presents a major disposal problem in most large cities, and the use of waste for electrical generation greatly reduces this burden for city officials. Under many of the contracts currently in effect, cities guarantee to supply the electric plant with a certain minimum amount of garbage. In most cities, however, this presents no difficulties. At present (1983), municipal waste-conversion plants are operating in a few U.S. cities, including St. Louis and Boston. Additional plants are planned for Dade County, Florida; Staten Island, New York; and Los Angeles.

4.7.2 Thermochemical Conversion

Solid fuels are useful for space heating and generation of electricity, but liquid and gaseous fuels are more versatile and easier to transport. For these reasons, a number of techniques have been developed for converting biomass to gas and liquid fuels. Pyrolysis and gasification methods produce a variety of fuels: low-, intermediate-, and high-energy hydrocarbon gases, hydrogen, methanol, oil, ammonia, and so on. Catalysts are widely used to achieve desired chemical conversions at higher efficiencies.

Conversion of wood to methanol is one important thermochemical conversion process currently being considered. The technology is well understood and has been widely used in the production of wood alcohol. The energy content of methanol is 45% as high as that of an equal volume of gasoline. Methanol is an odorless, explosive, poisonous alcohol which can be obtained from any substance containing hydrocarbons. It can be used as an automotive fuel in specially-converted engines, and for a variety of other uses.

The first step in methane production is converting biomass to a "synthesis" gas, a mixture of carbon monoxide and hydrogen. This synthesis gas is formed by incomplete combustion of the biomass in the presence of steam. The carbon in the material combines with water to produce CO and hydrogen. Then the synthesis gas is subjected to high pressures and temperatures (53 kg/cm² and 250°C, or 750 psi and 480°F). In the presence of a catalyst, methane is produced.

$$CO + 2H_2 \rightarrow CH_3OH$$

Historically, wood was used to produce charcoal; then a distillation process was used to produce methanol together with other chemicals, such as acetic acid and tars. But natural gas has been used extensively in the past 30 years as a methanol feedstock. The state of Maine plans to produce a methanol using 5 million acres of trees that are infested with spruce budworm. Other methanol projects are under consideration in West Virginia, California, and several other states.

4.7.3 Biochemical Conversions

When organic waste products break down in nature, yeasts, bacteria, and other microorganisms convert the complex organic molecules into simpler hydrocarbons and alcohols (Table 4.7.1). Biochemical conversions employ these same processes in a controlled situation to produce gas and alcohol fuels. Although many conversion techniques are practiced, the most common are production of methane gas and distillation of ethanol.

1. *Methane production* is a simple, rather easily controlled process. Under anaerobic conditions (no air supply), vegetable matter will decompose, giving off a mixture of methane and carbon dioxide. Methane constitutes about 60 to 70% of the resulting mixture, and the mixture has a heating value of approximately 700 Btu/ft³. A well-operated biogas project can produce up to 13,000 ft³ of gas per ton of waste. Besides methane, biogas conversion also produces a high-nitrogen slurry residue which has excellent qualities as an agricultural fertilizer.

In some cases, landfills have begun to ferment spontaneously, producing artificial gas deposits underground. A few landfill operators have begun to mine this source of methane by drilling into the site and drawing off the combustible gases. This method of methane production is quite haphazard and inefficient, but it illustrates the relative ease with which methane can be produced from organic wastes.

A more controlled conversion of waste matter to methane occurs in specially designed *methane digesters*. The technology for these devices is not difficult;

TABLE 4.7.1 Representative Alcohol Fuel Feedstocks

	Average Crop Yield (units/ acre)[a]	Alcohol Yield (gallons/ unit)	Alcohol Yield (gallons/ acre)
Simple sugar feedstocks:			
Sugar beets	21 tons	24.4	512
Sweet sorghum	20 tons	19	380
Sugarcane	36 tons	16.8	605
Grapes	5.4 tons	15.1	81.5
Molasses (50% sugar)	—	64 gal/ton	—
Apples	8 tons	14.4	115
Citrus (culls) (10% sugar)	—	12 gal/ton	—
Starch feedstocks:			
Corn	90 bu	2.5	225
Wheat	31 bu	2.5	81
Oats	56 bu	1.1	62
Barley	44 bu	2.1	92
Grain sorghum	56 bu	2.5	140
Potatoes	225 cwt	1.4	315
Sweet potatoes	120 cwt	1.7	204
Rice	2.2 tons	79.5	175
Rye	21 bu	2.2	46
Bakery waste (dry flour)	—	6.5 gal/100 lb	—
Jerusalem artichokes	20 tons	21	420

[a]bu, bushels; cwt, hundredweight.

Source: Reproduced with permission from *Making More Fuel Alcohol* (Butte, MT:, National Center for Appropriate Technology, 1980).

a single household can easily produce its own methane from garbage and household wastes. Millions of methane digesters have been sold in China and India. Use of these simple digesters is helping to reduce the direct burning of animal wastes, which has led to decreased soil fertility in many areas of the Third World.

On a somewhat larger scale, village-scale digesters are currently under development in India, and dairy farms throughout the world are beginning to look toward methane production as part of their manure-handling systems. The residues from these fermentation operations have excellent fertilizer value, so the agricultural benefits of manure spreading can be maintained together with the production of a useful fuel for heating and cooking.

BOX 4.7.1 METHANE PRODUCTION FROM HYBRID KELP: ONE LARGE-SCALE BIOMASS PROJECT

Some energy analysts believe that large amounts of biomass energy can be obtained through the cultivation of hybrid kelp (seaweed) farms in the ocean. The Gas Research Institute and the U.S. Department of Energy have sponsored a test program to test the feasibility of this idea. The process involves the construction of a strong yet light structure of PVC pipes submerged in the deep ocean. This structure forms a base to which the kelp plants attach themselves. Nitrogen-rich cold water is pumped from depths of approximately 2000 ft to the surface, where it provides nutrients for the kelp plants, which will later be processed to produce methanol.

Seaweed is regarded as an ideal resource for methane production because of its high moisture content and its lack of lignin (a woody cellulose structure that retards decomposition of plant matter). Seaweed farms do not require irrigation, and the fertilizer is readily available at a depth of a few thousand feet. It is hoped that ocean power (tides, waves, or ocean currents) will power the pumps that draw the nitrogen-rich water to the surface.

Although seaweed projects offer the possibility of enormous energy production, they have their drawbacks. Hybrid kelp grows well only in cooler climates, where ocean thermal power production is not feasible. The structures can be damaged by storms, and harvesting and transport of the kelp presents major logistical difficulties. Nonetheless, this project shows how large projects could produce the enormous amounts of energy which are needed if biomass is to provide an adequate replacement for natural gas.

2. *Ethanol production* using vegetable matter is a very old technology. Beer and wine have been produced for many centuries, and the technology for producing alcohol fuels is very similar. The treated cellulose waste materials are mixed into a liquid slurry, which is then treated with alcohol-producing yeasts. Under controlled conditions of temperature and atmosphere, the sugars in the liquid mixture are converted to ethanol and carbon dioxide. The resulting liquid can then be distilled to increase the alcohol concentration. Mixed in a ratio of 10% ethanol to 90% unleaded gasoline, this fuel is available as "gasohol" at many service stations across the United States. (Often, what is called "premium unleaded" gasoline is actually gasohol.)

4.7.4 Resource Availability

The commercial availability of biomass depends on the price it can fetch. At $20 per dry ton of biomass, up to 250 million tons per year could be made available, and up to 300 million tons per year within a decade. In the United States total organic waste material amounts to more than 1 billion tons per year of dry organic matter, much of which could be utilized to produce 1 quad of usable energy. The use of forestry residues, crop remains, animal refuse, and

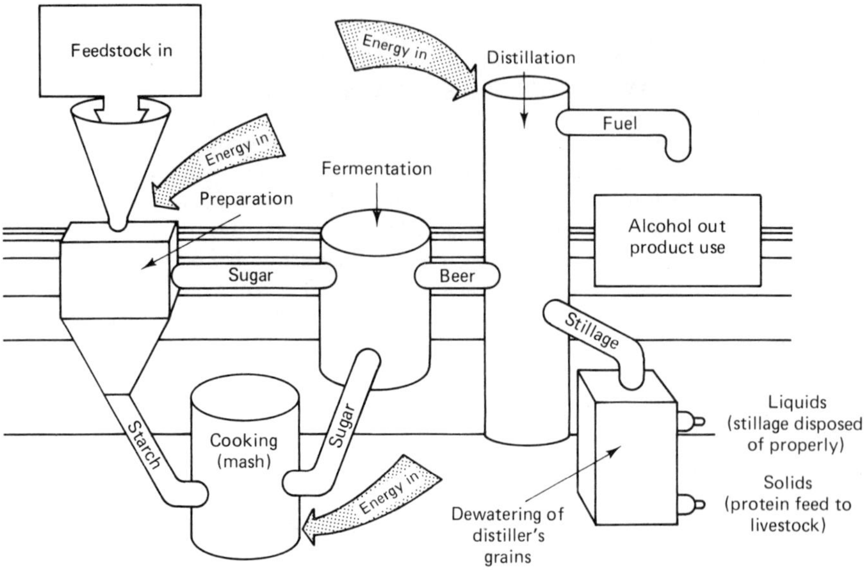

Figure 4.7.1 Fuel alcohol production process. [Reproduced with permission from *Fuel Alcohol Production* (Butte, MT: National Center for Appropriate Technology, 1981).]

city waste could provide even more energy for the nation. According to current projections, biomass will provide over 2 quads of energy by 1985 and over 3.25 quads by the turn of the century.

Electric power production from biomass is not likely to be significant in the national energy picture, but on a regional basis it may have more of an impact. There are currently "waste-to-electricity" plants operating in some of the nation's cities, and others are being built or planned. It is currently projected that by the year 2000, there will be 200 small (50-MW) electric power plants in the United States operating on wood chips, sugarcane stalks, city refuse, and other biomass materials.

Estimates are that alcohol-based fuels can replace up to 50,000 barrels of oil per day by 1985, 100,000 bbl/day by 1990, and 200,000 bbl/day by the turn of the century. Biomass alcohol can also displace a substantial amount of petroleum-based alcohol in chemical feedstocks. This use alone could account for 0.1 quad by the year 2000. Current estimates of alcohol distillation projects in the United States place total production at 115 million gallons per year.

4.7.5 Environmental and Economic Considerations

Biomass production is not an especially efficient way to convert solar energy. Most plants store only 1 or 2% of the sunlight that reaches their leaves. This means that large-scale biomass projects would require large areas of land.

Promoters of biomass conversion recommend the use of land which currently lies idle. This land could be used to produce rapidly growing energy crops such as sugarcane or hybrid trees. It is estimated that up to 20 quads of energy per year could be obtained by employing about 5 to 10% of the idle land in the United States. However, it is necessary to consider the costs and energy needs of irrigation, transportation, and processing of the vegetable matter produced by such a scheme.

Some environmentalists fear that widespread cultivation of biomass crops could take prime farmland out of production or impede wider cultivation of human foods. Others claim that only agricultural residues need be used for biomass conversion in prime agricultural regions. This issue is discussed more fully in Section 11.4.

In many cases, the environmental effects of biomass conversion are beneficial. Reduction of municipal wastes and replacement of fossil fuels are both possible with widespread biomass conversion. Use of the residual slurry can improve soil fertility. And if proper agricultural processes are used in the growing of biomass crops, air and water pollution can be minimized, soil erosion can be reduced, and increased habitat can be provided for field-dwelling animals.

Like most other new energy technologies, biomass conversion offers a great potential, but we do not know how successfully the technology will be developed. At present, biomass conversion appears to be cost-effective when practiced in conjunction with other agricultural operations. It remains to be seen whether biomass farming, either on land or in the oceans, will come into its own as an independent energy-production technology.

4.8 INDIRECT SOLAR: WIND ENERGY

Wind power, like water power, dates back thousands of years. Many cultures used wind to their advantage. The Egyptians built and used sailing ships as early as 2000 B.C. Chinese merchants reached India under sail about 2000 years later. The New World (America) was explored by the Vikings, and by Columbus and others, through the use of sailing ships. In fact, until the nineteenth century, the sailing ship remained the queen of the sea. But once steam-powered vessels were developed, sailing as a means of maritime transportation declined rapidly.

The use of wind power on land has been much more limited than the widespread use at sea. This has been due partially to economics and resource availability. When abundant wood, hydro, and fossil fuels have been available on land, development of wind power has not been a high priority. Other factors have also inhibited the development of wind power. The wind cannot be depended on to provide large amounts of power at all times when that power is needed. Also, it is impossible to store wind energy directly, like hydro energy behind a dam. Nonetheless, wind has been used on land as well as on the sea to provide energy for a number of useful purposes.

Figure 4.8.1 Multi-bladed windmill. (*Source:* U.S. Department of Agriculture.)

In the past, windmills, like watermills, were used for direct mechanical conversion (e.g., to grind wheat or pump water). The rotation of the large vanes about their horizontal axis was coupled to a grinding wheel or a pump. In the American Midwest, windmills were used to pump water for a number of years until widespread rural electrification made electric pumps easier to install and more economical to operate. Even today, hundreds of "prairie fan" windmills dot the farmlands of the Midwest. In Australia, too, small multivane wind machines were used to pump water from artesian wells.

Today, with the increasing use of electronics in every aspect of human life, the wind is used to generate electricity, a high-quality form of energy that can be used for nearly any purpose we wish. However, this creates a number of problems for the designers and manufacturers of wind power machines. Sophisticated controls and instrumentation are required to generate electrical energy at constant voltage and frequency, and in synchronization with the power being supplied by an electrical grid. This makes small wind generators impractical for the profitable generation of electricity from the wind. Many individual households have installed small-scale wind-generating systems to provide electric power for domestic use. But corporate efforts have been primarily directed toward the development of large-scale "wind farms," where a large number of wind generators are tied together to produce a substantial amount of electric power. Most of these wind farms have been located in the western regions of the United States, where prevailing winds are strong and predictable, but a few large-scale projects have also been undertaken in the eastern United States.

Commercial wind power production is still a relatively new development throughout most of the United States. Proponents of wind power believe, however, that its economic advantages will soon cause a rapid, large-scale expansion of wind-powered electric generation. The estimated potential for wind power in the United States is 5 quads. However, the U.S. Department of Energy projects a total production of only 0.1 quad for the year 1990 and 0.6 quad for the year 2000.

4.8.1 Design Criteria for Wind Power Installations

When a wind power system is being considered, a number of criteria should be taken into account. These criteria apply whether the proposed wind installation is a small household system or a large-scale wind farm. Of course, a site that is perfect for a one- or two-family installation might not be at all suitable for a municipal wind power project. A detailed design process should be carried out in each case.

1. *Preliminary feasibility study.* Power needs should be determined as precisely as possible. Often a projection based on past experience is the best that can be achieved. But as far as possible, the preliminary analysis should include a projection of future power needs.

2. *Economic analysis.* Costs of various wind systems should be compared, taking into account all costs of electrical production, storage, conversion, and distribution. Cost comparisons should be specific to the particular site. Alternatives should be considered, including the alternative of rejecting wind power for the site in question.

3. *Legal and environmental analysis.* Legal codes often restrict the building of high structures such as windmill towers. Also, power-line easements may have to be obtained if the power is to be transmitted over long distances. Noise can be a problem in built-up areas; neighbors often complain about the sound of wind generators.

4. *Site study.* Selecting a wind site is a long process in itself. Ideally, several sites should be studied and the best one selected. In a site study, wind speeds should be measured, at different altitudes if possible, for a period of at least one year. Seasonal data, time-of-day information, and periods of calm wind should all be recorded, together with the average wind speed. This is important, because the wind may not blow at the "average" speed very much of the time. In areas where periods of calm alternate with fierce gales, "average" wind speed is a meaningless concept.

Once the wind has been measured for a site, it is possible to calculate the actual power that can be obtained from the various wind systems under consideration. This is extremely important, because manufacturers' estimates of their wind machines' capabilities usually do not take into account the great variations between different sites.

5. *Wind system selection.* It is not enough to buy a high-quality wind generator, a good tower, and an excellent storage system. All the components

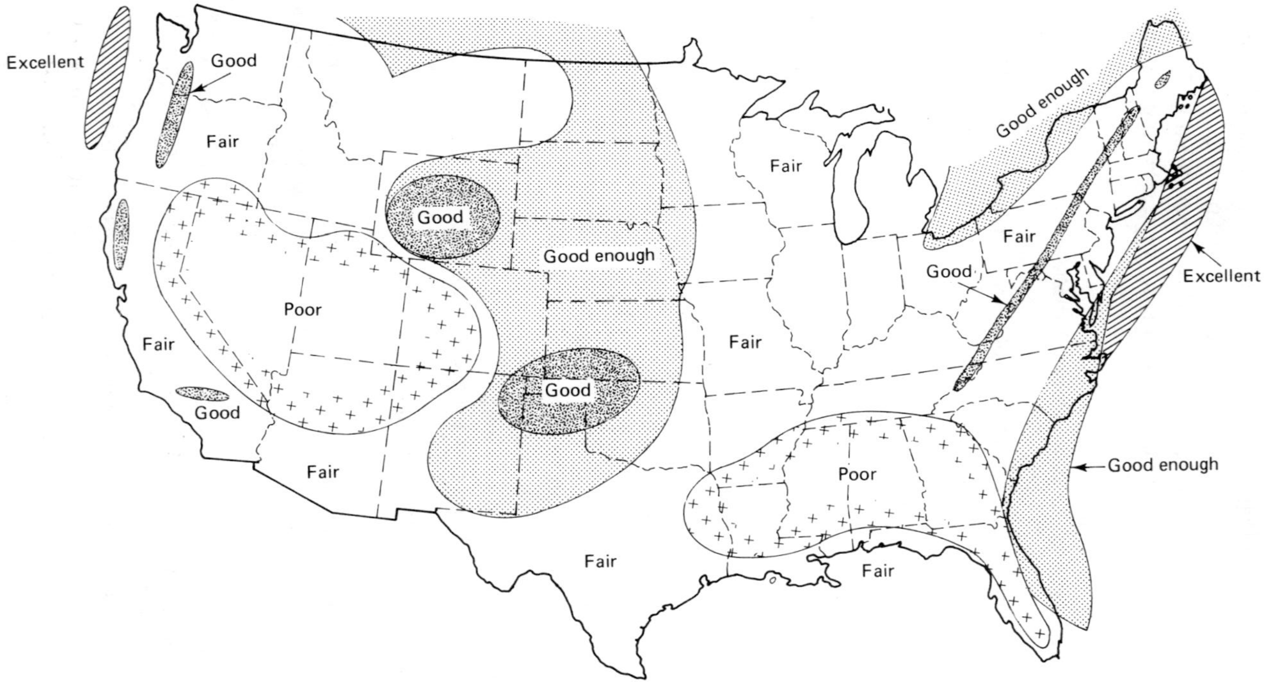

Figure 4.8.2 Potential of wind energy in the United States. (*Sources:* NSF; NASA; J. W. Reed, *Wind Climatology,* Sandia National Laboratories.)

of a wind energy system must work together. If components are not well matched, the weakest component will limit the performance of the rest of the system.

It must be emphasized that selecting and siting a wind system is a complex and integrated process—just like any other power plant siting, but more involved because of the additional constraints caused by the use of wind. The process is generally easier and more efficient if the wind's energy can be used *directly,* without going through a number of intermediate steps. For example, it does not make sense to use a windmill to produce electric power to drive a pump to pump water. Why not use the windmill to pump water directly? However, if the wind energy is to be used for pumping water, compressing air, and operating a band saw, it makes more sense to generate electricity, since electricity is a highly versatile form of energy and can readily be converted for a nearly unlimited number of applications.

4.8.2 Types of Wind Systems

As Figure 4.8.3 indicates, several different rotor types are available. Each type of rotor is best suited for a particular kind of task. For instance, if the wind power is to be converted directly to mechanical power (grinding grain, pumping water, etc.), the best choice is a low-speed, high-torque design such as a *Savonius* rotor (Figure 4.8.3a) or a *multivane fan* (Figure 4.8.3c). But if the wind is to be used for generating electric power, a high-speed, low-torque design should be used. The high speeds are necessary if the electricity produced is to have the same characteristics (frequency, voltage, etc.) as the electricity available from electric utility companies. *Darrieus* and *propeller-type* rotors are both suitable for electrical generation (see Figure 4.8.3b and d).

In addition to their speed and torque characteristics, rotors are classified by the orientation of their axes. Vertical-axis machines (*Savonius* and *Darrieus* rotors, for instance) rotate in a plane parallel to the ground, whereas horizontal-axis machines (the type most people think of as "windmills") rotate in a plane perpendicular to the ground. Each category has certain advantages, and advocates of each type can be found in any discussion of wind power.

When they are used for mechanical tasks, vertical-axis machines have the advantage that they can transmit power to the bottom of the tower without requiring a system of gears or belts. A horizontal-axis machine must convert the power from the horizontal axle to a belt or a vertical drive shaft in order to transmit the power to ground level. A second advantage of vertical-axis machines is that they do not have to swing back and forth to face into the wind. They respond equally well regardless of the wind direction. Finally, because of the lack of complex gearing and steering mechanisms, vertical-axis machines do not require as strong a supporting structure as do horizontal-axis machines.

The horizontal-axis windmill is an older design, and it can achieve much higher efficiency than the vertical-axis design. This advantage in efficiency compensates for the higher costs required in gearing, steering mechanisms, and tower

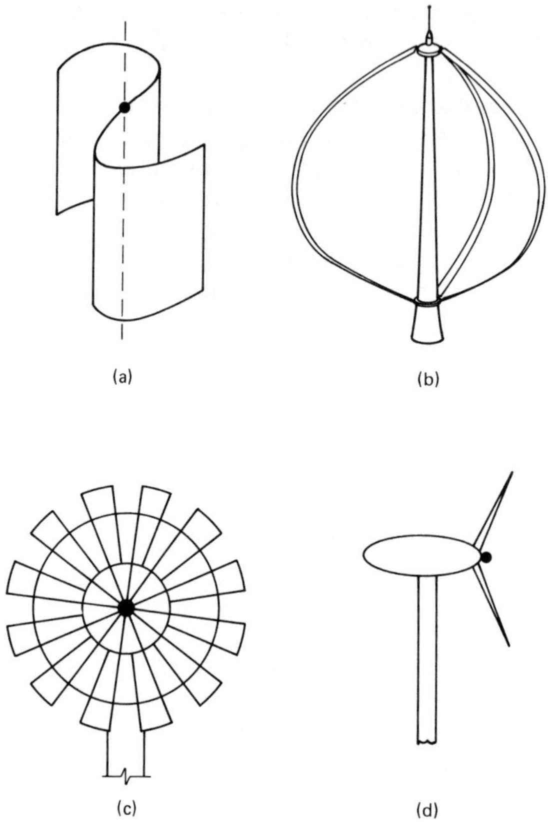

Figure 4.8.3 Various types of windmills. Vertical-axis wind machines: (a) simple Savonius-rotor, (b) Darrieus. Horizontal-axis wind machines: (c) multibladed horizontal-axis farm windmill, (d) two-bladed propeller type.

construction. Also, because the horizontal-axis machine is an older and more common design, used horizontal-axis machines are easier to find, and parts are more widely available for both new and used horizontal-axis wind generators.

4.8.3 Electrical Generation

After the rotor has converted the linear movement of the wind into rotary motion, the rotor is coupled to an electrical generator, either direct current or alternating current. (An ac generator is called an "alternator.") Systems of belts, gears, and pulleys are often used to couple the rotor to the generator, although in some cases the generator is coupled directly to the shaft of the rotor. Both alternating and direct current have their advantages, and both have their adherents.

Direct current is the type of electricity used in automotive electrical systems, and many small wind power systems use automobile generators to convert wind energy into electrical energy. Automobile parts are inexpensive, readily available, and easy to repair. But they operate at a low voltage, which causes a significant loss of efficiency due to the resistance of wiring systems.

A second advantage of direct current is that it can be stored in batteries. Again, automobile batteries can be used, but for proper long-term storage it is preferable to use costly, specially designed "deep cycle" batteries such as those used by telephone companies. Automobile batteries are designed to provide a large amount of power for a short period of time, but they are not designed to have a large amount of their energy drained from them. Frequent discharging and recharging will damage automobile batteries rapidly. Deep cycle batteries are better able to withstand the strain of being drained in periods of low wind.

Finally, advocates of direct-current systems point out that it is much easier for self-sufficient individuals to cut themselves off from utility companies altogether if they use direct-current systems with battery storage. Although this is true, it is unlikely that direct-current, battery storage systems can replace a substantial portion of the electric power now being produced in the United States. However, in remote areas where no utility grid is available, direct-current systems are certainly a reasonable alternative.

Alternating current is the type of electric power supplied by utility companies in the United States. Most home appliances are designed to operate on alternating current, although a few, such as incandescent light bulbs, can operate on direct or alternating current. If a wind system provides alternating current, it can be tied in with the utility's electric grid. This way, excess power can be sold to the utility when the wind machine produces more electricity than is needed, and power can be bought from the utility when the wind machine does not produce enough to supply the household's needs. This eliminates the need for another type of storage, because the power grid itself serves as the storage system.

But the convenience of tying into the utility power grid has its price. As was mentioned earlier, the electricity produced by the wind system must be precisely matched to the electricity supplied by the electric power grid. This requires the power output of the system to conform to very stringent power, frequency, and waveform requirements. Safety controls are also essential to ensure that utility repair crews will not be injured by the electrical output of individual wind systems. All these controls are expensive, and they cause a drop in the overall efficiency of the wind system. In many cases, the safety systems actually require the wind system to stop producing electricity altogether in case of grid failure.

4.8.4 Support Structures

In most cases, wind generators are located on high towers in order to place the generator system above nearby obstructions and to take advantage of the higher wind speeds at high altitudes. Even where no specific obstructions are

Figure 4.8.4 Improved design of a vertical-axis wind turbine developed at the Sandia National Laboratories. Notice the starter buckets and the air foil blades. The 15-ft blades drive into the wind and are capable of generating 3 hp at a 20-mph wind speed. (*Source:* U.S. Department of Energy.)

present, the higher wind speed at higher altitudes makes the tower a good investment. Adding another 20 ft of tower height is usually quite a lot cheaper than purchasing a more powerful wind generator to compensate for the lower wind speed at lower altitudes.

Like any moving fluid, wind is subject to eddying and surface obstructions. This creates two problems for wind systems: (1) at lower altitudes and near obstructions, the overall wind speed is lower; and (2) near irregular obstructions, the unpredictable wind currents can place great strain on generating equipment and support structures. These effects occur whether the wind generator is upwind or downwind from the obstruction, although the effect is stronger when the generator is downwind.

To avoid these undesirable wind effects, it is best to place the windmill at an elevation of 2.5 to 3 times the height of the highest obstruction within 2.5 miles. This can often be accomplished by locating the tower on a high point of ground. However, it is not advisable to place the tower on top of a building. Harmonic stresses, caused by the irregularities of wind speed and vibrations from the wind machine, can cause structural problems in the building, to say nothing of unpleasant noises and sensations for its occupants.

Towers are constructed in a variety of designs. Freestanding towers are the strongest, while guyed towers are the easiest to construct. Wind-generator manufacturers usually recommend certain types of towers for their products. However, tower construction should be regarded as a major construction project not to be undertaken lightly.

4.8.5 Special Situations

In a few special cases, wind speeds are unusually high and consistent. For example, along seacoasts, predictable land and sea breezes can provide a steady supply of wind power. In large flat areas, such as the Great Plains of the United States, winds are steady and relatively strong, although tornadoes can create tremendous destruction in these areas.

Mountains create substantial obstructions to the steady flow of moving air. This can create irregular winds of great intensity, placing severe stress on wind power machinery. A possible exception to this principle is the very tops of mountains, where wind speeds are quite high and few obstructions are present to create undesirable eddies. However, mountaintops are often wilderness areas, and the ecological systems on mountaintops are easily disrupted. Long-distance transmission lines cause additional disruptions in mountain wilderness areas.

4.8.6 Technical Considerations

The power, and therefore the energy derived from the wind depends on the wind speed. It is apparent to everyone that the greater the wind speed, the greater the amount of energy that can be obtained from the wind. But the relationship is not a simple one. In fact, the amount of power available from the wind is proportional to the *cube* of the wind speed. This means that power available from a 10-mph wind is eight times the power available from a 5-mph wind. Once this principle is understood, it can be seen that the increases in power from a relatively small change in wind speed, say from 10 mph to 12 mph, can be very significant.

$$10^3 = 1000$$

$$12^3 = 1728$$

Thus the power available from a 12-mph wind is over 1.7 times as great as the power available from a 10-mph wind.

Higher wind speeds naturally cause increased stress on the machinery. The force exerted on the blade surface increases with the square of the wind speed. Therefore, the machinery needs to be protected from high winds if the windmill is to survive a storm. At the time when the greatest power is available from the wind, the windmill must be shut down to prevent damage to the machinery. It is clear, then, that a windmill, in order to work effectively, must deliver as much power as possible over a wide range of wind speeds, not simply at high wind speeds.

BOX 4.8.1 COMPUTATION OF WIND POWER

Before calculating the power available from the wind, we need to establish a number of constants, including the following:

ρ = density of air, kg/m^3

D = diameter of the rotor, m

V = wind speed, m/s

C_p = power coefficient of the rotor (the portion of the wind's power that is converted to rotational power by the turbine)

E_g = efficiency of the generator expressed as a decimal (e.g., 0.6 rather than 60%)

Using these basic quantities, we can compute the following:

A (area swept by the rotor) = $\pi \times D^2/4$

power available from the wind = $\frac{1}{2} \times \rho \times A \times V^3$ (watts)

power harnessed by the rotor = power available $\times C_p$

= $\frac{1}{2} \times \rho \times A \times V^3 \times C_p$ (watts)

power output of the generator = power harnessed $\times E_g$

= $\frac{1}{2} \times \rho \times A \times V^3 \times C_p \times E_g$ (watts)

energy produced = power output of the generator × hours

= $\frac{1}{2} \times \rho \times A \times V^3 \times C_p \times E_g \times H/1000$ (kWh)

Note: The ideal coefficient of power (C_p) is 0.6, but this is never achieved in practice. The coefficient of power is influenced by the ratio of the tip speed to the wind speed, and also by the design of the rotor. For example, the American prairie fan windmill has a tip-speed ratio of about 0.8 to 1.0. This means that when the wind is blowing at a rate of 10 m/sec (about 22 mph), the tip of each blade is moving at 8 to 10 m/s. This windmill operates with a power coefficient of approximately 0.3. A high-speed two-blade rotor with a tip-speed ratio of 6 can achieve a C_p greater than 0.4.

BOX 4.8.2 ILLUSTRATIVE EXAMPLE

Suppose that a certain windmill has a rotor diameter of 10 m and a coefficient of power equal to 0.45. The generator operates with an efficiency of 0.85. The wind generator operates in a constant wind speed of 5 m/s for a period of 10 hours. (In fact, the wind speed is rarely constant for even 10 *seconds,* but for performance calculations it is best to assume a constant wind speed.) Under those conditions, we can supply the following values for the constants listed above.

ρ	=	1.3 kg/m^3 (this is a typical value)
D	=	10 m
V	=	5 m/s
C_p	=	0.45
E_g	=	0.85
power available from the wind	=	$\frac{1}{2} \times 1.3 \times 5^3 \times \pi/4 \times 10^2$
	=	6378 W = 6.38 kW
power converted by the rotor	=	$6.38 \times 0.45 = 2.87$ kW
power output of the generator	=	$2.87 \times 0.85 = 2.44$ kW
energy output over 10 hours	=	$2.44 \times 10 = 24.4$ kWh

4.8.7 Prospects for Wind Power

It should be pointed out that despite all the uncertainty and variability involved in power production using wind energy, wind appears to be a feasible source of commercial and domestic power. In some areas, especially remote areas far from power lines, wind power may be the only reliable source of power that is at all affordable.

Although construction costs for wind power systems are substantial, operating costs are very low, consisting only of routine maintenance and occasional repairs. The environmental problems created by wind systems are of a different type and magnitude from the problems caused by other types of power plants. The unsightliness and noise pollution caused by a wind power system are confined to the immediate area. Wind systems cause no air or water pollution and only negligible amounts of thermal pollution.

Wind power is a lively area of technology, and substantial amounts of research are being done in the area. A number of small manufacturers have sprung up to produce wind machines of widely varying size and efficiency. Wind power manufacturers and installers in the United States have profited from the tax benefits associated with the use of renewable energy systems. But it is still not clear whether wind power will become a source of electric power generation comparable to hydro power or thermal power plants.

In the meantime, wind power is still unmatched for specialized applications such as pumping water in remote rural locations far from any power lines. It seems likely that in the future, wind power will be most successful where the end use of the power needed is matched to the power-supply characteristics of wind energy systems.

TABLE 4.8.1 Wind-Energy-System Alternatives

Energy-Conversion Systems	Energy-Storage Systems
Mechanical pumps Air Water Refrigerant fluid Oil: hydraulic motors	Compressed air Heat pumps Pumped hydro Flywheels Compressed hydrogen
Electric generators Alternating current Direct current Direct current with inverters	Electric batteries Power-line connection
Direct heating Churning Friction Centrifugal pump	Hot water Flywheels
Sailboats on water	Direct conversion to useful kinetic energy

4.9 INDIRECT SOLAR: OCEAN THERMAL ENERGY

Ocean thermal energy conversion (OTEC) has been drawing considerable attention from energy experts during the past few years. Like most renewable forms of energy, OTEC is ultimately a form of solar energy. Since more than 70% of the earth's surface is covered by oceans, the majority of the solar energy reaching the earth falls on the surface of the oceans. Some of this solar energy is absorbed and stored in the form of small temperature differences between the ocean's surface and the lower depths. OTEC systems draw on this enormous reservoir of low-temperature solar energy.

Temperature differences in the ocean can be quite substantial. In tropical oceans, temperature differences up to 22 °C have been measured. The surface waters reach temperatures of 23 to 27 °C, compared to temperatures of 10 °C at a depth of 400 m or 4 °C at 1000 m. The regions between 10 degrees north and 10 degrees south latitude are ideal for OTEC plants because winds are low, there are few hurricanes, and ocean currents generally move at a speed below 1 knot. In this area, the mean incident solar radiation is about 215 W/m^2. Within this range of latitudes, the total ocean area exceeds 80 million square kilometers, enough to provide the world's entire electrical needs.

4.9.1 Basic OTEC Design

Several designs have been proposed for converting ocean thermal energy to electric power. Since the technology is still in the development stage, no one knows which design will ultimately be most economically feasible; however, all the designs follow the same basic principles. We will discuss one promising design here.

The key to OTEC design is a circulating fluid with a low boiling point (also called a *refrigerant*). Ammonia is one common refrigerant. This fluid is pressurized to about 50 to 75 kg/cm² (70 to 110 psi) and pumped through a series of heat exchangers. First, it is cooled by water from the ocean's depths and condensed into a liquid. This is done in a heat exchanger called a *condenser*. Then it is heated by water from the surface. This is done in another heat exchanger called an *evaporator*. The refrigerant boils, creating high-pressure vapor which drives a turbine to generate electricity. Finally, the spent vapor is cooled again in the condenser, and the process starts over.

In the design pictured in Figure 4.9.2, a floating platform is built in the ocean to support the power plant. This platform can be similar to a deep-water oil-drilling platform. The evaporators and condensers are submerged in the water. Cold water is pumped from the ocean's depths through tubes connected to the condensers. Warm water is drawn through inlets located just below the surface. The used water is returned to the ocean. If the electricity is to be trans-

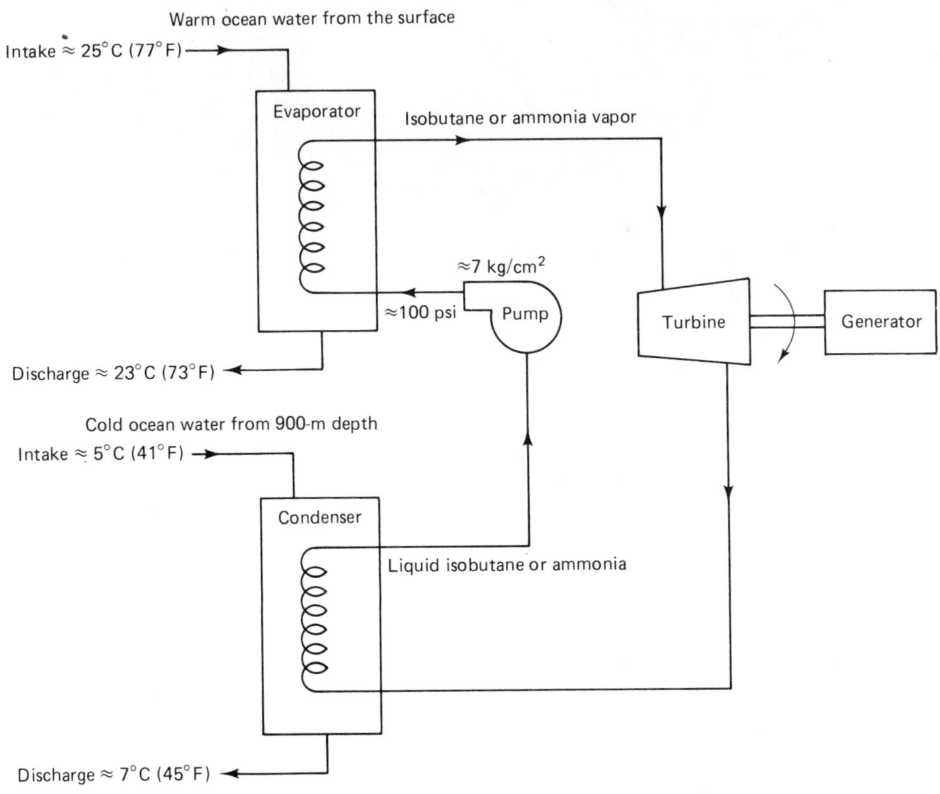

Figure 4.9.1 Ocean thermal energy conversion (OTEC) power plant.

Figure 4.9.2 Ocean thermal energy conversion (OTEC). Shown here is a model of one proposed structure which would use the small temperature differences in the ocean layers to drive a heat engine or turbine. The plans call for a 100-MW power plant 340 ft across and 17 stories high. The working fluid will be ammonia, and the engine will work on a temperature difference of 40°F. (*Source:* TRW.)

mitted to the land, a cable can be laid on the ocean floor. Otherwise, the power can be used on the platform to manufacture hydrogen or to extract chemicals and minerals from the seawater.

4.9.2 Performance Characteristics

Low-grade heat is a low-quality form of energy. It can be converted to a high-quality energy form such as electricity, but only at a low efficiency (see Chapter 5). For this reason, the efficiency of OTEC systems is quite low—approximately 2 to 3%—because of the relatively low range of temperatures involved. These low efficiencies raise the capital costs of OTEC systems: current projections place the construction costs at $1500 to $2000 dollars per kilowatt.

However, ocean thermal energy conversion can be a constant, uninterrupted source of energy—unlike solar heating or photovoltaic systems, whose effective operation is highly dependent on direct sunlight. The constancy and predictability of OTEC, together with its ready accessibility, may help offset the cost disadvantage caused by low efficiency.

4.9.3 Environmental Effects

Because OTEC does not create air pollution or emit toxic wastes, it is often considered to be an environmentally benign technology. But no one can be sure of the effects that would be produced by pumping massive quantities of cold water to the surface, together with the nutrients that tend to concentrate at lower depths. The carbon dioxide (CO_2) released to the atmosphere in this process could cause ecological disturbance. Changes in biological systems and ocean temperature patterns are possible. This means that the environmental effects of OTEC technology must be studied carefully before there is any rush to construct power plants.

4.9.4 Prospects for Development of OTEC

Because of the uneven distribution of ocean heat throughout the world, OTEC development favors tropical areas. Thus northern countries, such as Great Britain, will not be able to benefit from OTEC as easily as southern regions of the world. In the United States, OTEC potential will be limited to Hawaii and areas of the Gulf coast. Since the Gulf coast is already an energy-rich region, OTEC will not help to offset the imbalance in continental energy supplies. In Hawaii, however, other energy resources are not readily available and OTEC could become a significant source of electrical power. In fact, a small demonstration project off the Hawaiian coast recently produced electric power for an appreciable amount of time before mechanical problems forced the system to shut down. Hawaiian Electric has plans to buy 50 MW of OTEC power by the late 1980s.

OTEC plants can be commissioned rather quickly. The time from conception to completion is estimated at 5 to 10 years. This will help prevent the massive cost increases which have been experienced recently at power plants requiring longer construction times. In fact, OTEC appears to be more economically feasible than some of the more exotic technologies such as solar satellites.

Serious research in OTEC systems has been under way for the past few years. Lockheed, Bechtel, TRW, Global Marine Development, and United Engineers and Constructors are among the corporations involved in OTEC research projects sponsored by the National Science Foundation. The U.S. Department of Energy describes OTEC as a "promising" energy technology; however, DOE does not expect it to produce significant amounts of power before the year 2000. So, although OTEC is not currently a "mature" technology, its future prospects appear promising.

4.10 INDIRECT SOLAR: WAVE POWER

At present, there are no wave power plants in commercial operation, and therefore we have no way of knowing whether wave power is an economically feasible energy resource. But research into wave power seems promising, so we include wave power among the list of "futuristic" energy resources.

Waves are generated by wind blowing across the open ocean waters. Therefore, wave power can be considered an extension of wind power and, ultimately, of solar power. Wave power differs from hydro or tidal power because there is no translational movement of water; each individual molecule travels in a circular path, up and down, back and forth. Thus wave power cannot be used to turn a turbine. However, as each water molecule rises and falls, it gains and loses potential energy. This is the energy transfer that is tapped by most wave power machines being developed today.

4.10.1 Design of a Wave Energy Conversion Device

Of the many wave power devices that have been proposed in Japan and Great Britain, the most promising design developed so far is a floating device designed by Stephen Salter at Edinburgh University. It is a rocking boom with a cam-shaped cross section. The leading edge is pushed back and forth by the wave as it propagates across the water's surface. The circular shape of the trailing edge allows the boom to rotate freely with the wave's motion, creating very little disturbance in the backwater behind the boom. When the crest of the wave passes, the weight in the bottom of the boom causes the leading edge to rotate back to its original position, with the leading edge inclined toward the oncoming waves. This rocking movement can be used to generate electricity or mechanical energy. In principle, this device works beautifully, with up to 90% efficiency under ideal conditions. But in laboratory experiments simulating less than ideal conditions, the efficiency drops below 50%. (This is still considerably higher than the efficiency of a thermal power plant.)

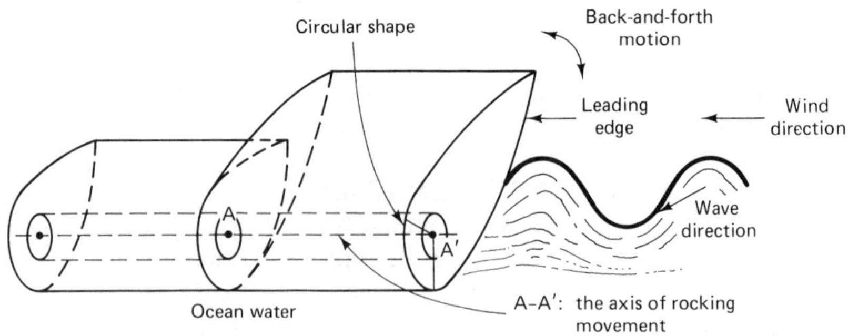

Figure 4.10.1 Stephen Salter's design for a wave-powered electric generating plant is made up of a large number of "ducks" or "rocking booms" connected in series.

Possible uses of wave power are not limited to the production of electricity. Other proposed applications include the following:

1. Production of hydrogen, which can be transported and burned as fuel.
2. Pressurization of water using a reversible pump. The pressurized water can be used to produce electrical or mechanical power.
3. Pumping seawater to extract valuable minerals, including uranium.

4.10.2 Obstacles to the Development of Wave Power

The movement of the waves remains the same as long as the wind velocity over the open ocean remains the same. However, as everyone knows who has lived by the sea, wind velocity is quite changeable, and so is the force of waves. So any wave power machine must be constructed to handle these fluctuations in wave motion. Other problems include the following:

1. It is difficult to synchronize mechanical motion to the relatively slow oscillations of the waves.
2. Getting power to the shore can be a difficult and expensive process.
3. Building strong, durable structures to hold the wave machines is a difficult and costly process.
4. Seawater is highly corrosive. Electrical equipment is especially susceptible to damage from seawater.

Finally, the economics of wave power are highly capital intensive. Current estimates of construction costs are much higher than the estimated costs of the tidal or OTEC systems.

4.10.3 Prospects for Wave Power

Despite these drawbacks, wave power has several advantages. Wave power is abundant and readily available over a large percentage of the earth's surface. Although no pilot plants have been constructed to determine the precise effects of wave power, it appears that environmental impacts would be low. Serious research work is currently being carried out to overcome the problems associated with wave power. In Great Britain and Japan, which both have an abundance of shoreline, significant work is being done toward harnessing wave power on a commercial scale. Potential power off the coast of England is estimated at 120,000 MW, of which approximately one-third can be harnessed. No significant research is being done in the United States, and the U.S. Department of Energy projections for the year 2000 do not list wave power as an energy resource of any consequence. But in island regions, wave power has the potential—in theory—of becoming a useful energy resource.

4.11 CONSERVATION: A RENEWABLE ENERGY RESOURCE

> Conservation . . . can be defined as the wise use of our natural environment: it is, in the final analysis, the highest form of national thrift—the prevention of waste and despoilment while preserving, improving and renewing the quality and usefulness of all our resouces.
>
> *President John F. Kennedy*
> *Conservation Message to Congress, 1962*

Conservation is different from the other energy resources we have discussed because strictly speaking, it is not an energy *resource* but rather a way of using other energy resources more effectively and more efficiently. However, in the highly industrialized, energy-intensive societies of the late twentieth century, conservation can offer as much or more energy, at much lower prices, and with less environmental degradation, than any other energy resource.

We have chosen to include conservation among the renewable energy resources because the energy saved by conservation is not "used up" in any sense. If a penny saved is a penny earned, then a Btu saved is one more Btu that will be available to future generations. Conversation is also renewable in the sense that many conservation measures, such as energy-efficiency improvements, offer the possibility of energy savings year after year. Some experts go one step further and say that renewable energy resources represent a form of conservation. In one sense they are correct: use of renewable resources conserves our precious nonrenewable resources. But in this section we use "conservation" to mean a reduction in the use of *all* energy resources, renewable or nonrenewable.

4.11.1 Deprivation or Innovation?

In many people's minds, conservation means living less comfortably, doing without many of the amenities of life, or even returning to the life of the past. In fact, President Reagan declared in a speech that conservation means "being too cold in the winter and too hot in the summer." This does not have to be the case. Certainly, it is possible to conserve by being too cold in the winter and too hot in the summer. But it is also possible, and really much easier, to save large amounts of energy by simply using energy more efficiently. In fact, many energy-efficiency improvements lead to a *greater* level of comfort at the same time that they save energy. Let us look at a few examples of how this works.

1. If a person lives in a drafty house with loose windows and doors and no insulation in the walls, the person needs to maintain the thermostat at a very high setting. This is true for two reasons:

a. The drafts caused by the air coming in around the doors and windows will make a person feel colder than will still air at the same temperature.
b. Radiated heat losses to cold walls will cause a person to feel colder than when in a room with the same air temperature and warmer walls.

In a properly insulated and weather-stripped house, these two problems are eliminated. So not only does the homeowner use less fuel to maintain the same temperature inside the house, but that temperature feels more comfortable.

2. In recent years, automobile manufacturers have worked very hard to improve the fuel-efficiency ratings of their cars. This has led to the development of front-wheel-drive automobiles with engines mounted in the front. Although the main incentive for this change was to reduce weight and thus to increase fuel economy, the change has also resulted in a more open, comfortable interior seating arrangement. There is no transmission and drive shaft bump in the middle of the seat, so seating can be arranged more comfortably. With more storage space in the passenger compartment and the trunk, the car can carry more luggage without being made any larger outside. Finally, the shorter lengths of many of the new cars makes them easier to park in tight spots. Some people complain that the smaller cars do not ride as smoothly as the older gas-guzzlers; but for short trips, there can be no doubt of the greater convenience of the new models. Even on longer trips, the money saved in lower fuel costs can be used to purchase other useful items for personal comfort.

3. Newer electronic systems use transistors, semiconductors, and integrated circuits rather than the old vacuum tubes and discrete circuits. This results in a dramatic decrease in power consumption. But that is not the only advantage. The use of integrated circuits allows electronic components to be much smaller and run much cooler than did the old vacuum-tube models. Solid-state electronic devices are also much more reliable than the older designs. In this case, conservation certainly does not mean deprivation; in fact, solid-state electronic devices were developed because of their great technological advantages, not for their energy-conservation potential.

These are only a few of the areas where energy savings improvements lead to a general increase in the quality of life. Readers will be able to supply other examples for themselves.

4.11.2 How Much Energy Can We Save?

A penny saved is a penny earned. But a gallon (or a kilocalorie) saved is more than a gallon earned. Besides money, labor, and time, a considerable amount of energy—mostly from nonrenewable resources—is required to recover oil, gas, or coal, to mine and process nuclear fuel, or to build a solar heating system or a wind turbine. Thus to be able to provide one unit of available energy, we actually have to come up with more than one unit of energy resources. But if we eliminate the need for that additional unit of energy resources, through the practice of intelligent conservation, we also save the energy needed to produce the resource. For this reason, projections of our potential for energy savings frequently understate the actual savings. In developing large-scale energy projections, these factors are usually taken into account. But when evaluating the merit of insulating a water heater or an attic, homeowners rarely take into account the secondary energy savings associated with conservation.

When we look at the national energy picture, we can see that there is a tremendous possibility for energy savings. Of the 79 quads of primary energy produced in the United States in 1979, 57% (45 quads) was simply lost, mostly through conversion inefficiencies. Of course, no energy-conversion system can be 100% efficient. But clearly there is room for more energy efficiency. As we will see in Chapter 5, it is inefficient to convert oil to electricity with an efficiency of 35%, and then convert that electricity to heat. Why not convert the oil directly to heat with an efficiency of 65% (in an older furnace) or 80% (in an efficient new furnace)? If the efficiency of energy conversions in the United States were increased from the present 43% to 53%, the United States could obtain the same amount of useful energy while reducing overall energy consumption to 63 quads. Looked at in another way: 79 quads of primary energy, used with an efficiency of 53%, could produce as much useful energy as 97 quads of primary energy used with an efficiency of 43%.

Although energy conversions represent a significant source of possible energy savings, they say nothing about the efficiency with which we *use* energy. For example, a furnace may burn oil with an efficiency of 80%, but if the house is poorly insulated, much of that oil will be used to heat the outdoors. A fluorescent light may convert electricity to light with a greater efficiency than an incandescent light, but if the fluorescent light is left on while no one is in the room, its useful efficiency is still zero. In the areas of end-use efficiency, tremendous savings are available. For example, space heating takes 16% of our national primary energy use, and automobiles take 13%. If all of the houses in the United States were insulated and weather-stripped so that they met current standards for thermal efficiency, and all the cars on the road met the 1982 fuel economy standard of 26 miles per gallon, the savings would be enough to eliminate all oil imports. Similar savings are available in nearly every area of energy use. Indeed, it has been estimated that with currently available conservation technology, U.S. energy use could be reduced to 56.8 quads per year, 80% of the current figure, with no reduction in standards of living. In fact, these figures assume a significant *increase* in the standard of living for many of the poorest members of American society.

When we add up the potential of energy savings from technological improvements and from changes in personal habits, we can see the potential for significant energy savings. For example, in European countries such as Sweden and West Germany, the standard of living is at least as high as in the United States, while energy use is 50% less. (Even in these countries, there is significant room for improvement.) It is entirely possible to reduce per capita energy consumption in the United States by 50% without causing any reduction in standards of living; in fact, these reductions in energy consumption could have many economic advantages, as we shall see in later chapters.

For the most dramatic proof of the conservation potential, we have only to look at the energy picture during the late 1970s. Between 1973 and 1978, U.S. energy savings from efficiency improvements were twice as great as the government's projections of future synfuel production. The efficiency improvements took only half as long as the projected synfuels schemes and cost less than one-tenth as much. Since 1979, U.S. energy consumption has declined from a high of 79 quads in 1979 to 71 quads in 1982 (DOE figures, excluding wood), primarily as a result of conservation.

4.11.3 Conservation Technology

Although we commonly think of conservation as a series of individual actions to save energy, it can also refer to technological innovations that enable us to use energy more efficiently. Here are a few conservation technologies that are helping to conserve energy today. Many other energy conserving technologies will be developed as incentives for conservation are increased in the future.

Cogeneration. When electricity is generated by a steam turbine, there is a considerable loss of waste heat. We are all familiar with the enormous cooling towers that are built at large nuclear and coal-fired electric power plants. These towers are designed to eliminate waste heat while minimizing the danger to the environment. But waste heat, especially in the form of steam or hot water, can also be used for other purposes, such as heating buildings or preheating water for industrial processes. Industrial cogeneration measures could increase electrical capacity by the equivalent of 100,000 MW and increase the efficiency of electrical power generation from 33% to 40% (in the case of steam cogeneration) or even 50% (gas and steam cogeneration).

Superinsulation. In central Canada and the northeastern United States, there is a growing interest in a type of building design that uses extremely thick insulation to essentially eliminate the need for heating and air-conditioning systems. The technology is rather sophisticated and requires the use of improved vapor barriers, moisture-control systems, and ventilation systems. Well-designed superinsulated buildings use essentially no energy for heating and cooling purposes. The necessary energy is provided by lights, appliances, and the occupants of the building.

Automotive technology. In 1973, the average new car built in the United States had a fuel economy rating of 12.75 mpg. Today that rating has nearly doubled. 1982 fuel economy standards require an average fuel economy of 26 mpg, and this figure rises to 27.5 mpg for 1985. Cars are now available in Japan with fuel economies of 70 mpg. As automotive technology continues to focus on fuel economy, even greater improvements can be expected.

Improved combustion systems. New furnaces and coal-burning systems are available which burn fuels with much higher efficiency than was available from earlier systems. Fluidized-beds, pulse burners, and wood chip burners are able to produce heat energy with a high degree of efficiency.

It is often said that conservation isn't "sexy"; that is, conservation lacks the kind of technological excitement and room for innovation represented by other areas of technology. But this view overlooks the significant developments in conservation technology that have occurred over the past decade. Today, energy conservation means more than turning down a thermostat.

4.11.4 Conservation on the Personal and Governmental Levels

When we think of conservation, we usually think of personal choices: driving a 40-mpg compact rather than a 12-mpg gas-guzzler, car pooling, turning down thermostats, turning off unneeded lights, insulating the attic, and installing a jacket on the water heater. There is a good reason for this view: since energy is used at a personal level, personal energy conservation can have an immediate and significant impact. Particularly when energy prices increase, as they did following the oil embargoes of 1973 and 1979, personal consumption can decline rapidly and dramatically.

But governments can facilitate energy conservation at the personal level, and can actually mandate conservation by businesses and governmental agencies. Here are a few governmental conservation measures which have stimulated energy savings on both the personal and corporate levels (see Section 9.5):

State and local energy auditing services

Community Action Program energy-conservation services

Automotive fuel economy standards

Building energy-efficiency standards

Government-mandated building temperature standards

Tax credits for installation of conservation equipment

Programs such as these make conservation more economically attractive and thus more likely to be done. Energy-conservation services for low-income citizens have an especially significant impact, since the poor tend to inhabit the most poorly insulated homes, drive the least energy-efficient cars, and spend the largest portion of their household budgets on energy.

4.11.5 Resource Recovery

The mining and processing of natural resources is a highly energy-intensive process. Glass production and smelting of metals require very high temperatures. Aluminum production uses enormous quantities of electrical energy. Many other resource extraction processes can be equally energy intensive, leading to a significant use of energy.

BOX 4.11.1 THE R'S OF CONSERVATION

*R*esource *R*ecovery
*R*ecycling and *R*euse of products
*R*eduction of energy consumption by:

- *R*enovating and *R*etrofitting existing structures
- *R*edesigning population centers
- *R*etooling industry for increased energy efficiency
- *R*apid transit (public transportation)
- *R*esidential units with high *R* values (see Chapter 10)

*R*egional energy planning
*R*elying on *R*enewable energy *R*esources whenever possible

When consumer products wear out and become useless, we lose the benefits of the energy that was consumed in their production. But we can recover the materials that were used in their manufacture, thus saving some of the energy that would have to be used to replace those natural resources.

In most countries of the world, resource recovery is simply taken for granted—as it was in the United States until very recently. Bottles are collected and melted down for the production of more bottles. Rags and cardboard are recycled to make paper products. Scrap metal is refabricated into new consumer products. In this way, the waste products of industrial societies can be mined for their valuable resources (see Table 4.11.1).

TABLE 4.11.1 Resource Recovery: Potential for Reduction in Energy and Environmental Costs

Industry	Energy Needed [Virgin Ore, (Btu/lb)]	Energy Savings (%)	Pollution Abatement Air (%)	Pollution Abatement Water (%)	Water Consumption Savings (%)
Paper	11,400	20–40	70	30	55
Glass	7,800	4–10	20	?	45
Steel	8,300	45–60	85	75	40
Aluminum	134,700	90–95	95	95	?

Sources: Denis Hayes, *Repairs, Reuse and Recycling,* Worldwatch Paper 23 (Washington, DC: Worldwatch Institute, 1982); Neil Seldman, "Resource Recovery," *Environment,* Vol. 25, No. 2 (1983), p. 43; Environmental Protection Agency, "First Report to the Congress," in A. Purcell, *Waste Watcher: A Citizen's Handbook for Conserving Energy and Resources* (New York: Technical Information Project, 1980); *A Guide to Recycling the Source Separation Way* (Indianapolis, IN: Indiana State Board of Health, 1979).

4.11.6 Advantages of Resource Recovery

The principal advantages of resource recovery are as follows.

1. *Economics.* The costs of producing materials from recycled products are less than the costs of production using virgin ores. Local, small-scale industries can be cost-competitive with large, centralized plants, even though they are dealing with smaller quantities of material.
2. *Transportation.* When recycled materials are used locally, transportation costs are reduced. In the past, many transporters charged higher fees for transport of recycled materials than for material produced from virgin ores. As this practice declines, the transportation advantages of recycled materials will become even more evident.
3. *Land use.* Land is not scarred by mines when recycled materials are used in place of new materials. Land for disposal sites and landfills is not used up as rapidly. Solid-waste management is made easier, producing lower costs for the community.
4. *Local benefits.* When resources are available locally, through resource recovery and recycling, funds tend to remain in the community rather than flowing to the outside for the purchase of raw materials. The money stays in the region, producing local jobs and stimulating the local economy.
5. *Environment.* When fewer raw materials are consumed and disposed of, there is less need to disrupt the land by mining and waste-disposal activities. Air and water pollution are also reduced.
6. *Social values.* Resource recovery and recycling promote values and social practices such as thrift, cooperation, awareness of systems, stewardship of the land, and providing for the future.

4.11.7 Prospects for Conservation

Many energy analysts believe that conservation is the most important of all energy resources, because it can be substituted for any other energy resource, renewable or nonrenewable. Nuclear power cannot be substituted readily for imported petroleum, because most petroleum is used for transportation or home heating, and at present nuclear-generated electricity is not a good choice for either of those purposes. However, conservation can easily be substituted for petroleum (in the form of better insulation or a higher-mileage car), for nuclear-generated electricity (in the form of a more efficient refrigerator or an insulating jacket for an electric water heater), or even for solar energy (a well-insulated house will require a smaller solar heating system). Because of this versatility, conservation will be an important part of any future energy policy.

Economically, conservation presents significant advantages over all other energy resources. Many energy-conservation measures cost nothing; others are quite inexpensive compared to energy-production alternatives. For equivalent amounts of energy, conservation measures cost roughly one-tenth as much as energy production facilities. This cost-effectiveness has led to dramatic increases in energy conservation in recent years. Energy conservation has become one of the fastest-growing industries in the world. Prospects for the future are even more promising, because the long-term trend in energy prices can only lead upward, despite short-term fluctuations due to temporary imbalances of supply and demand.

Finally, conservation is one of the few energy resources that is almost universally popular. Advocates of solar energy, nuclear energy, coal, and synthetic fuels all can agree on the significant benefits offered by energy conservation. This widespread acceptance, together with its many other benefits, makes conservation one of the world's major energy resources today and for the future.

STUDY QUESTIONS

Solar Energy

1. List the ways in which solar energy affects your everyday life. To what extent are you aware of these effects from day to day?
2. Classify active, passive, and hybrid solar energy systems. Is this a meaningful classification? Can you propose a better way of classifying systems?
3. Do you know of any houses in your neighborhood with solar water or space heating? Do they have active or passive systems? Prepare a simple sketch of the basic solar heating design.
4. Do all solar heating systems have thermal mass, or is it only part of passive solar energy systems? What is the importance of thermal mass in solar heating designs?
5. Why is there so much interest in photovoltaic cells despite their low efficiency (around 16% at the maximum)?
6. What are the main applications for photovoltaic cells? Do you know of any photovoltaic installations?
7. Which solar technologies are economically competitive with other energy technologies? Which solar technologies are most popular?

Hydro Energy

8. Why is hydro energy an indirect form of solar energy? Describe the energy exchanges in the water cycle.
9. Why do most hydroelectric installations use dams and reservoirs rather than using the flowing water directly?
10. Describe a typical hydroelectric system, identifying all the conversions from one form of energy to another. Assume that the electrical power is sent directly into the electric utility lines. Sketch a block diagram of the system.
11. Describe a pumped storage hydro system, identifying all the conversions from one energy form to another. Assume that the electric power is received and sent out on the same utility line.

12. Are there any hydroelectric systems near where you live? What type are they? Do they incorporate pumped storage? How is the energy sent out? How much electricity do they produce? How does this compare with the amount of electricity used in your community?
13. There is currently great interest in reviving hydro sites that were abandoned during the mid-twentieth century. Why were these sites abandoned, and why is there now a renewal of interest in them?

Tidal Energy

14. What are the similarities and differences between tidal and hydro energy resources?
15. Describe the functioning of a single-basin, two-way tidal electric power plant, listing all the energy conversions. Assume that all the electric power is fed into the utility grid.
16. Where are the significant tidal sites in the United States? Why haven't they been developed? What does this suggest about the future of tidal power in the United States?
17. Why is the Bay of Fundy considered a good tidal site? Looking at a world map, find some other spots that might share some of the same geographical characteristics. Are they considered good tidal sites?
18. Do you think tidal power will provide a significant contribution to the world's energy needs? Why or why not?

Wood

19. Wood heat is becoming increasingly popular in the United States, especially in rural areas. What are the causes of this renewed popularity?
20. Why doesn't the U.S. Department of Energy keep records of wood production and consumption as it does for other energy resources?
21. Is wood a significant energy resource? How does it compare to hydro or nuclear power in total energy produced? Would you call it a nationally important or a regionally important resource?
22. What are the desirable characteristics of a wood stove? Explain why a fireplace is so inefficient.
23. What are the health and environmental factors, if any, associated with wood energy?

Biomass

24. "The oldest and most reliable solar power plants are green plants." Comment on this statement.
25. Is municipal garbage or refuse a feasible source of energy? Could the garbage in your community be used to produce biogas or other energy resources? Are any such projects currently being carried out?
26. What are methanol, ethanol, and gasohol? How do they differ, how are they produced, and what are their different uses?
27. What are the environmental impacts of biomass energy production? How do they compare with the impacts of alternative energy resources?

Wind Energy

28. Why has interest in wind energy begun to reappear only during the past few years?
29. What are the characteristics of a good wind site? Identify and describe a good wind site in your neighborhood, if you can find one. What are its limitations or drawbacks?
30. Compare the advantages and drawbacks of ac and dc systems for wind power.
31. How does wind energy compare with other renewable energy resources in terms of cost, environmental impacts, and social effects?
32. Do you believe that wind energy will be a significant energy resource by the turn of the century? Support your opinion with factual information.

Ocean Thermal and Wave Energy

33. Why are ocean thermal and wave energy considered to be inefficient methods of producing electrical power? Can you think of any more efficient uses for these energy resources?
34. What environmental effects would you expect from widespread use of OTEC? From use of wave energy? Which of these technologies presents fewer environmental problems?
35. Which areas of the country are most likely to benefit from wave energy and OTEC? Do these areas currently consume large amounts of electrical power? What does this suggest about the future of these technologies?

Conservation

36. How significant is conservation in the national energy picture? What energy resources provide comparable amounts of energy?
37. Which sectors of the American economy offer the greatest potential for energy savings through conservation?
38. Can you think of any examples where conservation measures have caused adverse environmental effects? If so, how could those effects be minimized?
39. Visit a local industrial plant and find out what energy-conservation measures have been taken there within the past 10 years. Are further measures being planned? Why or why not?
40. Some economists have suggested that conservation is not as desirable as increased energy production because conservation limits overall growth in the economy while increased energy production contributes directly to growth. Do you support this statement? Give your reasons.
41. What personal conservation measures have you carried out within the past year? Do you expect to undertake any conservation projects in the near future? Describe them.
42. Although conservation does not produce any additional energy, it is often considered an energy resource. Do you feel that conservation should be classified this way? Explain your reasoning.
43. On page 201 readers are encouraged to give their own examples to show how energy efficiency can lead to a general increase in the quality of life. What examples can you think of?

PROBLEMS

Solar Energy

1. An active water heating system is required to heat 300 gallons per day. The incoming water temperature is 45 °F, and the water will be heated to 120 °F.
   (a) How much collector area will be needed under the following conditions?

   Maximum intensity of direct solar radiation, 300 Btu/ft^2
   Average intensity of solar radiation, 25% of maximum
   Efficiency of system, 35%

   (b) At a system cost of $35.00 per square foot of collector area, how much will this system cost?
2. In a properly designed passive solar space-heating system, 1 ft^2 of south-facing glass will save roughly the equivalent of 1 gallon of fuel oil per heating season. (This is approximately 100,000 Btu, taking into account the efficiency of the oil heating system.) If this is the case, how much south-facing glass would be required to provide 60% of the space-heating needs for a house with total seasonal heating requirements of 70 million Btu? How does this compare with the south-facing surface area of a typical house?
3. A passive solar heating system requires 500,000 Btu of energy to be stored in the thermal mass. Assuming that the temperature of the mass can change by 10 °F, what is the needed volume of thermal mass if the mass consists of:
   (a) Water, with a heat capacity of 62.4 Btu/ft^3?
   (b) Concrete with gravel, with a heat capacity of 22 Btu/ft^3?
4. Assume that the peak electrical power requirement in your house is 5000 W. What area of photovoltaic cells would you require to provide this amount of peak power, with a safety margin of 10%? Assume that the efficiency of the photovoltaic system is 10% and the intensity of solar radiation is 0.1 kW/m^2. How does this compare with the roof area of a typical house?
5. Look up the current price for photovoltaic arrays. At this price, what is the required investment to construct the system described in Problem 4?

Hydro Energy

6. A stream has an average flow of 0.1 m^3/sec and the available head is 2 m.
   (a) How much power (in kilowatts) is available in the stream?
   (b) At 85% efficiency, what will be the power output of the electrical system?
7. A hydro site has a head of 5 m, an average flow rate of 0.2 m^3/s, and a flow rate of 0.4 m^3/s in the spring (three months).
   (a) What is the annual energy potential of this site?
   (b) If the dam is built to limit spring flow to 0.2 m^3/s, what volume of water will be stored behind the dam during the spring?
   (c) Assuming that the average depth of the reservoir is 1.55 m, calculate the land area flooded by the hydro power site.
8. Hoover Dam has 17 generators with a combined capacity of 1345 MW. Assuming

a head of 225 m, calculate the water discharged through each turbine in 1 second. Assume an efficiency of 0.95 for the generator and 0.88 for the turbine.

Wood

9. Suppose that your house requires an average of 80 million Btu per heating season.
 (a) Using an airtight stove with an efficiency of 50%, how many cords of maple would you need to burn in one heating season?
 (b) How many cords would you need if the wood were basswood instead of maple?

10. The homeowner in Problem 9 is planning to build a woodshed large enough to hold a 2-year wood supply. Assuming that he can depend on a steady supply of maple, how large will the shed have to be? (Assume that the wood will be stacked to a height of 7 ft.)

11. A wood stove with a capacity of 50,000 Btu/h is installed in a house. If it is burned continuously for 24 hours on a very cold day, how many cords of maple wood will it need? (Assume that the stove has an efficiency of 50% and that the rated output is stated in terms of useful heat rather than wood consumption.)

Wind

12. What is the theoretically available power in a wind turbine with a swept area of 10 m^2 if the wind speed is:
 (a) 2 m/s,
 (b) 10 m/s,
 (c) 20 m/s,
 (d) 50 m/s?

13. If the coefficient of the rotor is 0.4, calculate the power harnessed by the rotor in Problem 12 at each of the listed speeds.

14. If the generator efficiency is 85%, calculate the electric power output of the wind system in Problem 12 at each of the listed wind speeds.

15. If the wind turbine in Problem 12 is used to pump water using an 80% efficient pump, calculate the work done in 1 hour at each of the different wind speeds.

Biomass

16. Sugarcane is planted on a 10-acre field with an average expected yield of 135 tons per acre. How much alcohol fuel can be produced per year? (One ton of sugarcane will produce roughly 20 gallons of alcohol.)

17. Current estimates of alcohol production from biomass is 115 million gallons per year. How much land area does this represent? Assume that the average piece of land produces two-thirds as much fuel alcohol per acre as the field in Problem 16.

Ocean Thermal

18. Calculate the theoretical maximum efficiency of an OTEC plant operating off the coast of Hawaii where the surface temperature is 25 °C and the water temperature at 100 m depth, where the condenser is located, is 4 °C. (*Hint:* Refer to Chapter 2 for the theoretical efficiency of a heat engine.)

19. If the OTEC plant in Problem 18 operates at 80% of the maximum theoretical efficiency, and if it cycles 1000 liters of water per minute, what is the output power of the plant in kilowatts?

Conservation

20. Here are some fuel-consumption figures for 1973 and 1983.

	1973	1983
Miles driven (average car)	10,000	9,000
Average fuel economy (MPG)	15	18
Number of cars (millions)	115	125

(a) How much was total automotive fuel consumption reduced between 1973 and 1983?

(b) If the 125 million automobiles in 1983 were driven the same distance at the same fuel economy as the 115 million automobiles in 1973, what would be their overall consumption of gasoline?

(c) What is the difference between the energy consumption figure in part (b) and the actual energy consumption in 1983? State your answer in terms of gallons saved and percentage reduction in gasoline use.

21. Suppose the fuel economy of an average car in 1983 were 27 mpg instead of 18 mpg.

(a) How many gallons would be saved compared to the actual 1983 fuel consumption?

(b) What percentage decrease would this represent?

22. In 1979, 79 quads of primary energy was used in the United States. Of this, 57% was lost in the transport and conversion process.

(a) What was the end-use efficiency of the system?

(b) If this efficiency could be improved by 25%, how much primary energy would be required to provide the same amount of end-use energy?

23. An average refrigerator sold in 1975 consumed 1600 kWh/yr. The California energy standards of 1976 called for 1400 kWh/yr. However, it is possible to manufacture a refrigerator that consumes only 450 kWh/yr. If there are 100 million refrigerators in the United States, calculate the annual energy savings if:

(a) The average energy consumption were equal to the California standard.

(b) The average energy consumption were equal to 450 kWh/yr.

24. An average new thermal power plant produces roughly 5.4 billion kWh/yr and costs 3 billion dollars to build.

(a) How many power plants could be eliminated under conditions (a) and (b) in Problem 23?

(b) How much capital investment would this represent?

FURTHER READINGS

BUNGAY, HENRY R. *Energy, the Biomass Option.* New York: John Wiley, 1981.

GIBBONS, JOHN H., and WILLIAM U. CHANDLER. *Energy: The Conservation Revolution.* New York: Plenum Press, 1981.

KENDALL, HENRY W., and STEVEN J. NADIS, eds. *Energy Strategies: Towards a Solar Future.* A report of the Union of Concerned Scientists. Cambridge, MA: Ballinger, 1980.

MARIER, DONALD. *Wind Power for the Homeowner.* Emmaus, PA: Rodale Press, 1981.

MAYCOCK, PAUL D., and EDWARD N. STIREWALT. *Photovoltaics: Sunlight to Electricity in One Step.* Andover, MA: Brick House, 1981.

MAZRIA, EDWARD. *The Passive Solar Energy Book.* Emmaus, PA: Rodale Press, 1979.

MCMULLAN, J. T., R. MORGAN, and R. B. MURRAY. *Energy Resources.* New York: John Wiley, 1977.

ROSS, DAVID. *Energy from the Waves.* Oxford: Pergamon Press, 1979.

STOBAUGH, ROBERT, and DANIEL YERGIN, eds. *Energy Future: Report of the Energy Project at the Harvard Business School.* New York: Random House, 1979.

WILLIAMS, ROBERT H., ed. *Toward a Solar Civilization.* Cambridge, MA: MIT Press, 1978.

5

Physical Analysis

5.1 THE EARTH'S ENERGY BUDGET

In Chapter 2, we learned that energy can neither be created nor destroyed. In other words, the total amount of energy in a closed system is a constant. But the earth is not a closed system. The earth is constantly receiving incoming energy from the sun and radiating energy to outer space. This means that the total amount of energy on the earth is not necessarily constant. In fact, some energy-conversion processes cause a net increase in the earth's stock of energy, some cause a net decrease, and some cause no change.

When solar radiation reaches the earth's system (which includes the atmosphere), some of the incoming solar radiation is immediately reflected back into space by the upper atmosphere. Some of it is absorbed by the atmosphere, raising the temperature of the atmosphere. The rest reaches the earth's surface. Of the energy that reaches the earth's surface, some is reflected back and some is absorbed, raising the temperature of the earth's surface. After this, many other energy transfers take place. Some of the heat energy is transferred to the atmosphere, some is stored in the form of evaporated water, some is stored in chemical and other forms, and some is radiated from the earth back into outer space.

In these thermal exchanges, the influx of solar radiation is generally balanced by the outflow of energy. It is easy to see why this is so. As the earth

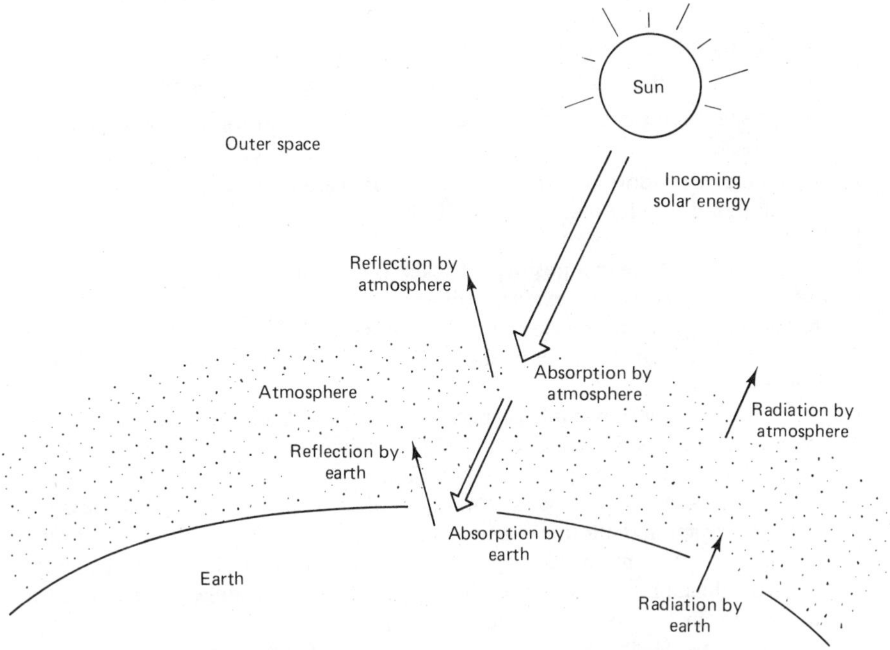

Figure 5.1.1 Earth's energy income.

becomes warmer, it radiates energy at a higher rate, because the temperature difference between the earth and outer space has increased. Conversely, as the earth cools down, it radiates energy at a lower rate. This process of *negative feedback* (see Box 5.1.1) keeps the energy exchange roughly in balance: that is, the amount of energy received from the sun is approximately equal to the amount of energy radiated from the earth. As we shall see, however, this balance can be upset, causing significant changes in the temperature of the earth, and thus in the earth's climate (see Chapter 7).

BOX 5.1.1 POSITIVE AND NEGATIVE FEEDBACK

Everyone has had the experience of attending a concert or lecture where the sound equipment began emitting an ear-splitting squeal. This squeal is a good example of *positive feedback*. The process of positive feedback is illustrated in Figure 5.1.2. The source sound is picked up by the microphone, amplified, and sent through the loudspeaker, where it is picked up again by the microphone, this time a bit louder than before. The louder signal is amplified and broadcast again, resulting in an even louder sound, which is picked up again, and so forth, until the sound reaches the maximum output of the amplifier–loudspeaker combination.

Theoretically, positive feedback has the following characteristics:

1. Conditions are met for the production of a certain effect.
2. Once the effect takes place, it causes changes in surrounding conditions.
3. These changed conditions are more favorable to the production of the effect.
4. The effect continues to occur with increased intensity as a result of the changed conditions.

It is easy to see that positive-feedback systems are very unstable and destructive. When a pattern of positive feedback develops, the processes rapidly reach the limits of the system, causing destructive breakdown. Interestingly enough, many of the processes that interest us in this text are examples of positive-feedback systems in action.

Fire is a good example of positive feedback. A high temperature, in the presence of combustible materials, results in rapid oxidation and a release of heat. This release of heat raises the temperature, thus causing more oxidation. The process continues to increase in intensity until the fuel is used up. A nuclear fission chain reaction is another example; in this case, a free neutron causes the division (fission) of an atomic nucleus. In the process of fission, more neutrons are released, which in turn cause more fission until there is no longer enough nuclear fuel to sustain fission.

Negative feedback, as the name implies, is the opposite of positive feedback. Consider, for example, a marble rolling inside a bowl. As the marble begins to travel up the side of the bowl, it encounters resistance from the force of gravity, which opposes its progress and sends it back toward the bottom of the bowl. Eventually, the marble comes to rest in the bottom of the bowl. Another example is a person taking a shower. If the water is too hot, the person adjusts the controls to lower the water temperature. If the water is then too cold, the person adjusts the controls to raise the temperature. The process continues until the water reaches the desired temperature.

Negative-feedback systems generally tend toward stability, although if there is a certain time lag between the input and the feedback, these systems, too, can go out of control. For example, imagine a shower with a very long pipe. The bather gets in the shower and turns on the water, but the water is too cold. The bather turns up the control, but the shower does not respond quickly. The impatient (and cold) bather turns up the control again. But now, when the water finally does warm up, it is too hot, so the unfortunate person turns down the control. Again, nothing happens right away, so the control is turned down again. But this makes the water even colder than it was before, causing a shriek of pain and a lunge for the control. Soon the shower temperature is completely out of control and the frustrated bather has given up in despair.

Negative feedback is important in maintaining the stability of natural systems. For example, when the population of a species increases, so does the population of predators that feed on that species. This brings the population of the species back down to the earlier level, and the lowered availability of food reduces the predator population to its earlier level.

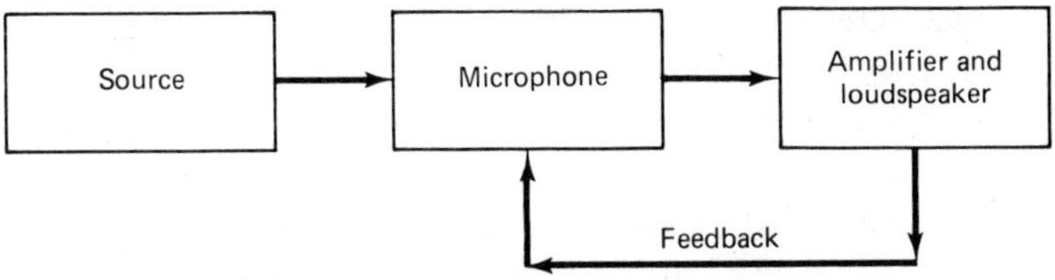

Figure 5.1.2 Positive feedback.

Besides the inflow of solar energy to the earth and the outflow of radiational energy from the earth, transformations occur which change energy from one form to another. For example, incoming solar energy is not all converted into heat. Some of the solar energy causes evaporation of water. Some is stored by green plants in the form of chemical energy. Some—a very small amount, to be sure—is converted into electricity by photovoltaic cells.

These primary transformations are often followed by further energy transformations. Some of the plant energy is burned as fuel in wood-burning stoves, some is converted into other forms of chemical energy by animals, and some is stored underground as fossil fuels. Some of these fossil fuels will be converted to heat energy by being burned. Others will be converted into plastics and other useful products. Some of the electricity produced by the photovoltaic cells will be converted into mechanical motion. Some will be converted into waste heat. Some will be converted into chemical energy in the form of battery storage.

Many of these energy transformations ultimately result in the production of waste heat. This raises the earth's temperature, which in turn increases the amount of energy radiated from the earth to outer space. So when energy is converted from any other form to waste heat, we can say with a fair degree of accuracy that it is being sent into outer space—a rather unproductive form of *space heating*. Of course, the energy can do a fair amount of useful work before it is ultimately radiated to the universe; that is the whole point of energy use. But when fossil fuels, for example, are converted to waste heat and radiated to outer space, the earth's total energy stock is decreased.

Other transformations will result in the storage of additional energy on the earth. For example, when solar energy is converted to chemical energy by green plants and stored as fossil fuels, that energy is not radiated back into space. It remains on the earth. Similarly, when solar energy causes the evaporation of water from the ocean and that water is carried to lakes and ponds at high altitudes, the energy remains on the earth in the form of potential (gravitational) energy. So the earth's total stock of energy is *increased* when solar energy is stored in some other form, such as chemical or potential energy.

Is an Increase in the Earth's Energy Desirable?

At first glance, we may tend to feel that any increase in the earth's supply of energy is desirable, and that any decrease in the earth's supply of energy is undesirable. However, the situation is not that simple. Suppose, for example, that the characteristics of the earth's atmosphere were changed so that the earth radiated less energy to outer space at a given temperature. This could be caused by the so-called *greenhouse effect*, in which carbon dioxide builds up in the atmosphere as a result of fossil fuels being burned. If this happened on a global scale, the earth's temperature would increase a few degrees. The total amount of the earth's energy stock would likewise increase. The effects of such a temperature increase would be devastating. Much of the world's landlocked polar ice would melt, causing a rise in the level of the oceans. A glance at a world map will show that even a rise of a few meters would be sufficient to flood many of the world's most important cities: New York, Los Angeles, Tokyo, Shanghai, Calcutta, Hong Kong—the list goes on and on.

On the other side, of course, the earth's supply of energy cannot decrease forever. After a certain point, the world's stored energy stocks would be totally depleted. But this does not imply that a small decrease in the world's stored supply of energy will necessarily be harmful in the long run. In developed countries, the conversion of fossil fuels to electricity has led to great improvements in medical technology, communications, and the quality of life. In some cases, a short-term depletion of the world's energy stores can even lead to a long-term reduction in the rate of resource depletion. For example, large quantities of fossil fuels are burned in the production of mineral wool for insulation, glass for solar collectors, and purified silicon for photovoltaic cells. Over many years, this will actually reduce the amount of energy that must be drawn from the earth's reserves of stored energy. As long as resources are used wisely, a decrease in the earth's total stock of energy is not necessarily an undesirable prospect.

5.2 SUSTAINABILITY: THE GOAL OF ENERGY ANALYSIS

In many discussions of energy policy, only the immediate energy situation is taken into account. As long as there is sufficient energy to get us through the next 20 years, we feel that we are safe. Many analysts project energy use to the year 2000. It is assumed that if we can last until then, our energy problems will be over. But the year 2000 is not very far away—less than one-fourth of an average person's lifetime. It will be small comfort to us if we make it through the year 2000, only to run out of energy five years later.

The best energy policy will be one that offers us the possibility of a *sustained* energy supply: one that we will be able to count on for generations into the future. In developing a policy of energy sustainability, we need to understand patterns of energy consumption, concepts of efficiency, and the optimum use of different energy resources. Once we understand these basic principles, we will be more successful in matching our energy supplies with the functions we hope to accomplish in our use of energy. We will also be able to see how sustainability in energy supplies is related to other issues, such as population control, deforestation and soil conservation, conservation, and development of renewable energy resources.

5.3 FINITE RESOURCES AND PATTERNS OF DEMAND GROWTH

Since early Greek times, it has been generally accepted that the earth is round. Despite popular representations to the contrary, medieval astronomy was based on the assumption that the earth was a sphere surrounded by other rotating spheres. When Magellan completed the first successful circumnavigation of the earth, no cherished theories were upset.

Today, we know the size and the mass of the earth very precisely. What is more important, we have begun to realize that the earth is limited, both in its capacity to supply us with raw materials and in its capacity to absorb environmental abuse. This means that a policy of using up the earth's natural resources is not a sustainable one. Although the earth is very large, so is our energy appetite. During the past few decades, we have used up significant portions of the earth's stock of oil and natural gas. As we continue to consume fossil fuels and other natural resources, we are coming ever closer to the point where our levels of energy use can no longer be sustained.

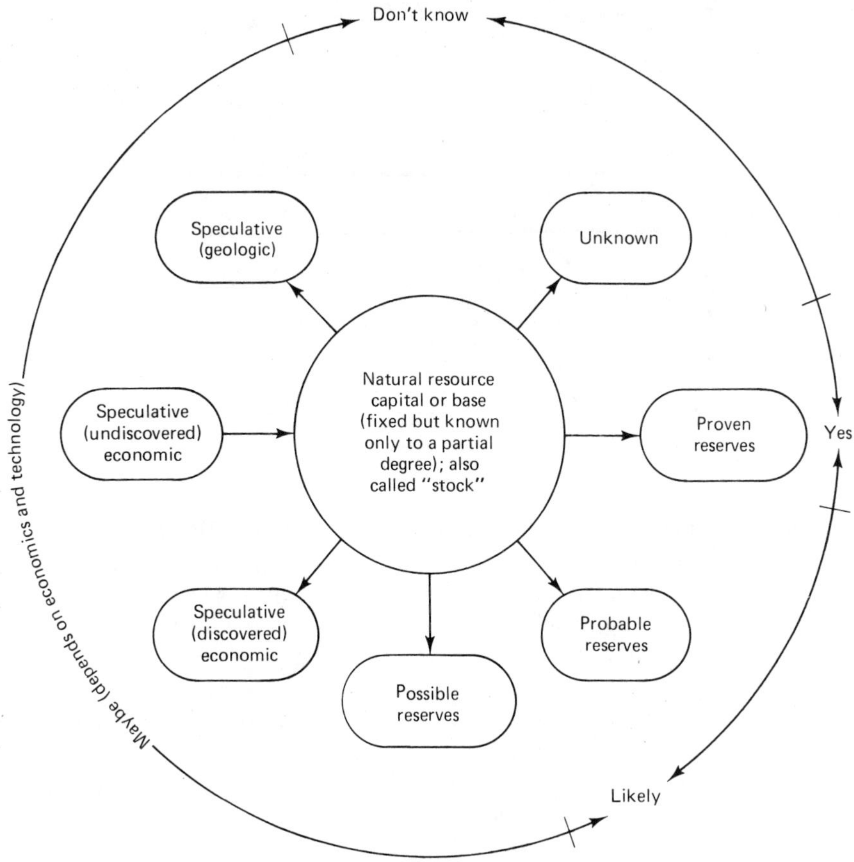

Figure 5.3.1 Round-earth theory of resource base.

Estimates of the earth's total energy reserves vary greatly depending on the person who is doing the estimating. Some claim that we have enough gas and oil for a few decades, whereas others claim that major new deposits remain to be discovered. But all agree that the earth's total deposits of energy resources are finite, and that the world's supply of nonrenewable energy resources cannot be replenished except over a period of many millions of years. Although we have no direct control over the earth's total supply of energy resources, we do have some control over our own patterns of energy demand. So let us look at the ways in which our demand for energy changes with time, and how those changes affect our energy future.

5.3.1 Constant Demand

In certain situations, energy demand remains essentially constant from year to year. For example, a gas furnace will tend to use the same amount of energy from year to year as long as the climate and the insulating qualities of the house remain essentially the same. In graphic form, this pattern of energy demand can be represented as shown in Figure 5.3.2.

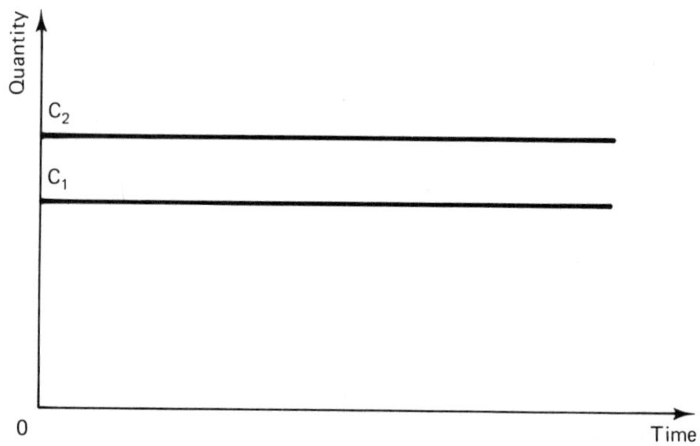

Figure 5.3.2 Constant demands graph.

Suppose that each year this gas furnace used 100,000 ft^3 of natural gas. Now suppose that there were 1 million such houses in the world, each using natural gas at the same rate, and that this was the world's sole use of natural gas. (This is necessarily a simplification for purposes of illustrating a principle.) This means that altogether, the 1 million houses would be using 100,000,000,000 (1 hundred billion, or 10^{11}) ft^3 of gas per year. Suppose further that the world's supply of natural gas amounted to 8,000,000,000,000,000 (eight quadrillion, or 8×10^{15}) ft^3. To find how long the world's supply of natural gas would last under these conditions, we divide the supply by the amount used per year. This yields the following:

$$\frac{\text{world supply}}{\text{annual use}} = \frac{8 \times 10^{15}}{10^{11}} = 8 \times 10^4 = 80{,}000 \text{ yr}$$

Of course, demand for energy is rarely constant. For one thing, the world's population increases from year to year. Other factors also contribute to increases in energy demand. But the precise *nature* of the increase in energy demand is not necessarily simple. Let us consider a few categories of increasing demand.

5.3.2 Linearly Increasing Demand

If demand for energy increases by a certain amount each year, the demand is said to be *linearly increasing*. A look at Figure 5.3.3 will show why this name is appropriate. In the example stated above, suppose that 100,000 new houses are built each year, each using 100,000 ft^3 of natural gas per year. This will increase the demand for natural gas by a constant amount each year. Graphically, the situation can be represented as shown in Figure 5.3.3. Let us assume that the world's total supply of natural gas is still fixed at 8×10^{15} ft^3. Now it becomes somewhat more complex to calculate how long the world's supply of natural gas will last. In the first year, the amount of natural gas used will be the same as in the preceding example: 100 billion cubic feet (BCF). But the next year, the amount of gas used will increase by 10 BCF. Thus the second year, the amount of gas used will equal 110 BCF. In the third year, another 10 BCF will be added to the amount used during the second year. This raises the gas consumption to 120 BCF for the third year. Similarly, gas consumption during the fourth year equals 130 BCF, 140 BCF during the fifth year, and so on.

If demand for natural gas increases by this constant amount every year, the world's supply of natural gas will not last nearly as long. In fact, the natural gas will be used up in 12,650 years. This is a long time, but not nearly as long as 80,000 years. Clearly, if energy demand is constantly increasing, our energy resources will not last as long as they would if demand remained constant.

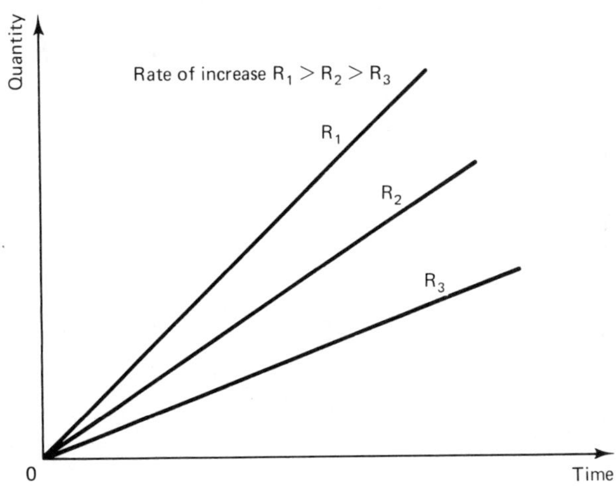

Figure 5.3.3 Linearly increasing demands graph.

5.3.3 Exponentially Increasing Demand

Unfortunately, the situation presented above, with demand increasing by a constant amount, may be unrealistically optimistic. In fact, the *growth* in energy demand may increase from year to year. This situation can be represented graphically as shown in Figure 5.3.4. Exponential growth is a difficult concept to understand, but our simple example may help make the situation clearer. In the previous case, we specified that 100,000 new houses would be built each year, each house using 100,000 ft^3 of natural gas per year. But suppose, instead, that the number of houses increases by *10% per year.* In this situation, let us see what happens during each of the first few years.

During the first year, natural gas use is still equal to 100 BCF. During the second year, this increases by 10%, or 0.1×100 BCF. Thus the *increase* during the second year equals 10 BCF, just as in the earlier example. This means that during the second year, the total amount of natural gas use equals 110 billion cubic feet.

$$
\begin{aligned}
\text{first-year natural gas use} &= 100 \text{ BCF} \\
\text{second-year increase} &= 0.1 \times 100 \text{ BCF} \\
&= 10 \text{ BCF} \\
\text{second-year use} &= \text{first-year use} + \text{second-year increase} \\
&= 100 \text{ BCF} + 10 \text{ BCF} \\
&= 110 \text{ BCF}
\end{aligned}
$$

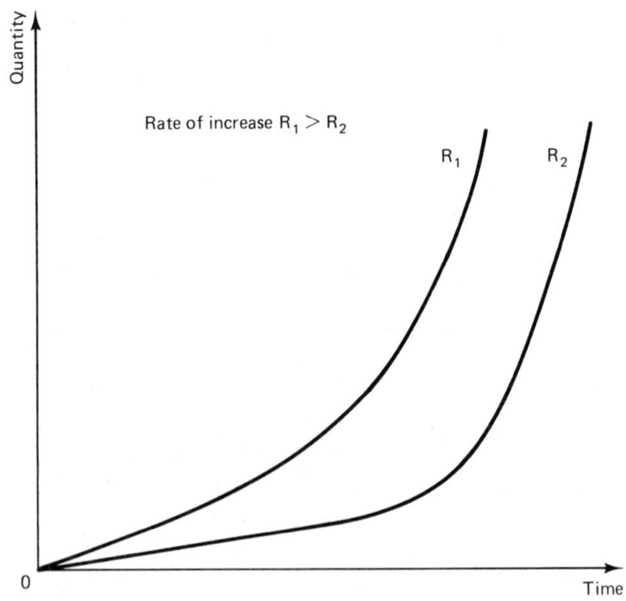

Figure 5.3.4 Exponentially increasing demands graph.

But now consider what happens during the *third* year. In this case, the increase will be greater than it was during the second year.

$$\begin{aligned}\text{second-year natural gas use} &= 110 \text{ BCF}\\ \text{third-year increase} &= 0.1 \times \text{second-year use}\\ &= 0.1 \times 110 \text{ BCF}\\ &= 11 \text{ BCF}\\ \text{third-year use} &= \text{second-year use} + \text{third-year increase}\\ &= 121 \text{ BCF}\end{aligned}$$

The *change* in demand will continue to increase in each succeeding year as long as the exponential growth pattern holds true. Here are the demand increases for the next few years of the progression (rounded off to the nearest 100 million cubic feet):

$$\begin{aligned}\text{fourth-year increase} &= 13.3 \text{ BCF}\\ \text{fifth-year increase} &= 14.7 \text{ BCF}\\ \text{tenth-year increase} &= 23.5 \text{ BCF}\\ \text{20th-year increase} &= 61.1 \text{ BCF}\\ \text{26th-year increase} &= 108.3 \text{ BCF}\end{aligned}$$

Note that during the 26th year, the *increase* in natural gas use is larger than the *total* use in natural gas during the first year. In this case, if we go through all the calculations, we find that the world's supply of natural gas will be used up in only 120 years. This is a far cry from 80,000 or even 12,000 years.

BOX 5.3.1 ESTIMATING DOUBLING TIMES

Calculating the effects of exponential growth can be a complex process, requiring a clear head and a sophisticated calculator. But for quick estimates, the concept of *doubling time* can simplify the process.

The doubling time is the number of years required for a quantity to double in value, assuming that the quantity is growing by a certain percentage each year. For example, in the problem above, the doubling time is the number of years it will take for the annual rate of natural gas use to reach 200 billion cubic feet per year.

For rough estimating purposes, where the values are not known precisely, it is sufficient to come within 10% of the exact doubling time. In these applications, the following formula will yield reasonable results for growth rates under 20% per year:

$$\text{doubling time} = \frac{70}{R}$$

where R is the rate of increase in percent per year.

In the example above, $R = 10$, so we reach the following value:

$$\text{doubling time} = \frac{70}{10} = 7 \text{ yr}$$

For more precise applications, such as financial analysis, errors of 10% may not be acceptable. When growth rates are very large (above 20%), the simple estimating method is not sufficiently accurate. In these cases, it is necessary to calculate the doubling time using the following method:

$$\text{doubling time} = \frac{\log 2}{\log (1 + R/100)}$$

where R is the rate of increase in percent.

In the example above, we can substitute the following values:

$$R = 10$$

$$\text{doubling time} = \frac{\log 2}{\log (1 + 10/100)} = \frac{\log 2}{\log 1.1} = 7.27 \text{ yr}$$

The following table lists doubling times for several representative growth rates, calculated according to this method.

Doubling Times and Exponential Growth Rates

Annual Rate of Growth (%)	Approximate Doubling Time (yr)
1.0	69
1.5	46
2.0	35
2.5	28
3.0	23
3.5	20
4.0	17.5
5.0	14
6.0	11.5
7.0	10
8.0	9
9.0	8
10.0	7
15.0	5
20.0	3.8

5.3.4 Decreasing Demand

A fourth possibility exists: the possibility that demand for energy will actually *decrease* with time. For example, if a house is insulated, its use of natural gas for heating will decrease. Suppose that through a variety of technological improvements, the total demand for natural gas to heat a house could be made to *decrease* each year. Let us start with the same situation as before: 1 million houses, each using 100,000 cubic feet per year of natural gas. Let us assume, however, that each year total demand for natural gas *decreases* by 10,000 ft^3 per year. This is *linearly decreasing* demand (see Figure 5.3.5). In this case, though, we find that after 10 years the demand for natural gas becomes *negative*; that is, we are putting more natural gas into the ground than we are taking out. This is clearly an impossible situation.

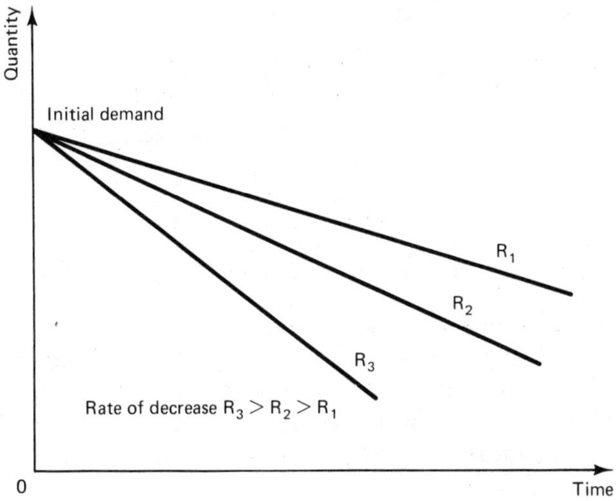

Figure 5.3.5 Linearly decreasing demands graph.

A more likely situation is one where demand decreases *exponentially*. In this case, demand for natural gas might decrease by 2% per year. If this rate of decline continued for 1200 years, natural gas use per household would be less than 1 ft^3 of natural gas per year—a plausible value, if passive solar heating technology continues to improve.

5.3.5 Limitations of Exponential Growth

Making projections on the basis of exponential growth can often lead to absurd situations. For example, let us project the population of the earth based on the assumption that the world's population will increase by 2% per year. (At present, the earth's population is growing at approximately this rate.) Beginning with the current world population of 4.5 billion people, we find that after 560 years the world's population will have increased to approximately 14,000,000,000,000,000, or fourteen quadrillion people.

This is a very large number. If this population were distributed evenly

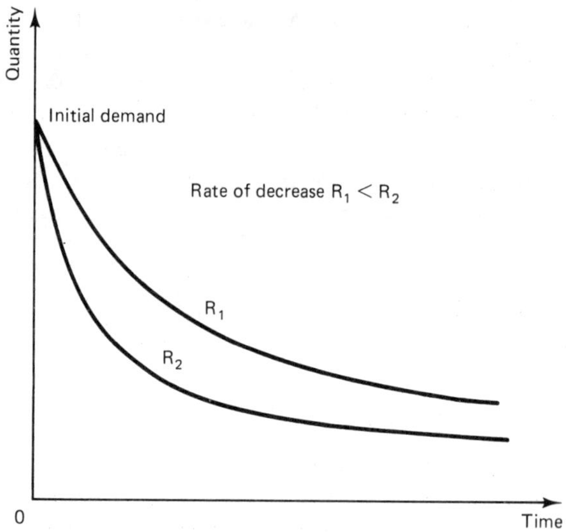

Figure 5.3.6 Exponentially decreasing demands graph.

over the surface of the earth, including the oceans, each person would have less than 4 ft² of living space—that is, each person would be confined to an area roughly 2 ft square. If the world's population increased at a rate of 4% per year, this same situation would come about in a matter of 380 years, and if the population increased at 8% per year, it would happen after a mere 200 years.

Needless to say, this situation is not likely to occur in reality. The world's food supply would be exhausted long before the population reached this level, and thus the world's population would no longer continue to increase at all; in fact, a catastrophic population decline would be the most likely result.

No exponential growth can continue indefinitely. This is a natural, scientific principle. The end of exponential growth can happen suddenly or gradually. If the growth continues beyond the limits of the system, the collapse will be sudden and complete. But the more common pattern is for the growth pattern to change gradually as the limits are approached, resulting in an S-shaped curve (see Figure 5.3.7).

Current growth in population and energy-use patterns are based on a world where overcrowding in some regions of the world has been balanced by ready availability of open land in other places, and where food and energy resources have been abundant and relatively inexpensive. High population growth in some countries has been balanced by lower growth in other regions of the world. But population and energy growth patterns can be expected to change significantly as we approach the limits of the earth's finite resources. It remains to be seen whether this change will be gradual and smooth, or sudden and catastrophic. The choices we make now may very well determine which course we will follow.

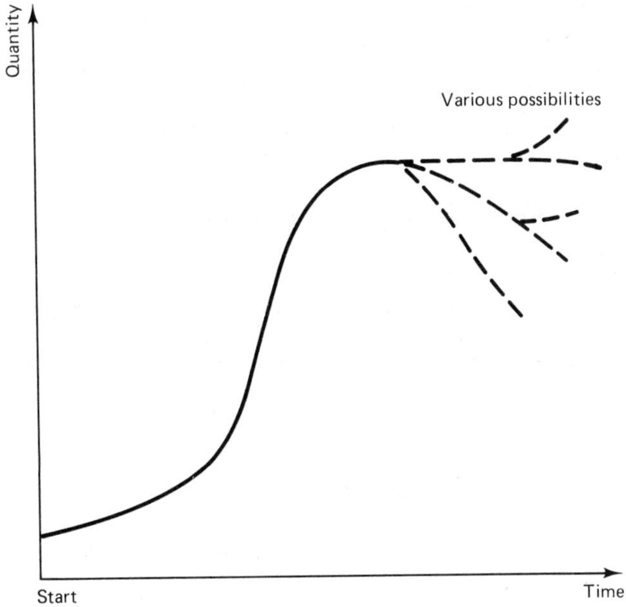

Figure 5.3.7 Ends to exponential growth.

5.4 FACTORS DETERMINING FUTURE ENERGY DEMAND

Energy policy analysts argue incessantly over the question of how energy demand is changing. Projections range all the way from exponential growth to exponential decline. All these projections are based on current trends in energy use. Here are some of the factors that are generally taken into consideration in demand projections or forecasts:

5.4.1 Population Growth

The world's population has grown at an exponential rate since the beginnings of the human race. Thomas Malthus, the eighteenth-century English philosopher, observed this exponential growth in human population and argued that a world shortage of natural resources was an inevitable result of the trend. Recently, with improvements in medicine, agriculture, and other technologies, the rate of population growth has increased; instead of growing at a rate of 1% per year, as it did for many thousands of years, the world's population is now growing at a rate of nearly 2% per year. This rate of growth alone is likely to cause a large increase in global energy consumption. But as we shall see, world energy use has grown even more rapidly than the world's population. Although population growth contributes significantly to increases in energy demand, it is not the only important factor.

5.4.2 Standard of Living

Energy use appears to be related to standard of living, although the precise nature of the relationship has recently been questioned by energy analysts. Consider the listing of energy use in developed and less developed countries given in Table 5.4.1.

TABLE 5.4.1 Energy Use per Capita in Selected Countries (MBtu/yr)[a]

Developed Countries		Less Developed Countries	
Country	Energy per Capita	Country	Energy per Capita
United States	320	Mexico	50
Canada	260	Jamaica	37
Sweden	165	Brazil	21
USSR	160	Zambia	14
West Germany	160	Ghana	5.5
Switzerland	160	China	2.5
Japan	120	Ethiopia	1.0

[a]MBtu = 1 million Btu.

Sources: Arthur M. Bueche, "Energy: An International Appraisal," *Energy in America: Fifteen Views* (Los Angeles: University of Southern California Press, 1980); Phillip G. LeBel, *Energy, Economics and Technology* (Baltimore: Johns Hopkins University Press, 1982).

This chart indicates two important patterns of energy consumption. The first is the difference between the developed countries of the world and the less developed countries. Japan, which has the lowest personal energy use of the developed countries, has a per capita energy consumption more than twice as high as Mexico, with the highest personal energy use of the less developed countries. Thus we can see that there is a clear connection between energy use and levels of industrial development.

A bit of analysis will show why this is so. Many industrial processes are heavy users of energy: production of steel, aluminum, plastics, and ceramics requires large inputs of energy. Also, personal consumption of energy is higher in developed countries, where private automobiles, electric appliances, and other energy-consuming devices are widely owned and used. As industrialization spreads and more people own energy-using consumer products, energy use is likely to increase.

But the differences between developed and less developed countries are less striking than the differences between individual countries in each category. For example, among the developed countries, the United States uses approximately twice as much energy per capita as any other developed country with the exception of Canada. Sweden, Switzerland, and West Germany have standards of living at least as high as the United States and similar levels of industrialization; thus other factors also appear to be at work. We will look at these factors in greater detail in Chapter 8.

5.4.3 Technological Progress

In recent years, technological developments have led to an increase in efficiency of energy use. In the most highly developed countries, energy use per capita has increased at a slower rate than before; and in a few cases, energy use has even decreased. A variety of factors can be cited to account for this change:

1. *Replacement of outmoded industrial equipment.* New industrial equipment tends to be much more efficient than the outmoded equipment which it replaces. For example, new furnaces and boilers use much less energy than those which were being manufactured just a few years ago. The same is true of many electrically powered machines. These efficiency improvements lead to reduced energy consumption in the industrial sector.
2. *Improved energy efficiency of consumer products.* Everyone is familiar with the dramatic improvements in automobile gas mileage that have occurred during the past few years. As new, fuel-efficient automobiles have begun to replace old gas-guzzlers, demand for gasoline has started to decline. Similar efficiency improvements are anticipated in refrigerators, televisions, and a variety of other consumer products.
3. *Technological breakthroughs.* In the electronics industry, energy use has decreased dramatically during the past 30 years. Integrated circuits use less than one-tenth as much energy as the vacuum-tube circuitry which they replaced. This has led to a tremendous reduction in energy requirements for telecommunications, computers, and electronic controls. Similar technological breakthroughs in other fields may lead to even greater energy savings.
4. *Economic effects.* As energy prices increase, energy use declines. The precise nature of this relationship is poorly understood, but economic factors have a powerful effect on growth in energy demand (see Chapter 6). In recent years, prices for electricity and petroleum products have increased faster than the general rate of inflation. This, in turn, has led to a slowing of demand for energy in developed countries.

Considering all these factors, we can see that projecting energy use is not an easy task. The difficulty of making accurate projections can best be illustrated if we look at some of the energy-use projections that have been made over the past 10 years and notice how the conclusions have changed. Looking at the projections shown in Table 5.4.2, we can see that the trend has been toward

TABLE 5.4.2 Projections of U.S. Energy Use for the Year 2000

Energy Projection Study	Year of Study	Projected Consumption[a] (quads) High	Middle	Low
Federal Power Commission (FPC)	1972	190		
Atomic Energy Commission (AEC)	1972		160	
Amory Lovins/Friends of the Earth (FOE)	1972			125
Edison Electric Institute (EEI)	1974	160		
Energy Research and Development Administration	1974		140	
Ford Foundation Energy Policy Project	1974			100
Edison Electric Institute (EEI)	1976	140		
Energy Research and Development Administration	1976		124	
Amory Lovins/*Soft Energy Paths*	1976			75
Electric Power Research Institute (EPRI)	1977–1978	159		
Committee on Nuclear and Alternate Energy Systems, National Academy of Science	1977–1978		124	
Steinhart, *Pathway to Energy Sufficiency*	1977–1978			33
Exxon Corporation	1980	105		
Department of Energy (DOE)	1981		100	
Solar Energy Research Institute/DOE	1980			57

[a]The figures quoted here are base line or middle-range projections. Most of the studies included several alternative assumptions and projections.

Source: Adapted with permission from John S. Steinhart et al., *Pathway to Energy Sufficiency: The 2050 Study* (San Francisco: Friends of the Earth, 1979).

predicting more modest increases in demand. This makes sense if we look at recent trends in energy prices. Until recently, real energy costs—especially electrical costs—were declining rapidly. But price increases, especially for electrical power, seem to have turned the trend around.

Even today, many energy analysts assume that demand growth will increase exponentially; the only question is whether energy is expected to grow by 7% per year (the traditional assumption until the mid-1970s), 4% (the figure used by most electric utilities for proposing construction projects), or 2.8% (the "low demand growth" figure used by many public utility boards). In fact, during 1980, 1981, and 1982, energy use in the United States actually *decreased,* from 79 quads in 1979 to 76 in 1980, 74 in 1981, and 71 in 1982 (DOE figures, excluding wood). In the light of these figures, we believe that it is unrealistic to cling to a projection of exponential energy growth. More fine-grained analytical techniques are available which should enable researchers to look more deeply into patterns of energy exploration, production, and use.

5.5 EFFICIENCY

It is clear that if we are determined to get what we desire, for as long as possible, at the lowest possible cost, we should use the most efficient procedures possible (see Table 5.5.1). But a concern for efficiency must be tempered by an awareness of other factors, such as human fallibility, cost-effectiveness, environmental effects, and social constraints. In principle, the concept of efficiency can be extended to include economics (get the most benefits per dollar spent), social effects (provide the greatest social benefits for a given expenditure), and environmental consequences (accomplish the task with the least environmental degradation). In other words, the concept of efficiency can be extended to include nearly every category of analysis. For the present, however, we will be concerned with the concept of physical efficiency: that is, the ratio of useful energy output to useful energy input.

Let us review what has already been established in Chapter 2. In a multistep system, the system efficiency is equal to the product of the step efficiencies.

$$E_{\text{system}} = E_{\text{step 1}} \times E_{\text{step 2}} \times \cdots \times E_{\text{step } n}$$

Whenever we assess the relative efficiency of various energy alternatives, we need to consider the efficiency of the entire conversion, beginning with the energy resource and concluding with the end use. For example, let us look at the overall efficiency of a stereo system powered by electricity from a hydroelectric power plant. Here are the step efficiencies:

$$\begin{aligned}
\text{Power plant: } E_p &= 0.9 \\
\text{Transmission: } E_t &= 0.9 \\
\text{Stereo: } E_s &= 0.6 \\
E_{\text{system}} &= E_p \times E_t \times E_s \\
&= 0.9 \times 0.9 \times 0.6 = 0.49
\end{aligned}$$

This means that approximately 49% of the energy contained in the water is actually transformed into stereo sound.

In analyzing the efficiency of energy systems, it is important to consider the entire process of energy production, distribution, and consumption. If any of these areas are omitted from the analysis, the results will be seriously flawed. For example, electric resistance heating is a 100% efficient process; that is, all the electrical energy that passes through the heater unit is converted to useful heat energy. Looking only at the consumption section of this energy system, it appears that resistance electric heating is the best possible choice because of its high efficiency.

But let us look at the overall system efficiency of the electric heating system, beginning with the original resource. Since most electricity is produced by fossil-fuel plants, we will look at the efficiency of this method of power production.

TABLE 5.5.1 Energy Efficiency of Selected Devices, Processes, and Systems

Device or Process Name	Nature of Work or Conversion	Approximate Efficiency (%) Low	High	Comments
Resource production				
Coal	Mining to processing to transport	55	61	The efficiency figures for extraction account for the percent of resource left unrecovered
Natural gas	Wells to refining to transport	65	67	
Uranium	Mining to refining to transport	50	55	
Oil	Production to refining to transport	28	40	
Wood	Harvesting to transport to splitting	55	65	
Diesel engine	Chemical to thermal to mechanical	25	38	Efficiency depends on the model, make, size, operating conditions, temperature device condition, controls, loading conditions, etc.
Gas turbine	Chemical to thermal to mechanical	25	42	
Gasoline engine: automobile	Chemical to thermal to mechanical	24	30	
Steam engine	Chemical to thermal to mechanical	10	16	
Steam turbine	Chemical to thermal to mechanical	20	45	
Water turbine	Linear (mech.) to rotational (mech.)	65	95	
Windmills, propellers	Linear (mech.) to rotational (mech.)	25	50	
Electric generators	Mechanical to electrical	80	99	
Dry cell battery	Chemical to electrical	85	90	
Fuel cells	Chemical to electrical	50	60	
Storage battery	Chemical to electrical	65	75	
Photovoltaic cells	Radiant to electrical	10	16	
Thermocouples	Thermal to electrical	5	15	
Home gas furnace	Chemical to useful heat	70	85	Efficiency depends on proper maintenance and tune-up, operating temperature, design and construction, proper sizing, etc.
Home oil furnace	Chemical to useful heat	55	75	
Electric heat	Electrical to useful heat	95	100	
Steam boiler, large	Chemical to useful heat	75	88	
Wood stoves, airtight	Chemical to useful heat	35	65	

Wood stoves, non-airtight	Chemical to useful heat	20	40	
Wood fireplaces	Chemical to useful heat	-5	15	Depending on outside temperature
Electric motors, large	Electrical to mechanical	90	95	Depending on design,
Electric motors, small	Electrical to mechanical	55	70	construction, and conditions
Incandescent lamps	Electrical to light (radiant)	4	7	
Fluorescent	Electrical to light (radiant)	15	21	
High-intensity lamps	Electrical to light (radiant)	25	33	
Lasers	Electrical to light (radiant)	25	35	
Electric transformers	Electrical to electrical	85	99	
Rectifiers	Electrical (ac) to electrical (dc)	55	65	
Electric transmission	Electrical to electrical	80	100	Depending on distance
Gears	Mechanical to mechanical	80	98	Depending on proper
Chains and sprockets	Mechanical to mechanical	94	99	lubrication, operating
Bearings	Mechanical to mechanical	94	99	conditions, etc.
Belts and pulleys	Mechanical to mechanical	94	99	
Coal conversion to gas	Chemical to chemical	60	95	
Coal liquefaction	Chemical to chemical	60	65	
Synthetic fuels	Chemical to chemical	20	30	Tar sands, shale oil, etc.
Hydro power plants	Gravitational to electrical	80	90	
Pumped storage plants	Electrical to potential to electrical	60	66	
Fossil-fuel plants	Chemical to thermal to mechanical to electrical	35	40	
Nuclear power plants	Nuclear to heat to mechanical to electrical	30	31	
Magnetohydrodynamics	Chemical to heat to electrical	45	50	Under development
Photovoltaic cells	Solar to electrical	1	12	
Automobiles	Chemical to thermal to mechanical	5	10	Depending on many factors
Heat pumps	Electric to mechanical to heat transfer	200	300	Depending on temperature difference

Sources: Earl Cook, *Man, Energy, Society* (San Francisco: W. H. Freeman, 1976); John M. Fowler, *Energy and the Environment* (New York: McGraw-Hill, 1975); Claude M. Summers, "The Conversion of Energy," *Scientific American,* September, 1971.

$$\begin{aligned}
&\text{efficiency of coal transport, } E_t &= 0.98\\
&\text{efficiency of the power plant, } E_p &= 0.38\\
&\text{efficiency of transmission, } E_d &= 0.91\\
&\text{efficiency of electric heating, } E_h &= 1.00
\end{aligned}$$

Combining all these step efficiencies, we find that the overall system efficiency is only 33.8%—nowhere near 100%.

$$\begin{aligned}
\text{system efficiency} &= E_m \times E_t \times E_p \times E_d \times E_h\\
&= 0.98 \times 0.38 \times 0.91 \times 1.00\\
&= 0.338
\end{aligned}$$

Now let us look at an alternative system, a fuel oil furnace with an operating efficiency of 70%. At first glance, the fuel oil furnace appears to be less efficient than the electric heater. But let us look at the total system efficiency. Again, we will begin with the basic energy resource, in this case crude oil.

$$\begin{aligned}
&\text{efficiency of refining, } E_r &= 0.88\\
&\text{efficiency of transport, } E_t &= 0.95\\
&\text{efficiency of combustion, } E_c &= 0.70
\end{aligned}$$

Combining all these step efficiencies, we find that the oil system actually has a much higher efficiency than the electric system: approximately 58%.

But even an oil furnace cannot match the overall efficiency ratings of the newest natural gas combustion systems. The very newest natural gas furnaces are capable of combustion efficiencies on the order of 92%, and natural gas refining and transport are more efficient than oil refining and transport.

$$\begin{aligned}
&\text{efficiency of refining, } E_r &= 0.97\\
&\text{efficiency of distribution, } E_d &= 0.95\\
&\text{efficiency of combustion, } E_c &= 0.92\\
&\text{system efficiency} &= 0.97 \times 0.95 \times 0.92\\
& &= 0.848 = 84.8\%
\end{aligned}$$

Strictly speaking, even natural gas is not the most efficient heating system if we look at where all these energy resources come from. All fossil fuels are a form of stored solar energy, so in principle we should also consider the efficiency of conversion from solar energy to fossil fuels.

The process of conversion from solar energy to fossil fuels is fairly well understood. First the solar energy must be converted to plant materials. This process is relatively inefficient. The highest conversion efficiency that has been observed is 8%, in the case of sugarcane. In general, the process is even less efficient, with typical levels around 1%.

Once the solar energy has been stored in the form of plant matter, the dead vegetation is compressed and transformed under the weight of sedimentary rock deposits. During this process, a large percentage of the vegetation is lost due to decay. This introduces further reductions in efficiency.

When these processes of conversion are taken into account, we find that in each of the cases described above, the overall efficiency of conversion from sunlight to home heating is only a tiny fraction of 1%. But because most of those conversions have taken place by natural processes over many centuries, we tend to discount the enormous losses that have taken place before the fossil fuels are mined.

Although efficiency is an important concept in energy analysis, it is not the only important consideration. For example, the efficiency of photovoltaic cells, which convert sunlight into electricity, can range anywhere from 2 to 3% all the way to 14%. Clearly, at 2%, the cells are not very efficient. But that does not mean that only cells with a 14% efficiency are worth our while. In order to achieve such high efficiencies, manufacturers must use extremely precise manufacturing techniques, which add to the costs of production. In the final analysis, it may be better to settle for a somewhat less efficient cell, but one that produces electricity at a lower price. As we shall see in later chapters, a large number of economic, environmental, and social factors enter into the equation as well.

5.6 EFFICIENCY AND ENERGY QUALITY

So far, we have been speaking of energy efficiency in terms of the percentage of energy transferred for useful purposes. But this concept of efficiency is incomplete. For example, if we heat a house with resistance heating using electricity from a hydro plant, the overall system efficiency will be over 80%. This would seem to suggest that in areas with abundant hydro potential, electric resistance heating is preferable to solar heating with an efficiency of 50%, but that in areas with fewer hydro resources, solar heating would be preferable. Clearly, something is missing from our analysis, and that something is the concept of energy *quality*.

We have stated that in every energy transfer, a certain amount of useful energy is transformed from a useful to a less useful form. This means that different forms of energy have different levels of quality. *High-quality* energy is in a highly useful form, and *low-quality* energy is in a relatively useless form.

Let us illustrate the difference between high-quality and low-quality energy. Imagine two streams. One stream is 120 ft wide and 1 in. deep. The other stream is 10 ft wide and 1 ft deep. Clearly, it is easier to harness power from the deeper stream, although both streams contain the same amount of water and the same amount of energy. In a similar way, all energy sources can be classified according to the ease with which they can be used. For example, electricity is the most useful form of energy because it can easily be converted to a variety of other forms. Gasoline is also a high-quality form of energy. High-temperature heat is fairly useful, but low-temperature heat is not very useful at all.

In determining the usefulness of an energy resource, three factors need to be taken into account:

1. *Density of energy.* How concentrated is the energy? If the energy is in a form that is insufficiently concentrated, we will not be able to use it to do very much useful work. Low-temperature heat, poor-quality oil shale, and thin seams of coal are all examples of resources that lack sufficient density to be high-quality energy resources.
2. *Convertibility.* How easy is it to convert the energy to other forms? Electrical energy can be converted to other forms with a high degree of ease and efficiency, whereas other forms, such as low-temperature heat, cannot be converted very easily or efficiently.
3. *Controllability.* How well can the energy be controlled? The density of the energy in a hydrogen bomb is very high, and in principle the energy can be converted to electrical energy with a high degree of efficiency. But the energy cannot be controlled at all, so its quality is not as high as more easily controlled energy forms.

Based on these three criteria, we can classify a variety of energy forms and energy resources according to a simple grading system (see Table 5.6.1). Then we will see how these classifications can be used to help us use energy resources as efficiently as possible. It is important to note that the lower grades of energy can be quite useful for some purposes. There is no need to use very

TABLE 5.6.1 Energy Grades and Energy Resources

Energy Grade	Energy Type
A Fancy	Electricity Lasers
A	Hydro energy Oil, natural gas, coal Nuclear energy
B	Biomass (wood, refuse, peat, paper) Geothermal energy
C	Sunlight Wind Low-head hydro, tidal, etc.
D	Room-temperature heat Low-temperature geothermal energy Wave power, ocean thermal (OTEC)
F	Freezing cold

high-quality energy for simple tasks such as heating a building. In fact, as we shall see, it makes good physical and economic sense to use the *lowest-quality energy* that will perform any given task, rather than using up high-quality energy resources unnecessarily.

BOX 5.6.1 HOW USEFUL IS USEFUL?

Let us look at representative energy resources from our various energy grades and see how each resource can be used.

Grade D. Low-temperature geothermal heat can be used to reduce heat loss in underground buildings. In conjunction with a heat pump, it can provide home heating.

Grade C. Sunlight can be used for home heating or water heating at a maximum efficiency of around 50%. It can produce electricity at an efficiency of around 10%. When concentrated, it can produce moderate temperature heat.

Grade B. Biomass or fuel oil can be used for home heating at an efficiency of approximately 50 to 85%. It can produce electricity at an efficiency of about 40%. When burned, it produces fairly high temperature heat.

Grade A. Hydro power can be used directly for mechanical work such as turning, grinding, or pumping. It can be used to generate electricity with an efficiency near 90%. When converted to electricity, hydro power can provide very high temperature heat with a high degree of efficiency.

Grade A Fancy. Electricity can be used for home heating at an efficiency of essentially 100%. If used to drive a heat pump, it will draw even more heat into the house. Electricity can provide light and very high temperature heat, and when used in conjunction with electronic devices, it can be used to handle information in very sophisticated ways, almost like the human brain.

When we use high-grade energy to do low-grade tasks, we are not using that enegy to its fullest potential. For example, suppose that we use electricity to heat a living room to 68 °F. This is a waste of a resource that is capable of doing much more sophisticated tasks, such as providing lighting, driving electric motors, or communicating over long distances. It is also important to realize that electricity is capable of performing those other tasks *and* providing heat to the living space at the same time. Suppose, for example, that we use electrical energy to operate a television set. This is a high-level use of electricity. But the energy used in the television set is ultimately converted to heat energy. So the electrical energy is still converted to heat, but the energy performs other useful functions along the way. Clearly, it makes more sense to use the electricity to operate a television *and* to supply a bit of heat to the house, rather than using it only to heat the house.

5.7 ENERGY QUALITY AND THE LAWS OF THERMODYNAMICS

Now that we have a good idea what high- and low-quality energy are, we are ready to restate the *second law of thermodynamics*. In Chapter 2 we stated the law in the following terms:

> In any energy transaction, some energy is converted from useful to useless forms.

In terms of energy quality we can state the law as follows:

> In all processes that convert energy to useful work, some energy is converted from a higher- to a lower-quality form.

This means that whenever energy is used, the energy loses some of its ability to do useful work in the future. Thus, when we say that we are using up energy, we really mean that we are using up the total amount of energy *available for doing useful work*.

BOX 5.7.1 MEASURING ENERGY QUALITY

In dealing with energy systems, it is useful to have a precise way of measuring the decline of energy quality that occurs in any energy transaction. To achieve this purpose, physicists have developed the concept of *entropy*. Stated as simply as possible, entropy represents energy that is not available to do useful work. This leads us to a third statement of the second law of thermodynamics:

> In any energy transaction, the total amount of entropy in the universe is increased.

In other words, entropy always increases.

As is the case with potential energy, it is rather difficult to measure or define *absolute* entropy. What we can do, however, is measure the changes in entropy of a system. Changes in the entropy of a system can be calculated using the following formula:

$$dS = \frac{dQ}{T_1} - \frac{dQ}{T_2}$$

where dS = change in entropy of the system

dQ = amount of energy transferred from the higher-energy region to the lower-energy region of the system

T_1 = temperature of the lower-energy region

T_2 = temperature of the higher-energy region

T_1 and T_2 are measured in *degrees Kelvin* (Systeme International) or *degrees Rankine* (English system).

5.7.1 Physical Description of a Heat-Transfer Process

If we consider the physical processes involved in a simple heat transfer, it is easy to see that the amount of useful energy in a system will always decrease. Consider, for example, two containers of water, one at 0 °C and the other at 100 °C. (Both containers hold liquid water; no ice or steam.) If we place the two containers in contact with each other, the temperature of the hotter container drops and the temperature of the cooler container increases until the two are the same temperature: 50 °C. This process is so obvious that it seems almost unnecessary to think about it. But it illustrates an important feature of energy transactions.

At the beginning of the heat transfer described above, the average temperature of the two containers is equal to 50 °C, the same as the average temperature at the end of the process. This means that the total amount of energy contained in the system is the same before and after the transfer. But a highly significant change has taken place. When the two containers are at widely different temperatures, it is possible to obtain useful work from the system by exploiting the difference in temperature between the two containers, by using thermocouples or a heat engine such as the one in an OTEC (ocean thermal) power plant. But after the transfer of heat, it is no longer possible to get any useful work out of the system because there is no temperature difference to drive the process.

The irreversible nature of heat transfer is easiest to understand if we look at the process on the molecular level. When a molecule of water interacts with a molecule of the wall of its container, the two molecules exchange their energy.

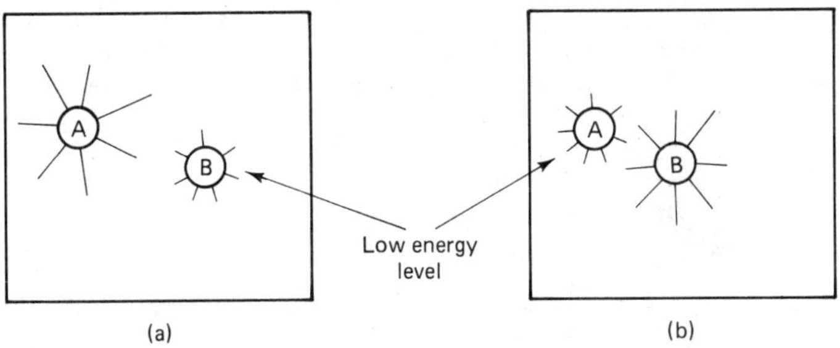

Figure 5.7.1 Molecular energy exchange: (a) before exchange; (b) after exchange. Before the exchange, molecule A has higher energy; after the exchange, molecule B has higher energy.

Similarly, the walls of the two containers exchange energy. Indirectly, then, we can say that the molecules in the two containers exchange their energy.

Now let us consider what happens as the energy transfer process continues. At the beginning of the heat transfer described above, the average energy of a molecule in the hotter container is much higher than the average energy of a molecule in the cooler container. If an average molecule from the hotter container transfers its energy to the cooler container at the same time that an average molecule from the cooler container transfers its energy to the hotter container, the cooler container gains energy and the hotter container loses energy. Multiplying this effect by all the molecules in each container, we can see that on the average, more energy will be transferred to the cooler container from the warmer container. This will reduce the temperature of the warmer container and increase the temperature of the cooler container, bringing the average molecular energy levels closer together. As the two containers reach the same temperature, the average energy levels of the molecules in the two containers are the same, which means that on the average, no energy will be transferred between the two containers.

We can see, based on this example, that energy transfer is a one-way street. Because energy naturally flows from a higher to a lower energy level (e.g., from a higher- to a lower-temperature region), energy-transfer processes are not reversible. Each time an energy transfer takes place, less useful energy remains in the system as a whole.

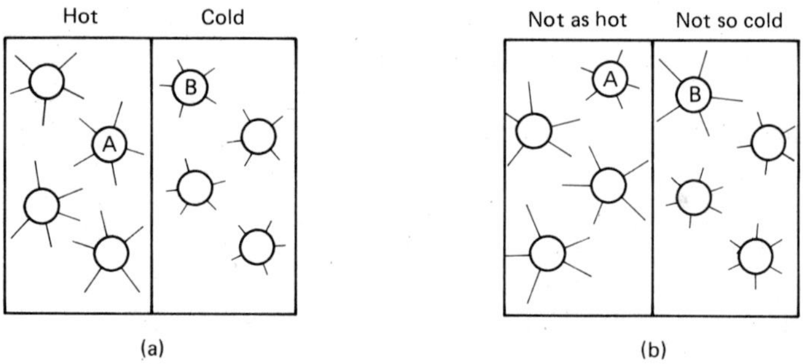

Figure 5.7.2 Molecular energy exchange between two containers: (a) before exchange; (b) during exchange. Molecules A and B have exchanged some energy.

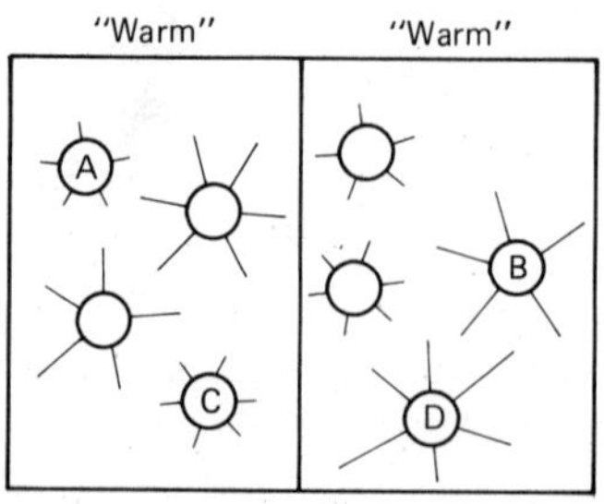

Figure 5.7.3 Energy exchange between two containers having equal temperatures; no net energy transfer occurs after thermal equilibrium.

5.8 THE LAW OF DIMINISHING RETURNS

It is generally accepted that by putting more work into any project, we can gain more benefits. This is true whether we are talking about studying for a test, practicing the piano, or exploring for oil. But most of us have also noticed that after a certain point, it seems that we get less in return for our additional effort. The first hour of studying seems to do us the most good, the second less good, and so on, until the extra effort no longer does any good at all. The same principle applies in many other areas of human endeavor, including energy production and use. This principle is called the *law of diminishing returns.*

Applications of the principle of diminishing returns fall into three general categories: (1) limitations imposed by theoretical limits of efficiency, (2) limitations imposed by finite supplies of resources, and (3) losses caused by increases in power or speed. Let us see how these categories apply to some real-life situations.

5.8.1 Theoretical Limits of Efficiency

Electrical production. The first electricity-producing steam turbines were quite inefficient—about 9% efficient. But as turbine design has become more sophisticated, the efficiencies have increased. Today, a typical steam turbine has an efficiency of approximately 39%. These improvements were accomplished quite rapidly and without a particularly significant increase in costs. It would be possible to achieve even higher efficiencies today, but only by investing very large amounts of money in highly sophisticated engineering processes. So as we approach the theoretical limits of turbine efficiency, the law of diminishing returns places a severe economic penalty on further improvements in efficiency.

Energy conservation. One technique for reducing energy use in a house is to add insulation to the walls, ceilings, and floors. But not all the insulation will save the same amount of energy. Each time we double the thickness of the insulation, we reduce the amount of transmitted heat by one-half. Clearly, a doubling from 1 to 2 in. will produce dramatic energy savings. But as the insulation becomes thicker, two things happen:

1. More insulation is required to cut the transmitted heat losses. For example, if the house already has 1 ft of insulation, we need to add another foot to cut the transmitted heat losses in half.
2. As the thickness of the insulation increases, the total amount of heat loss decreases; thus an additional 50% reduction in heat loss represents a smaller energy savings.

Ultimately (say, when the insulating value of the walls exceeds *R*-50, or when the insulation in the ceiling exceeds *R*-80), the cost of added insulation greatly exceeds the savings offered by the insulation. Once again, as we approach the theoretical limit of efficiency, the law of diminishing returns comes into play.

5.8.2 Finite Supplies of Resources

Oil exploration. When petroleum was first used as an energy resource, it was available in pools on the surface of the earth. Very little energy was required to obtain supplies of oil. As the surface pools were used up, more energy was required to drill oil wells. Today, most of the easily available oil has been pumped out of the ground, and producers have turned to secondary recovery techniques (see Section 3.3.4). Clearly, it takes more energy than it did 50 years ago to produce the same amount of oil.

In theory, this process could continue forever: ever-deeper oil wells could be drilled and more complex and costly techniques could be used to extract the oil from the wells. But beyond a certain point, the amount of energy required to produce the oil can actually exceed the energy value of the oil. At that point, further oil production represents a net loss of energy.

5.8.3 Increased Speed and Power

Automobile travel. Suppose that an automobile is traveling at a steady 80 km/h (50 mph). If there were no frictional forces acting on the car, it could continue to travel at 80 km/h with no additional input of energy. But the drag of the air and the additional resistance of the road surface will require the expenditure of some energy to maintain a constant speed.

Now let us consider what happens if the car is traveling at 120 km/h (75 mph). The air resistance increases with the *square* of the velocity. This means that as we increase the speed of the car by 50% (going from 80 to 120 km/h), the air resistance increases by 125%. This is a very significant increase. Other components, such as the resistance of the tires, will not increase as rapidly as the air resistance, but it is still true that to maintain a higher speed, more energy will be required to cover the same distance. Thus, as we increase our energy input, and thus our speed, the return (in this case the distance traveled per gallon of gasoline) decreases.

BOX 5.8.1 SPAGHETTI DINNER AND DIMINISHING RETURNS

Suppose that you go to an excellent Italian restaurant and order a bowl of spaghetti. After a short wait, you are presented with a fork, a spoon, a napkin, and a bowl of spaghetti.

You take your first bite, and immediately your taste buds are flooded with the flavor of the sauce. All you have to do is dig in and you will get some on your fork.

After a while, you notice that the strands of spaghetti are getting a bit shorter than they were before, and you get less spaghetti per forkful. Eventually, your fork becomes useless, so you use your spoon to pick up the bits and ends of spaghetti together with the sauce that is left in the bowl.

Eventually, even your spoon cannot pick up enough spaghetti sauce to make it worthwhile. You tip your bowl sidewise to consolidate the remaining sauce so that you can reach it with your spoon. This produces one or two respectable spoonfuls, but then you are back to ground zero again.

At this point, if you are in polite company, you will probably stop. But if your hunger overcomes your sense of courtesy, you grab the bowl with both hands and lick it with your tongue. You get some benefit from this "enhanced recovery" technique, but even all that hard work will not get you all the molecules of the precious sauce.

By now, you have reached the point where you are spending more energy collecting the "resource" than the spaghetti sauce can deliver to your body's metabolic system. Fortunately, you are in a restaurant and can order another bowl of spaghetti.

Now imagine that the bowl of spaghetti has turned into an oil well. At first, the oil is near the surface and can be pumped out quickly and easily. But after a fair amount of pumping, production drops and you need to use secondary recovery techniques, injecting water or steam. But even after secondary recovery, you have recovered only one-third of the oil in the reservoir. To recover more oil, a tertiary or "enhanced" recovery process must be used. Detergents and carbon dioxide are pumped into the well and extracted, together with dissolved petroleum, two or three years later.

If more oil recovery is desired, more energy must be spent on drilling deeper and horizontally along the bottom of the rock formation to drain some of the residual oil into an artificially created reservoir.

In principle, you can go on spending more energy and money to recover additional bits of residual oil from the formation, although 100% recovery is impossible, no matter how much energy is spent on recovery. In more practical terms, the financial and energy costs of the recovery process will eventually exceed the benefits derived from the enhanced recovery methods. At this point, the oil well is no longer useful. But unlike the situation in an Italian restaurant, there comes a time when we can no longer simply order another oil well.

5.9 END-USE ANALYSIS

In developing a sustainable energy policy, the most difficult task is deciding which questions to ask. Are we concerned primarily with abstractions such as energy quality, energy efficiency, and demand projections, or should we be looking at the availability of specific energy resources, the uses to which energy resources will be put, and the technologies that will enable us to make the conversions? In the final analysis, our choice of analytical methods makes more difference than any other single factor.

In all the talk about system efficiency, energy quality, and patterns of demand, it is easy to overlook what the energy is being used for. What people need is the things that energy can do for them: driving motors and engines, keeping houses warm, providing entertainment. As long as the end result is the same, we do not ask how it was achieved. So before we decide how to produce the energy we need, we will do well to ask ourselves why we need the energy in the first place. Then we can look at alternative ways of achieving the same result (heating the house, getting to work, cooking supper) and choose between these alternatives. End-use analysis tells us that if we have a choice between two alternative ways of achieving the same result, we should choose the method where the type of energy provided is well matched to the task for which it is required.

If we look at U.S. energy consumption from an end-use point of view, we find that a very large percentage of our energy use consists of creating relatively low temperature heat. Once we have discounted the energy that is simply lost in conversion processes, we find that 35% of the remainder is used to provide heat at temperatures below 100 °C. Another 6% is used to provide temperatures between 100 and 200 °C. These temperatures can easily be achieved by solar collectors and other renewable energy resources. However, only a tiny percentage of low-temperature heating needs are supplied by renewable resources. Most low-temperature heat is produced by electricity or fossil fuels. Clearly, a high-quality (grade A) resource is being used to provide for a low-quality (grade C) energy need.

One of the most attractive features of end-use analysis is that it does not limit the analysis to energy-supply options. If increased insulation will provide the desired result—in this case a warmer house—it will be considered a valid alternative to increasing the supply of heat. For example, consider the following situation, conceived by Amory Lovins.

A homeowner is taking a bath, but the plug in the bathtub drain is defective and the water keeps running out. To keep the tub full, the homeowner must leave the faucets on, which rapidly depletes the supply of hot water. At this point, the homeowner has two choices: buy a larger water heater or invest in a simple 59-cent plug. Clearly, the latter choice makes much more sense.

In a similar way, not all our energy problems need to be solved by increased production of energy resources, renewable or nonrenewable. If the same objective can be achieved through energy-conservation techniques, there is no need for the increased use of energy resources. They can stay in the ground, or in the forest, to be used in other situations where there is no substitute.

In recent years, end-use analysis has been accepted by increasing numbers of policymakers. Conventional projection techniques have failed to predict changes in demand patterns during recent years, while studies based on end-use analysis have been somewhat more successful. End-use analysis also permits policymakers to look toward a sustainable energy future based on renewable resources, whereas conventional analytical methods call for ever-increasing use of nonrenewable resources, including nuclear power. In a time of rising interest rates and strained governmental budgets, end-use solutions often seem more workable than large increases in energy supply.

But perhaps the most attractive feature of end-use analysis is the degree to which it can be integrated with a consideration of other energy issues:

economics, environmental impacts, social effects, and political consequences. Energy use is not something that happens "out there"; it is part of the everyday life of every person on the planet. As we choose our analytical tools, we must recognize that essential connection. If we do not, we will merely be engaged in a futile intellectual exercise.

STUDY QUESTIONS

1. List examples of each of the following:

Zero growth
Constant (linear) growth
Exponential growth
Exponential decrease

Which types of growth are most likely to be sustainable for long periods of time?

2. How has your own use of energy changed over the years? Try to estimate your overall energy consumption at five-year intervals, beginning at birth.

3. Take an inventory of your current energy consumption in terms of energy quality. To what extent are you using high-quality energy to perform low-quality tasks?

4. Why is it easy to convert from higher-quality energy forms to low-quality forms, but very difficult to go the other way? Explain this principle in terms of order and disorder.

5. Based on your own experience, give three examples of the law of diminishing returns.

6. Make a list of some of the end uses for energy in your life (examples: drying hair, reproducing recorded music, transportation to and from work or school, etc.). What energy resources are you using to perform these functions? What other energy resources might perform the same functions equally effectively?

7. Explain what is meant by end-use analysis and relate this idea to the concept of sustainability.

PROBLEMS

1. Suppose that the population of the earth in 1975 was 4 billion and that the population doubles every 35 years. Assume that 75% of the earth's surface is covered with water and that half the land area is uninhabitable.

(a) What is the average density per square mile?
(b) What will it be in 2010? In 2045?
(c) When will you begin to feel crowded?

2. Using an almanac, atlas, or statistical abstract, determine the population density of New Jersey, California, and Alaska.

(a) Which of these states is closest to the population density of the earth in 1975? (See Problem 1.)
(b) Which is closest to the density in 2010? In 2045?

3. Which method of home heating gives more useful energy per unit of input energy?

    Electric resistance heat, assuming that the electricity was produced in a nuclear power plant

    Oil heat, assuming a furnace efficiency of 65%

    A heat pump that delivers 3 kWh of heat energy for every kilowatthour of electric energy produced in a nuclear plant,

   Use the high-efficiency figures from Table 5.5.1.

4. Calculate the net efficiency of a gas heating system given that the production of natural gas is 67% efficient, the transport of natural gas is 97% efficient, and the furnace efficiency is 80%.

5. Calculate the net efficiency of
   (a) Incandescent lighting.
   (b) Fluorescent lighting.
   (c) Electric water heater.

   The step efficiencies are:

    Production of coal, 60%

    Transportation of coal, 96%

    Generation of electricity, 34%

    Electrical transmission, 90%

    Incandescent lamp, 5%

    Fluorescent light, 20%

    Electric water heater, 92%

6. Production of crude oil is 45% efficient. The process of refinement that gives commercial gasoline is 87% efficient. Transportation of gasoline is 95% efficient. The automobile engine efficiency is about 28%, the transmission of power from the engine to the axle is about 75% efficient, and the rolling efficiency of the wheels is about 60%. Estimate **(a)** the net efficiency of automobile transport and **(b)** the energy lost per gallon of crude oil drilled.

FURTHER READINGS

COMMONER, BARRY. *The Closing Circle: Nature, Man, and Technology.* New York: Alfred A. Knopf, 1972.

COOK, EARL. *Man, Energy, Society.* San Francisco: W. H. Freeman, 1976.

EHRLICH, PAUL R., and ANNE H. EHRLICH. *Population, Resources, and Environment.* San Francisco: W. H. Freeman, 1970.

GOSWAMI, AMIT. *The Concepts of Physics.* Lexington, MA: D. C. Heath, 1979.

LEBEL, PHILLIP G. *Energy Economics and Technology.* Baltimore: Johns Hopkins University Press, 1982.

LOVINS, AMORY B. *Soft Energy Paths: Toward a Durable Peace.* New York: Harper & Row, 1977.

6

Economic Analysis

Most of us simply take for granted that gasoline, home-heating fuel, electricity, wood, and other energy resources are available to us. In fact, those resources are only made available through the functioning of economic processes. For public utilities as well as private energy companies, economic analysis is a major component of every energy project.

Economics is a large, complex subject. Entire books have been written on the economics of public utilities and energy supply systems. But the basic principles are not especially hard to understand; it is the applications of those principles that lead to the greatest complexities. In this chapter, without going into a great many calculations, we present the general principles of energy economics. Using these principles, we then take a look at the operations of the energy industry, with special emphasis on regulated investor-owned utility systems.

6.1 PRICES AND THE MARKETPLACE

In a free and open market, prices are determined by the process of supply and demand (Fig. 6.1.1). Very simply, the process is based on four principles:

1. If a product is offered for sale at a high price, people are eager to produce the product, but few are willing to buy it.
2. If a product is offered for sale at a low price, people are eager to buy the product but few are willing to produce it.

3. If there is a surplus of a product, the price will go down.
4. If there is a shortage of a product, the price will go up.

When the price of a product is high, the high production creates a surplus of the product. This drives down the price, which increases people's willingness to buy. At the same time, the falling price causes production to decrease. Soon the surplus is eliminated. If this process goes too far, a shortage will be created, and prices will begin to rise again. This cyclical process can result in wild price fluctuations. But in the long run, prices will tend to stabilize at an *equilibrium level* where production equals people's willingness to buy.

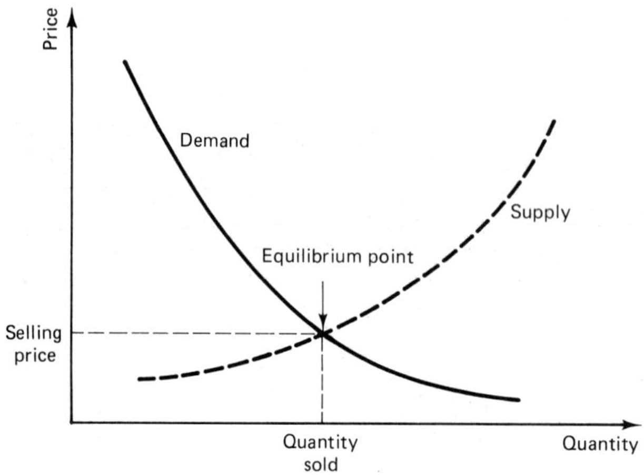

Figure 6.1.1 Supply and demand curves.

BOX 6.1.1 A PARABLE

In 1956, Mary Jones operated a gas station at the corner of Main Street and Washington Ave. in Smalltown, Ohio. She sold *Zowie* gasoline. The petroleum shortage was unheard of, and Mary sold gasoline for 26.9 cents per gallon. Across the street, Al Smith operated a *Rocket* station. He also charged 26.9 cents per gallon. Both Mary and Al knew that in the city 20 miles away, gasoline sold for 23.9 cents per gallon. But most of their customers were willing to pay more for the convenience of buying gas close to home.

One day, young Steve Johnson decided to open a gas station. He considered locating in the city, but finally he decided to open his station in Smalltown because prices were higher and he could make more money for each gallon of gasoline he sold. He knew that Mary and Al had loyal customers, so he set his price at 25.9 cents per gallon, figuring that some customers would be attracted by his lower prices. His plan seemed to work, and soon quite a few customers were coming to Steve's station.

Two weeks later, another station opened just outside of town selling

Tightwad gas for 24.9 cents per gallon. Soon most of the customers in Smalltown were stopping at the new station. Mary and Al lowered their prices to 24.9 cents per gallon, and Steve lowered his price to 23.9. But the new station immediately lowered its price to 22.9 cents, and a gas price war was on. What had happened in Smalltown was this: the supply of gasoline had suddenly increased, but demand was unchanged. The same number of customers could choose between twice as many gas stations. This meant that prices had to move down until supply equaled demand.

Within a week, prices at all four stations in Smalltown were down to 17.9 cents per gallon. Drivers from the city heard about the low prices in Smalltown and started driving over to fill their tanks. The stations had as much business as they could handle, but they were all losing money. In terms of economic theory, the increase in demand for gasoline was a natural result of the lower prices in Smalltown. Lower prices nearly always create an increase in demand.

Finally, Steve gave up and closed his station. Now there were only three stations in Smalltown. Then Al Smith decided to raise his prices to 18.9 cents per gallon. He had noticed that customers were coming from the city where gasoline cost 23.9 cents per gallon, and he figured they would consider gas a bargain at 18.9 cents. Sure enough, his business kept up. Soon all the stations were charging 18.9 cents, then 19.9, and so on.

What happened here was that the lower prices caused a decrease in the supply of gasoline. In this case, the supply decreased when Steve Johnson sold his station. Then the prices increased naturally as a result of the lower supply.

When prices reached 21.9 cents per gallon, people stopped driving over from the city, and the rise in prices became slower. Finally, prices leveled off at 23.9 cents per gallon. There were now three stations in Smalltown, where there had been only two before. The prices in Smalltown had stabilized at the same level as prices in the city. In economic terms, the price increase caused a decline in demand for the product. After a certain point, demand had dropped enough so that the supply and demand were in *equilibrium*, and the prices stabilized at that level.

The last we heard of Steve Johnson, he had opened a gas station in Suburban Grove, where gasoline was selling for $1.25.9 per gallon.

As Box 6.1.1 shows, energy prices respond to the forces of supply and demand, as do the prices of any other commodities. But other factors besides supply and demand have a powerful effect on energy prices. Here are a few:

- Costs of production and distribution
- Prices of competing fuels
- Incentive pricing to create demand
- International price-fixing
- Government regulation
- Tax policies
- Nontax government subsidies

In the rest of this section, we consider the marketplace forces that affect energy prices. Section 6.2 deals with governmental actions.

6.1.1 Costs of Production and Distribution

In many cases, energy costs are set by the cost of developing a resource. This is especially true when the purchaser cannot choose freely between a range of energy options. For example, since the early days of the space program, photovoltaic cells have been used to power satellites. Competing fuels could not be considered, because the satellites had to operate for long periods of time with no maintenance or fuel deliveries from outside. So photovoltaic cells were used, despite an extremely high cost: $500 per peak watt, several hundred times as much as conventional sources of electric power.

The extremely high price of photovoltaic cells was caused by the high costs of production. Some of the more expensive parts of the project included basic research, design and engineering, production of highly refined silicon, assembly, testing, and installation of photovoltaic arrays. As photovoltaic technology has developed, the costs of basic research and design have dropped, and increased production has caused production costs to fall significantly. This, in turn, has caused prices of photovoltaic arrays to drop proportionately. In 1984, prices were approximately $8.00 per peak watt, still three or four times as high as conventional energy sources, but less than one-fiftieth what they were 20 years earlier.

Although production costs have a particularly striking effect on the prices of experimental or unconventional energy resources, prices of more conventional forms of energy are also strongly influenced by production costs. For example, electric power prices are much lower in areas with a high percentage of hydroelectric generating capacity than in areas where most electric power is produced by thermal power plants. This is because the costs of operating a thermal power plant are much higher than the costs of operating a hydroelectric plant.

Distribution costs also affect the price of energy. Fuel costs are generally higher in remote areas than in more densely populated places. This is largely due to the added transportation costs. Distribution costs are also an important factor in the price difference between natural gas and petroleum. Natural gas is distributed entirely by pipelines, whereas petroleum is usually distributed to customers by trucks. Since pipelines are much more efficient than truck transport, the overall distribution costs of natural gas are somewhat lower.

6.1.2 Prices of Competing Fuels

Sometimes the price of an energy resource is determined less by the costs of production than by the prices of competing fuels. For example, natural gas is currently very inexpensive to produce in the United States and the price of natural gas has traditionally been lower than the price of oil (see Table 6.1.1).

In the 1970s, as oil prices increased, they exerted an upward pressure on natural gas prices. For a while, natural gas prices were held down by government price controls. But after the price controls began to be lifted in the late 1970s and early 1980s, natural gas prices increased sharply. During 1982–1983, some areas of the country experienced natural gas price increases of up to 70%.

Natural gas is still very cheap to produce, but most analysts believe that prices for natural gas will increase until they are roughly equivalent to oil prices. The reason for this is clear: customers are not likely to abandon natural gas as an energy resource unless they are able to replace it with a cheaper fuel. At present, no cheaper fuel exists (except, perhaps, firewood in heavily forested areas), so the price of natural gas can rise without causing a massive switch to other fuels. Thus the high price of competing fuels is directly causing increases in the price of natural gas.

TABLE 6.1.1 Average Cost of Energy (September 1982)

Form	Average Cost per Unit	Energy Equivalence	Cost per MBtu
Crude oil, wholesale	$33.0/bbl	0.132 MBtu/gal	$ 5.69
No. 2 heating oil, retail	1.148/gal	0.135 MBtu/gal	8.31
Gasoline, retail	1.295/gal	0.131 MBtu/gal	10.36
Diesel, retail	0.985/gal	0.135 MBtu/gal	7.14
Natural gas, retail	5.56/1000 ft^3	1.031 MBtu/1000 ft^3	5.39
Coal, wholesale at plant	24.46/ton	22.5 MBtu/ton	1.65
Wood, retail	85/cord	19.5 MBtu/cord	4.36
Electricity, residential	0.0718/kWh	3413 Btu/kWh	21.03
Electricity, industrial	0.0525/kWh	3413 Btu/kWh	15.38
Propane, retail	0.495/gal	4.01 MBtu/bbl	5.07
Butane, retail	0.776/gal	0.0955 MBtu/gal	7.95

Source: U.S. Department of Energy, *Monthly Energy Review,* December 1982.

The effect of competing fuel prices works both ways, of course. During the late 1970s and early 1980s, large numbers of homeowners switched from fuel oil to natural gas and wood because of the lower prices of natural gas and wood. This, in turn, helped reduce demand for petroleum, which eventually led to lower prices for heating oil. Heating oil prices are still (1984) much higher than natural gas and wood prices for equivalent amounts of energy. But without the competition of lower-priced alternative fuels, it is certain that petroleum prices would be much higher than they are today.

6.1.3 Incentive Pricing

Sometimes prices are set at an artificially low level in order to produce an increase in demand for a product. In principle, the increase in demand and the resulting higher production will result in greater production efficiency, a reduction in costs, and greater profitability for the producer. Thus, by setting prices at an artificially low level, producers hope to lower their costs to reach those levels.

Incentive pricing has frequently been used in energy production. Perhaps the most familiar example is the "turnkey" contracts that were offered to electric utilities by manufacturers of nuclear power plants during the 1960s. Under these contracts, utilities were guaranteed a completed, operable power plant at a fixed price. (All they had to do was "turn the key.") The reactor vendors did not expect to make a profit on these early plants. Instead, they hoped to create a market for nuclear power plants so that they would be able to make a profit on later power plants. The reactor manufacturers believed that increased production experience would lower costs, enabling them to produce nuclear reactors at a low price.

In practice, costs did not decline as rapidly as reactor vendors had hoped, and the market for nuclear power reactors disappeared during the late 1970s. But during the 1960s and the early 1970s, the incentive pricing practiced by the reactor manufacturers paid off in a substantial number of reactor orders from utility companies.

Incentive pricing of energy resources is an economic gamble. If it pays off, production capacity develops rapidly and energy supplies grow at an increased rate. But when incentive pricing does not produce the expected result, as in the case of nuclear power, the result is a temporary distortion of the energy marketplace followed by a traumatic series of economic readjustments.

6.1.4 International Price-Fixing

In the early 1970s, oil prices rose dramatically because of the efforts of an international cartel of oil-producing countries. This organization, called OPEC, consisted of a number of Middle Eastern nations and kingdoms, together with a few oil-producing countries in Asia and Latin America. The organization had existed for a number of years before 1970, but it was not able to influence prices significantly because of a worldwide surplus of petroleum. But in the 1970s, a worldwide shortage developed, which enabled the OPEC countries to raise their prices repeatedly, from a price of \$1.20 per barrel in 1969 to a high of \$34 a barrel in 1981.

Although the role of OPEC has been well documented, other international factors contributed to the success of OPEC's price-fixing. The reduction in United States oil production, the tremendous success of the automobile, the decline in automobile gas mileage during the 1960s, and the switch from coal to oil as a fuel for electric power plants all contributed to the international situation that permitted oil prices to rise dramatically.

After 1981, overproduction and reduced demand (both natural consequences of high prices) caused the world market for oil to soften, and oil prices declined somewhat. In 1982, plans were developed for the OPEC countries to reduce oil production in an attempt to eliminate the temporary oversupply and thus create the conditions for resumed price increases. So far, the strategy has failed, although increased demand for petroleum could bring about a renewal of OPEC's power. (A further discussion of OPEC is presented in Chapter 9.)

During the 1960s and 1970s, an attempt was made to create an international uranium cartel (UPEC) similar to OPEC. In 1972, several uranium mining firms met in Johannesburg, South Africa, to take steps toward assuring adequate profits in the uranium business. Their efforts met with some success as nuclear capacity increased and long-term uranium fuel contracts expanded to absorb world production of uranium. A severe shortage of uranium fuel was projected, based on planned nuclear plant construction for the 1970s and 1980s. As a consequence of the projected shortages, the international cartel was able to raise prices for refined uranium oxide (yellowcake) from $7 per pound in 1973 to $45 per pound in 1978. The Westinghouse Corporation was forced to default on a number of long-term contracts to supply uranium fuel at a fixed price of $8 per pound, and utilities had to scramble to ensure their own future supplies of uranium. The stage seemed set for the uranium cartel to imitate the great successes of OPEC.

But then, as construction of nuclear power plants slowed, the shortage of uranium fuel evaporated and prices began to decline. By 1982, uranium prices had dropped below $20 per pound, and the power of the uranium cartel appeared greatly diminished.

In summary, international price-fixing has been a significant element in world energy pricing, but only to the extent that world energy supply and demand have favored such practices. The power of international cartels has been shown to fade rather quickly when decreases in demand result in a temporary oversupply of energy resources.

6.2 GOVERNMENT ACTION AND ENERGY PRICES

In the early part of the twentieth century, the American oil industry was totally dominated by one company, Standard Oil, under the control of John D. Rockefeller. In an effort to eliminate this monopolistic condition, the U.S. Supreme Court ordered the breakup of the Standard Oil Trust into a number of smaller (but still quite large) oil companies. This marked the beginning of the government's control of the energy industry. In this section we look at some of the ways in which government control affects the price of energy.

6.2.1 Price Controls

For a number of years, gasoline and other petroleum products cost three or four times as much in Europe as they cost in the United States. Even today, gasoline prices are lower in the United States than in Europe, and much lower in Venezuela than in the United States. These price differences are largely the result of governmental policies. In the United States, the prices of domestically produced oil and gas were set by government regulations for many years. This eventually led to a situation where domestically produced oil and gas could only be sold at a price far below the world market price. As predicted by the principle of supply and demand, this led to a reduction in U.S. production of oil and gas, and thus to increased reliance on imported supplies.

During the late 1970s, in an attempt to stimulate domestic production, oil prices were decontrolled. This allowed prices for most domestic oil to rise until they reached the world market level. This meant a substantial price increase for older U.S.-produced oil. To compensate for the possible inequities caused by such a sudden increase in oil prices, the U.S. government imposed a special tax on the "windfall profits" expected from the price decontrol. Critics of the tax pointed out that while decontrol was intended to increase domestic production, the effect of the new tax would be to reduce production. They further pointed out that certain categories of oil, especially Alaskan oil, were exempt from the tax. However, the windfall profits tax survived a court challenge and is firmly in place as an element of U.S. energy policy.

6.2.2 Tax Policies

As every business manager knows, tax policy has a tremendous effect on business operations, including prices. Favorable tax treatment can give a company the edge on its competition, open up markets that might otherwise have been unavailable, and provide the company with investment capital. Unfavorable tax treatment can penalize a company seriously, even driving it out of business. In the United States, major industrial groups maintain staffs of lobbyists in Washington. These lobbyists try to influence tax legislation as well as government regulations, and often they succeed in producing significant tax benefits for their employers.

Perhaps the greatest energy tax subsidy in American history was the oil depletion tax allowance. Because oil is a nonrenewable resource, oil companies reduce their stock of marketable resources every time they sell a barrel of petroleum. To offset this loss, or *depletion,* of energy resources, the energy companies used to be allowed a tax credit equivalent to 27½% of their gross income. Clearly, this was an enormous tax credit, equivalent to a subsidized price cut of 27½%. This giant tax break, together with governmental price controls, maintained U.S. oil prices at an artificially low level for a number of years in the 1960s and early 1970s. The oil depletion allowance was finally repealed in 1975. But in August 1981, a new set of tax incentives was instituted which greatly benefited oil producers. The windfall profits tax was cut in half for many categories of oil, and new tax credits were instituted for private investors in oil-drilling schemes. Over five years, the total tax benefits were expected to total $11.8 billion, with larger benefits after 1986.

In 1978, a major tax subsidy was provided for purchasers of conservation and renewable energy equipment. Tax credits ranging from 15 to 40% of the purchase price (up to a certain established limit) were offered to businesses and householders in an attempt to encourage conservation and the development of renewable energy resources. This represents an effective price reduction of 15 to 40% on the products covered by the tax subsidy—a major subsidy for conservation and renewable energy resources.

Take the example of solar hot water heaters. According to the Arizona Solar Energy Commission, a well-designed hot water system for a family of four costs about $2700. The federal tax subsidy of 40% ($1080) plus the Arizona state tax credit of $945 makes the actual cost to the family $675. Savings on electric bills could further amount to about $325 per year—a 45% rate of return on the original investment. Thus in two and a half years this system will have paid for itself. Nearly 75,000 Arizona households and businesses were expected to install solar hot water heaters by the end of 1983.

6.2.3 Nontax Government Subsidies

Governments influence energy supplies and prices not only through tax policies, but also through direct and indirect subsidies. For years, the U.S. Department of Energy and its predecessors have provided subsidies for research and development of energy resources (see Chapter 9). Table 6.2.1 shows the amounts of money allocated to various energy resources in recent years.

Besides providing financial assistance for research and development, the federal government provides indirect subsidies in several ways. For a number of years, energy companies have been able to explore for oil, gas, uranium, and coal on government-owned land. The companies are required to pay the government a leasing fee, but no royalties are required to be paid for the energy resources that are extracted from the government-owned lands. This is quite different from the situation in other countries and states such as Alaska, where oil companies are required to pay substantial severance royalties to the governments that control the resources.

A second type of indirect subsidy is provided to the nuclear power industry by the U.S. government in the form of fuel enrichment services. Because of the dangers associated with enriched uranium fuel, private companies are not allowed to operate fuel-enrichment plants. Instead, the federal government provides enrichment services at a low price. The government can afford to offer these low prices partly because civilian nuclear fuel is produced as a by-product of enrichment for nuclear weapons production.

Yet another indirect subsidy is provided to the nuclear industry in the form of the Price-Anderson Act. Under this law, utility companies are liable for only a maximum of $560,000,000 in case of a nuclear accident. This means that if an accident caused property damage of $2 billion (not an unrealistic figure for an accident near a major city), the utility company would be liable for less than one-third of the damages actually caused. So far, no accidents have occurred in which damage to the public exceeded the limit of the Price-Anderson Act. But insurance costs for utility companies are much lower as a result of the Act. In fact, private industry was unwilling to participate in the development of nuclear power until the government instituted the indirect subsidy of

TABLE 6.2.1 United States Government Expenditures for Energy Research and Development (millions of current dollars)

Category	1963	1969	1972	1974	1976	1978	1980	1982
Nuclear	235 (71%)	309 (86%)	375 (72%)	1034 (80%)	753 (59%)	1378 (54%)	1254 (36%)	1530 (63%)
Fossil	51 (16)	37 (10)	101 (19)	110 (9)	323 (26)	684 (27)	743 (22)	430 (18)
Solar	—	—	—	45 (3)	93 (7)	368 (14)	507 (15)	257 (11)
Geothermal	—	—	—	—	31 (2)	106 (4)	112 (3)	55 (2)
Conservation	43 (13)	15 (4)	49 (9)	105 (8)	66 (5)	31 (1)	827 (24)	145 (6)
Total	329 (100)	361 (100)	525 (100)	1294 (100)	1266 (100)	2567 (100)	3443 (100)	2417 (100)

Note: Percentage figures may not add up because of losses due to rounding.

Source: Adapted from Phillip G. LeBel, *Energy Economics and Technology* (Baltimore: Johns Hopkins University Press, 1982). 1982 figures are from Environmental and Energy Study Conference Fact Sheet, Nov. 21, 1983.

the Price-Anderson Act in 1957. Without the Act, the nuclear industry would not exist today.

Recently, the Department of Energy has proposed a research grant to help finance the cleanup of the nuclear power plant accident at Three Mile Island in Pennsylvania. This would be an indirect subsidy to the utility that operates the nuclear plant, because it would shift some of the costs of the cleanup from the utility to the federal government.

Renewable energy resources have also benefited from direct and indirect subsidies. Perhaps the best known subsidies are the water projects of the Far West. Strictly speaking, these projects are not simply energy projects. They provide water for irrigation, transportation, and recreation, as well as drinking water for many western cities. But they also provide an enormous government subsidy for the development of hydroelectric power.

Solar energy has also benefited from government subsidies. The National Aeronautics and Space Administration (NASA) purchased large numbers of photovoltaic cells to power satellites and space stations. These government contracts underwrote much of the basic research and development effort for photovoltaics and have helped reduce the price of photovoltaic cells from $500 per peak watt in the late 1960s to $8 today.

Government subsidies and tax policies have long been the subject of a major national debate. Proponents of the various energy resources often accuse the government of undermining free market principles, and many call for a reduction in government subsidies and controls. But the principle of tax and other subsidies is firmly established as a governmental policy tool. It seems likely that governmental subsidies will continue to exert a strong influence on energy development in the United States.

6.3 PUBLIC UTILITIES AND ENERGY PRICES

Public utilities, such as telephone and electric power companies, are defined as *regulated monopolies.* This means that they do not compete directly with other companies that provide similar services, because such competition would lead to an undesirable duplication of facilities. For example, suppose that a city were served by two (or three or four) electric power systems. Each system would need to acquire land for its transmission lines, which would result in a proliferation of unsightly and expensive systems of cables, poles, and conduits. Also, no single company would be responsible to ensure continuity of electrical service. As companies went in and out of business, customers could be subjected to serious inconvenience. To avoid these problems, local or state governments grant one company the exclusive right to provide services to a community.

Because of the monopolistic structure that is established by such a system, prices cannot be set by the normal processes of supply and demand. If given a free hand, utilities could raise their prices at will and threaten to cut off any customers who refused to pay the high prices. To avoid this problem, prices are set by state governmental agencies with names such as *public service commission, public utility board,* or *state utility commission.* These state agencies evaluate an electric company's costs and operating procedures and set the prices that the utilities may charge. If a utility wants a price increase, it must receive the approval of the agency before raising prices to the customers.

6.3.1 Factors Affecting Rate Decisions

Different public service boards use somewhat different criteria for establishing the prices that utilities are allowed to charge their customers. But a few general principles are followed in nearly all cases.

Rate base. Utilities are guaranteed a rate of return on their investments. But not all investments are eligible for this rate of return. For example, if a utility decided to invest in a tennis shoe factory, it would not be permitted to include the cost of the factory in its rate base. In general, the rate base is limited to investments that help the utility provide energy services to its customers.

In some cases, public service boards exclude investments that the utilities want to include in the rate base. For example, after the nuclear accident at Three Mile Island, both the damaged nuclear plant and another plant at the same site were dropped from the rate base because they were not providing any electric power to the utility's customers. The utility objected to the exclusion of the undamaged power plant, but the public service board argued that the plant was "neither used nor useful" and thus was ineligible to be included in the rate base. The utility is currently making a strong effort to restart the undamaged power plant in order to have it included in the rate base.

It is important to understand that only *capital investments* are eligible to be included in a utility's rate base. Operating expenses are not included. This means that if a utility is given the choice between a facility with high capital costs and low operating costs (such as a nuclear power plant) or one with lower capital costs and higher operating costs (such as a coal-fired plant), the utility can earn a higher profit by building the facility with the higher capital costs. Theoretically, this could lead utilities to choose uneconomical investments in order to maximize their profits, and some critics of the energy industry claim that this has been the case with nuclear power plants. However, in order to receive approval to invest in an energy facility, the utility must first convince the public service board that its investment is financially sound. Thus, in principle, the public is protected against such unethical practices.

Return on investments. Once the utility's rate base has been determined, the guaranteed return on investment can be computed. This guaranteed return is equal to a set percentage of the rate base, typically between 12 and 17%. This rate of return is necessary in order for the utility to pay interest to its bondholders and dividends to its stockholders (see pp. 274-276).

Operating costs. In addition to its guaranteed return on investment, the utility is allowed to recover its operating costs from the ratepayers. These operating expenses include fuel costs, salaries, taxes, operating and maintenance expenses, and miscellaneous other costs. In some cases utilities are allowed to include costs of advertising and promotion materials in their operating costs, whereas in other cases those costs must be absorbed by the stockholders.

Utilities are not permitted to earn a profit on their operating costs, but critics charge that this principle is not always followed. For example, utilities are permitted to pass along the costs of fuel to their customers. Critics charge that utilities pay unreasonably high fuel costs because they know that those fuel costs will be automatically passed along to their customers. In many cases, the provider of the fuel and the utility are both owned by the same holding company, so the high fuel costs are used as a method for increasing profits at the expense of ratepayers.

Another practice that has been criticized is the use of "phantom taxes" in figuring utility rates. These are taxes that the utility is permitted to include in its operating costs even though the taxes are never paid. In some cases, the utility uses two different accounting methods for computing taxes: one method produces high tax figures, which are billed to the ratepayers as operating costs; and the other method yields lower tax figures, which are the basis of actual tax payments.

As these practices have become more widely publicized, public utility boards have begun to scrutinize utility operating costs more carefully. No one suggests that utilities should be prohibited from recovering their costs of operation; but neither should utilities be able to use their operating costs as a way of exceeding their legally established rates of return.

Setting utility rates. After the rate base, the return on investment, and the operating costs have been determined, the utility adds those figures together to determine the amount of income that can legally be obtained from ratepayers. Then the utility estimates its sales of electric power and establishes its electric rates at a level that will allow it to recover the legal amount of income. If the utility's calculations show that the rates must be raised or lowered in order to match the guaranteed level of income, the utility requests the public utility board to grant a rate change.

In the past, utility costs decreased steadily, and utilities adjusted their rates downward. Needless to say, these rate changes did not generate much public opposition. But during the 1970s, the situation turned around, and utilities began asking for large rate increases. Many of these rate increases became highly controversial. Public advocates began to question the wisdom of utility investments, and operating costs were scrutinized more carefully. Today, when a utility requests a rate increase, it can expect to go through a lengthy series of hearings and respond to many questions from intervenors.

6.4 UTILITY RATE STRUCTURES

The price people pay for energy has a significant impact on their energy use. Yet many commercial customers have little understanding of the components of their bills. In fact, most customers believe that electricity is sold much like any other commodity, although this is not the case. In fact, electric rates vary widely, not only from one region of the country to another, but even between two customers of the same utility. The electric utility rate structure reflects the particular nature of electricity as an energy resource.

Electricity is a very perishable form of energy. Once it is generated, it must be used quickly or it is lost forever. But demand for electricity fluctuates widely from season to season, and even from one hour to the next. This creates a problem for utilities. They must be prepared to provide enough power to supply the highest demand of the entire year. This means a large generating capacity. But during most of the year, the demand for power is much lower than the "peak." So much of the utility's generating capacity sits idle most of the time. The generating facilities cost money whether they are operating or idle. In fact, for many types of generators, operating costs are only a small fraction of total costs. The utility must find a way to recover the costs of these generating facilities that operate only part of the time.

In response to this complex situation, utilities have developed a number of factors to be used in calculating electric bills. The most important of these are the following: (1) energy consumption, (2) "time-of-day" charges, (3) demand for power, and (4) power factor. We will examine each of these in turn.

6.4.1 Energy Consumption

This refers to the total amount of energy used by the customer. The higher the consumption, the greater the energy costs, because it takes more money to produce and deliver more energy. But the situation is not quite that simple. Consider a power plant with a capacity of 50 MW. If the plant is not used at all, only the operating costs are saved. The fixed costs must still be paid. If the plant is operating at 50% capacity, there will be additional operating costs, but the fixed costs will not change. Thus the cost per unit of energy generated will decrease for the plant operating at 50% of capacity. Following this principle, we can see that a plant operating at 100% of capacity has the lowest cost per unit of energy (see Figure 6.8.2). So it is in the utilities' interest to keep their generators running as much as possible.

To ensure that their generating capacity will be used as fully as possible, many utilities use a *declining block* rate structure for energy costs. A typical declining block structure is shown in Table 6.4.1. This type of rate encourages customers to take full advantage of the utility's generating capacity. However, it does not encourage conservation; in fact, it indirectly discourages conservation because it reduces the cost benefit of conservation measures. This type of

TABLE 6.4.1 Declining Block Rate Structure for Energy Charges

Energy Consumption	Rate Charged per kWh
All categories	Flat rate for electric service, typically called *customer charge;* assessed regardless of the amount of energy used
0–500 (kWh/month	Highest rate
500–1000	Slightly lower rate
1000–4000	Lower rate
4000–10,000	Still lower rate
Over 10,000	Lowest rate

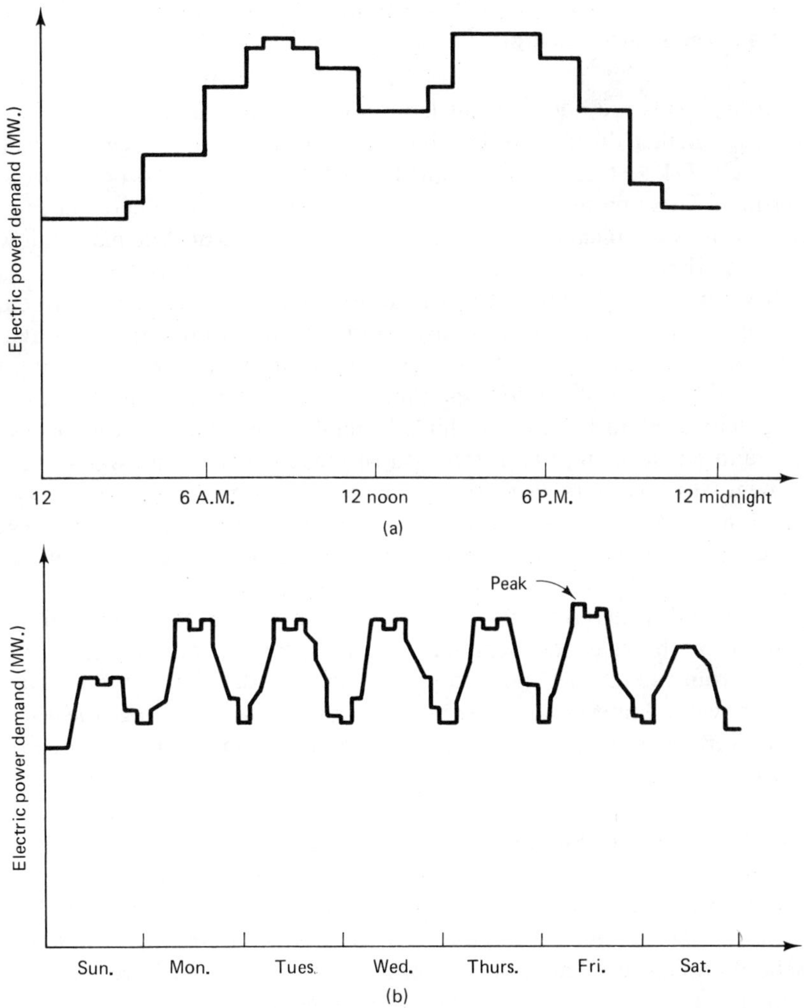

Figure 6.4.1 (a) Typical daily demand curve; (b) weekly demand curve.

rate structure is called *inverted* because rates decline as usage increases. (Note: Readers may encounter other names for the rate structures.)

In some parts of the country, inverted rate structures are being replaced by *progressive* rate structures, where higher energy consumption means higher costs per kilowatthour. Some energy analysts claim that this misrepresents the utilities' actual costs, because higher energy use often allows utilities to reduce their costs per kilowatthour. But progressive rate structures, by encouraging conservation, may reduce the necessity for installing new, more expensive generating capacity. Thus many economists believe that progressive rate structures more accurately reflect the total economic impact of increased energy consumption.

6.4.2 Time-of-Day Charges

Utility loads are not constant from hour to hour. Electrical use is generally much higher during the day than at night. The lowest demand occurs late at night (11 P.M. to 5 A.M.), and the highest use tends to occur around suppertime. These patterns will vary from one region to another, but Figure 6.4.1 illustrates a typical demand curve. (Note that electrical demand is lower on Sundays than on weekdays.)

It would be in the interest of the utility company to reduce the fluctuations in usage, or at least to discourage consumption during hours of peak demand. For this reason, some utilities install time-of-day metering capacity. By measuring the electric use at different times of day, the utilities can charge more for electricity used during hours of high demand. Some utilities are going a step further and encouraging the installation of electric space- and water-heating equipment which can draw electric current only during off-peak hours. These installations usually purchase electric energy at very low rates. Some of the rate differentials necessitated by variations in energy demand are shown in Table 6.4.2.

Time-of-day metering and off-peak heating systems allow the utilities to shift some of the electrical demand from the peak periods to the off-peak periods. From the utilities' point of view, the ideal demand curve would appear as shown in Figure 6.4.2. Any billing methods that encourage the leveling off of power demand will permit the utilities to make more efficient use of their generating capacity.

6.4.3 Demand for Power

Consider two customers, each using equal amounts of energy, say 800 kWh per month. One customer uses electricity to operate electric lights and a few household appliances. This customer uses electricity at a fairly constant rate. The other customer uses electricity only on weekends, but when he does, he uses the electricity to power large machinery, including an electric welding

TABLE 6.4.2 Electric Rates for a Winter-Peak Utility[a]

Rate Type	Minimum Customer Charge (per month)	Peak Season Rate (Jan.–Apr.)	Off-Peak Season Rate (May–Dec.)	Comments
Rate 1 residential service	$5.78	3.371 cents/kWh for first 200 kWh 11.799 cents/kWh for use over 200 kWh	3.371 cents/kWh for first 200 kWh 4.944 cents/kWh for use over 200 kWh	No customer-operated or customer-owned generation allowed without permission
Rate 2 general service	9.10	13.5 cents/kWh if no demand charges With demand meters: first 10 kW of peak at \$10.56/kW; additional at \$15.45/kW; plus 4.5 cents/kWh	5.169 cents/kWh if no demand charges With demand meters: first 10 kW of peak at \$2.75/kW; additional at \$7.65/kW; plus 3.371 cents/kWh	Three-phase service at additional \$5.96 per month If peak demand is more than 100 kW, then 60% of that value for next 11 months
Rate 3 off-peak water heating	3.33	3.821 cents/kWh	2.585 cents/kWh	Must have a storage capacity of 50 gal
Rate 11 residential off-peak service	8.43	Off-peak: 3.371 cents/kWh Peak: 20.676 cents/kWh (Peak hours are: 4 hr between 6 A.M. and 12 noon 4 hr between 4 P.M. and 10 P.M.)	3.371 cents/kWh	No customer-owned or customer-operated generation allowed unless granted by special contract

[a]These rates were in effect for the customers of Central Vermont Public Service Corporation, Rutland, Vt. Temporary surcharge of 5% in peak season and 10% in off-peak season. Rates as of December 1982; requests for rate hike pending with the Public Service Board. Based on CVPS pamphlets CV858-1, -2, -3, -11; available for the general public.

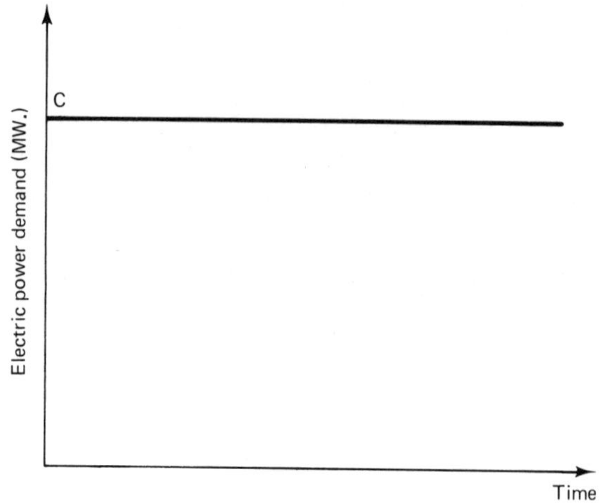

Figure 6.4.2 Flat demand curve.

machine. Both customers use the same total amount of electricity over a month's time, but the customer with the heavy machinery uses much more electric power.

It is clear that the customer with higher power requirements puts a greater burden on the utility, because the utility has to maintain more generating capacity to provide the needed power. To account for this problem, utilities often measure *power demand,* especially for large industrial and commercial users. A meter is installed which measures the power demand during critical periods of the day. A fee is then levied on the customer, based on the highest power demand for the month.

In some cases, the utility's demand charge reflects the highest demand measured during the entire year. This means that an instant of high demand, for whatever reason, can mean an entire year of high costs. In other cases the utility will apply a *ratcheting* factor, say 75%, to this maximum demand figure. Thus the customer will pay either the actual maximum demand for the month, or 75% of the annual maximum demand, whichever is higher.

It is worth pointing out that in many cases, large users pay more for their power demand than for actual energy consumption, especially in months of low energy consumption. Consider the following example. During the summer, a company has a measured demand of 1000 kW. This high demand is caused largely by air conditioning. In the fall, its energy demand falls to 300 kW. But because of the "ratcheting" factor, the company still pays demand charges based on 750 kW (see Figure 6.4.3). Of course, during periods of heavy demand, the energy use will be higher, and energy costs will exceed demand charges. But this example illustrates how costly it can be for a utility customer to experience a brief surge of demand during a period of peak usage.

Measured demand: 300 kW
"Ratchet" demand: 750 kW
Demand charge: 750 kW × $15.00 = $1125
Energy use: 20,000 kWh
Energy cost per kWh: 4 cents
Energy charges: $800

Figure 6.4.3 Sample August electric bill.

6.4.4 Power Factor

Electric power has two components: voltage and current. Both voltage and current alternate between a positive and a negative maximum. It is not necessary for the maximum voltage and the maximum current to occur at the same time. Due to the characteristics of electrical devices, the maximum voltage and maximum current can be offset, as shown in Figure 6.4.4.

Simply speaking, power equals the product of voltage and current.

$$\text{power (watts)} = \text{voltage (volts)} \times \text{current (amperes)}$$

At any given time, a power meter will measure this value. However, this power figure is misleading when applied in situations where the maximum voltage and maximum current do not coincide. In cases like this, the power being used at any given moment will be less than the product of the maximum voltage and the maximum current. But the electric utility must be capable of delivering the maximum voltage and the maximum current. For this reason, utilities calculate the *power factor,* which accounts for these differences.

$$\text{power factor} = \frac{\text{power use (meter reading)}}{\text{maximum voltage} \times \text{maximum current}}$$

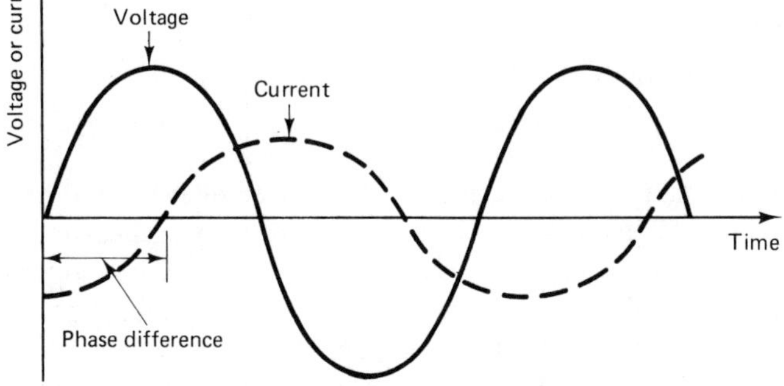

Figure 6.4.4 Voltage and current out of phase.

If a user's power factor is 85 to 90% or above, utilities will usually not assess a penalty. But if the power factor drops lower, they will usually assess a fixed charge for each percentage point below the minimum acceptable power factor. The customer can avoid this extra charge by installing a storage "bank" (usually a set of capacitors) to handle these fluctuations during the power cycle.

6.4.5 Other Factors in Utility Bills

In addition to the factors described above, a typical utility bill also contains other charges, such as *fuel adjustment* charges. A fuel adjustment charge allows the utility to cover any increases in fuel prices that occur during the billing period. Without this provision, utilities would be required to request a rate increase every time fuel costs increased or decreased. Rate hearings are long, complex affairs, and fuel prices are quite volatile. For this reason, utilities and public service boards try to reduce the number of hearings through the use of fuel adjustment clauses.

In some areas utility bills may also include a charge for *construction work in progress* (CWIP). These charges are assessed to pay for the construction of future generating capacity. CWIP is a highly controversial practice, and it has been made illegal in some states (see Section 6.6.4).

6.4.6 The Debate over Utility Rate Reform

Some critics have charged that present electric rate structures are designed to reflect the needs of the utility companies—namely, to maximize their profits—rather than to reflect the needs of the public. The critics point out that public utility boards are usually staffed by former energy industry employees, and they back up their case by pointing out some of the most inequitable features of utility rate structures:

Inverted rate structures
Fuel adjustment clauses
Phantom taxes
"Ratcheted" demand charges
Inclusion of promotional expenses in utility costs

In recent years, many of these criticisms have been heeded, and rate structures have been changed. Inverted block rates are rapidly disappearing, and utilities are sometimes in the forefront of encouraging conservation. In many regions of the country, so-called *social rate structures* are in effect. Low users (below 200 kWh per month for residential users) are charged very low rates, and higher rates apply to larger users. In this way, conservation can be encouraged without penalizing low-income families.

These changes are likely to continue as public service boards become more responsive to concerns of the general public. Utility rate reform is only one sign of a more widespread change in the way our society deals with the problems of energy supply and demand.

6.5 ECONOMIC ASSESSMENT OF ENERGY PROJECTS

Like any other business venture, energy projects must make good economic sense. That is, they must be advantageous to the entrepreneur. In theory, this means that as long as the project can generate more money than it costs to build and operate, the project is worth undertaking. But in actual practice, matters are not quite so simple. Determining the economic feasibility of a project is quite an involved process involving a wide range of variables.

An energy project can be considered economically justifiable if *all* the following conditions are fulfilled:

1. The benefits of the project exceeds its costs.
2. The project is more profitable than alternative methods of achieving the same objectives.
3. The project is on the proper scale to achieve optimum economic results.
4. Capital is available to complete the project.
5. Each separable segment or part of the project is economically justifiable, meeting conditions 1 through 4.

Let us examine these conditions one at a time.

1. *The benefits must exceed the costs.* In the simplest interpretation, this means that the project should be profitable. But an energy project can also involve other less obvious costs and benefits, which may be somewhat more difficult to quantify. It is important not to overlook these intangible factors.

For years, the U.S. Army Corps of Engineers has been using a dollar figure to estimate the "recreational benefits" of water projects. But in assessing costs, such intangible factors have not been widely considered. In traditional economic analysis, intangible factors such as ecological, social, and political costs were described as *externalities;* that is, the costs did not have to be paid by the company doing the analysis. As a result, these costs were not usually included in economic assessments of proposed energy projects.

An analysis that excludes "externalities" not only conceals the possibility of social and ecological harm, but it can also conceal long-term financial costs that may eventually jeopardize a project's economic success. For example, if a company is building an electric power plant, it may elect not to include pollution controls, regarding them as unnecessary, wasteful, and expensive obstacles to short-term profitability. But after a few years of operation, the plant might cause such severe environmental damage that state or local legislators would compel the company to add pollution control equipment, or perhaps even close down altogether. This can seriously affect long-term profitability. Similarly, safety-related equipment, which may seem expensive in the short term, can cut down on the probability of future lawsuits. So it seems prudent to look beyond immediate profits and consider an analysis that includes externalities.

It is easy to understand why many companies do not consider such matters as social and environmental costs. Because such factors as environmental quality or human health cannot be quantified, there is no established method of including them in economic assessments. How is it possible to assign a dollar value to a clear-running stream or a healthy child?

The problems are not likely to disappear overnight. But serious attempts are being made to integrate external factors into economic decision making. For example, the National Environmental Policy Act (NEPA) requires all new energy projects to meet certain environmental standards as a condition of operation. Violators are subject to fines, lawsuits, and other economic penalties. This prevents entrepreneurs from overlooking environmental factors as a cost of doing business. Similar methods are being used to ensure that public and worker safety are being protected: the federal Occupational Safety and Health Act establishes workplace safety standards, and class-action lawsuits have enabled citizens to bring claims against companies that have endangered public safety. In recent years, some states and municipalities have begun to hold companies liable for the social effects of their operations: for example, a corporation may be required to assist in the building of schools, roads, and sewer systems to serve the additional employees brought to work at an energy facility. Many of these ideas are still experimental: no doubt some will be found ineffective and others will be modified. But thoughtful economic planners will realize that the cost of doing business includes a number of factors that may not have been included in traditional economic assessments.

2. *The proposed energy project must be more profitable than other ways of achieving the same objectives.* In other words, no other way should exist that would achieve the same objectives at a lower cost. But again, as in figuring costs, there may be substantial disagreement over the factors to be included in the analysis. How broadly should we define our "objectives?" What about secondary benefits of proposed alternatives? Finally, who should perform the comparisons?

Consider the following example. A certain community determines that it will need 100 MW of electrical power within five years. The power can be generated either by a fossil-fuel plant or by a hydro project. Proponents of the fossil-fuel plant show that capital costs will be lower for the fossil-fuel plant, but advocates of the hydro plant counter that the hydro plant will create a small lake, thus providing recreation for local residents and attracting tourists to the region. They claim that this extra benefit makes their project superior to the fossil-fuel plant. Finally, a third group appears, with figures indicating that current demand for electricity can be reduced by 100 MW through the imposition of strict energy consevation standards, at essentially no cost. Who is right?

If we define our objective very narrowly (i.e., providing 100 MW of electric power at the lowest possible cost), the proponents of the fossil-fuel plant are right. But proponents of the hydro plant are also right when they claim that the additional benefits of their project should be taken into account. Finally, the proponents of conservation are also right when they claim that they can meet the community's needs for power at a lower cost than either of the other two projects. So in a sense, all three are right. The problem is that all three groups define the problem somewhat differently. Only by comparing the three different assessments of the community's needs, and perhaps by conducting an independent study, can planners be sure that they have chosen the best alternative.

3. *The project must be on the proper scale to achieve optimum economic results.* For example, it would be absurd to build an 850-MW thermal power plant in a region where electrical demand was only 75 MW. The plant could operate at low power in order to provide electricity at the level desired, but the costs would be very high. This is because the *construction* costs of the 850-MW plant are much higher than for a smaller, 75-MW plant. The interest charges on those construction costs would raise the cost per unit of electricity dramatically. Power plants must be built to scale.

But it is not enough to build the power plant to the correct scale. The plant should also be built at the correct scale for the energy resource being used. For example, although nuclear reactors can be built in a range of sizes, it would not be economically feasible to construct a nuclear power plant to produce 25 kW of electricity. The costs of licensing, safety equipment, and personnel would be so high that the electricity would be greatly overpriced. On this small scale, low-head hydro, a diesel generator, or perhaps wind energy might be a better buy. On the other hand, low-head hydro or wind energy is not so well suited to very large scale production of electricity (500 to 1000 MW). The additional expense of control circuitry and interconnecting grids would raise the costs per kilowatt far above the rate for larger installations. For such large-demand situations, large-scale hydro projects or thermal power plants appear to be more economically feasible.

4. *Capital must be available to complete the project.* A favorable cost/benefit ratio, an appropriate choice of resources, and a proper design are not enough to ensure economic feasibility. The economic resources must be available to build the project. This capital can be raised in a number of ways, which will be discussed later in this chapter. However the capital is raised, enough capital must be available to carry out the complete project. In the case of a wind farm, a bank of photovoltaics, or an ocean thermal (OTEC) installation, the project can be partially completed and then revenues from the partial project can be used to raise additional capital. But in large-scale hydro projects or thermal power plants, a major portion of the project must be completed before any power (and thus revenue) can be generated.

A lack of sufficient capital can cause a planned energy installation to be scrapped even when the consequences are economically damaging. For example, in 1982 and 1983, the Washington Public Power Supply System (WPPSS, sometimes called WHOOPS by its critics) was forced to abandon construction on four out of five proposed nuclear power plants because of a lack of available capital. The economic consequences have been very severe. In 1983, WPPSS failed to make interest payments on $2.5 billion of bonds.

5. Not only must the complete project meet conditions 1 through 4, but *each separable segment of the project must also be economically justifiable.* For example, suppose that a utility plans to build a "wind farm" electric power installation. In order to store electrical energy during periods of low demand, the developers propose a battery storage facility. However, they discover that they can construct a pumped storage hydro system at a lower cost than the battery storage facility. This means that the project as originally proposed is no longer economically justifiable. It should be modified to include the pumped storage rather than the battery storage system.

Projects cannot always be modified in this way. Sometimes a segment cannot be considered independently. Consider, for example, a multipurpose hydroelectric facility. A river is to be dammed, creating a reservoir that will be used for flood control, irrigation, recreation, freshwater fishing, and a city water supply. Turbines and generators will be installed below the dam, providing electric power. Critics of the project point out that a thermal power plant could provide the same amount of power at a lower cost. But a switch to a thermal power plant would eliminate all the other benefits of the hydro project, requiring another reliable source of drinking water, water for irrigation, and so on. Thus the electrical component of the project is not really separable, and the hydro project may still be economically feasible.

Figure 6.5.1 Aerial view of Hoover Dam and Lake Mead, Boulder Canyon Project in Arizona–Nevada. This multipurpose project, completed in 1935, provides flood protection, irrigation, water storage, recreation, and of course, electricity. (*Source:* Bureau of Reclamation, U.S. Department of the Interior.)

6.6 CAPITAL FORMATION FOR ENERGY PROJECTS

Except for a few low cost or no cost conservation measures, every energy project requires a fairly large amount of money to be invested before any energy can be derived from the project. Depending on the type of energy project, the money may be tied up for 4 to 15 years before the energy facility produces any power—or any income, since the plant must be in operation before its power can be sold (see Table 6.6.1). In some cases (generally nuclear plants), construction has gone on for more than a decade; and in a growing number of cases, projects have been abandoned without producing any power. For these reasons, capital must be available, not only for a short time, but up to 15 years and perhaps longer.

TABLE 6.6.1 Typical Construction (Project Completion) Periods[a]

Type of Project	Minimum Time (yr)	Maximum Time (yr)
Coal production		
Strip mining	2	4
Underground mining	3	5
Oil and gas production		
Existing field expansion	1	3
New field	3	10
Uranium production from new mines	7	10
Power plant projects		
Coal	5	8
Nuclear	9	12
Hydro	5	8
Wind	2	4
Oil	5	7

[a]These are typical time spans in years for project completions. A construction period is defined here as the time lag between the "green light" for construction work and the "ribbon-cutting" ceremony.

Source: Data from "A National Plan for Energy Research and Development and Demonstration: Creating Energy Choices for the Future," U.S. Energy Research and Development Administration, April 1976.

The amount of money involved in an energy project can be quite considerable. An average nuclear plant begun in 1972 took six years to build at a cost of $170 million. However, a plant begun in 1982 is likely to take 12 to 14 years and cost anywhere from $1.5 billion to $3.7 billion. Boston Edison's Pilgrim II plant in Plymouth, Massachusetts, was proposed in 1972 at a projected completion cost of $402 million. The plant was finally abandoned in 1981, when the startup date was pushed back to 1990 and the estimated completion cost reached $4 billion. So if an electric utility plans to construct a 1000-MW nuclear plant—a typical size—the utility must be prepared to raise several billion

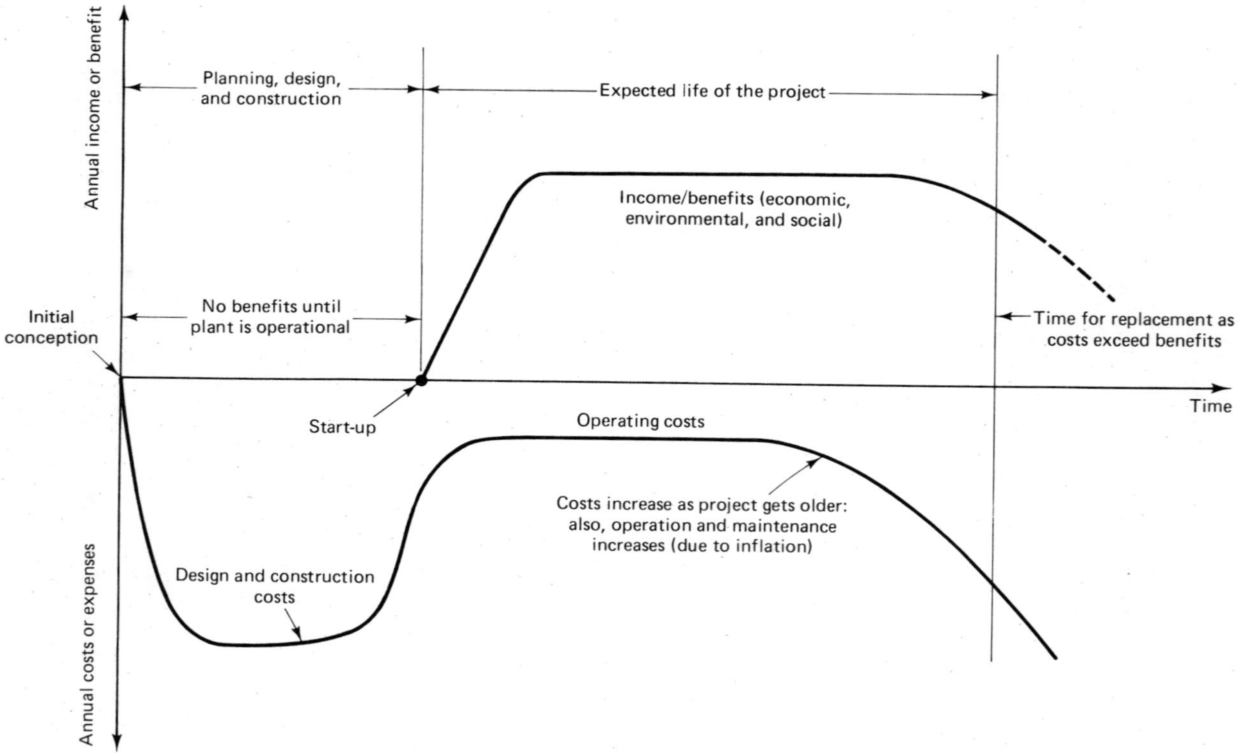

Figure 6.6.1 Cost and benefit curves for a typical energy project.

dollars. Where does the money come from? How does the utility persuade investors to provide these huge amounts of money? What do investors get in return for their money? All these are important aspects of financing energy projects.

In a few cases, companies have been able to save their profits from current operations to finance future energy projects. For example, many oil companies use a share of their income for exploration and drilling. But more commonly, the capital must be raised from outside sources. There are three basic techniques used for capital formation: (1) borrowing, (2) sales of equities, and (3) rate increases. We will examine each of these methods in some detail.

6.6.1 Borrowing

Borrowing is the most common method of raising large amounts of capital for energy projects. If all the money for a project can be borrowed from a bank or other financial institution, the money can be available almost immediately. Other methods of borrowing, such as bond sales, require somewhat more time, but they are still quite effective ways of raising designated sums of money quickly. Since energy companies usually want to begin construction as soon as possible, borrowing is by far the most common method of raising capital.

Although the different methods of borrowing differ in certain particulars, all methods of borrowing share the following characteristics:

1. The money that is borrowed must be used for a purpose that is specified in the lending agreement.
2. The borrower must pay back the money borrowed (the *principal*), together with *interest* on the money.
3. The loan is repaid according to a predetermined schedule. If the borrower fails to meet the repayment schedule, the lender can seize property (in this case the energy project) in order to recover the money that was lent.
4. If the company goes bankrupt, the lenders will be repaid before any of the company's assets are distributed to proprietors or shareholders.

The first three conditions will be familiar to anyone who has borrowed money to buy a house or car. The fourth condition is particular to business situations, where claims on company assets follow a legally determined *order of precedence*.

Loans from banks. When a large business, such as a public utility, borrows money, it pays a somewhat lower interest rate than everyday citizens. Most banks post a *prime lending rate,* which is the rate of interest they charge their "best" corporate customers. Public utilities usually fall into the category of "best" customers; in fact, they are sometimes able to borrow money below the prime interest rate. So utilities that borrow directly from banks enjoy the benefit of relatively low interest rates.

However, banks do not usually offer business loans at fixed interest rates. The usual practice is for interest rates to be reevaluated every year or six months.

In this way, the bank can offer low interest rates without running the risk of losing money if interest rates rise. Recently, for example, prime interest rates rose dramatically, from a level around 9% in 1979 to more than 20% in 1981, then fell back to 10.5% in 1983. When the rates were high, banks had to pay high rates of interest to attract deposits. To cover their increased costs, the banks raised the interest rates that they charged corporate customers. At the same time, holders of residential mortgages continued to pay very low rates—sometimes as low as 7% per year—because interest rates on those loans were fixed by legal agreement.

In general, energy companies do not like to borrow money from banks and other financial institutions. This is true for a number of reasons, including the following:

1. Energy companies, especially electric utilities, often have to raise immense amounts of money—billions of dollars—and banks are often not willing to lend such huge sums.
2. Banks do not usually lend money for extremely long periods of time, such as 30 or 40 years. Because of the long periods of time required for the construction of energy projects, this presents difficulties for energy companies.
3. Borrowers do not have the use of all the borrowed money for the full period of a bank loan. The amount borrowed must be paid off gradually throughout the period of the loan.

Bonds. A corporate bond is simply a loan agreement between a corporation and the bondholder. The bondholder may be an individual person, a bank, a pension fund, a union, or a university. The corporation is the borrower, and the bondholder is the lender. Under the terms of a bond agreement, the lender gives the corporation a certain amount of money, called the *principal.* (Bonds are usually sold in *principal amounts* of $1000 or $10,000.) In turn, the corporation agrees to pay a fixed rate of interest (called the *coupon*) for a certain number of years. At the end of that period, the corporation agrees to return the money which was borrowed.

Under most bond agreements, all of the principal is available to the corporation for the entire period of the loan. The corporation pays only the interest during the period of the loan and pays back all the principal at the end of the lending period. In many cases, too, the corporation reserves the right to retire the bonds early by paying back the principal before the period of the loan has been completed. This is usually done if interest rates go down and the company can sell new bonds at a lower rate. However, in a period of rising interest rates, the corporation usually keeps the principal for the full period of the loan. This is a definite advantage over bank loans, because it gives the corporation a greater degree of flexibility.

Bond Ratings. Because thousands of corporations sell bonds to the general public, there are many categories of bonds. Financial service corporations called *rating services* evaluate the quality of the various bonds that are offered to the public. For example, if the rating services are certain that the corporation will be able to meet all its interest payments and repay the principal, they will give the company's bonds a high rating (A or AA, depending on the rating service). If the company's ability to repay is not so certain, the rating service may give the bond a lower rating. When this happens, investors are reluctant to buy the company's bonds and the company must offer a higher rate of interest to attract investors. So it is in the utility's financial interest to maintain a favorable bond rating.

Sometimes, if a utility's bond rating falls too low, it will be unable to sell any bonds at all, and energy projects may have to be abandoned. In 1983, the bond rating of the Washington Public Power Supply System (WPPSS) dropped so low that the organization was unable to sell enough bonds to finance its projects. This led to the abandonment of three planned nuclear power plants and seriously jeopardized a fourth. Ultimately, WPPSS failed to make payments on some of its bonds. So a drop in a company's bond ratings can have a very serious effect on energy projects.

Electric utilities are especially dependent on bonds as a method of raising capital. On the average, 40% of all corporate bonds are offered by utility companies.

6.6.2 Appropriations

When the energy company is a public utility, such as a municipal electricity company, capital for energy projects can be raised through the appropriations process. (Because taxpayers are the ultimate source of capital for these projects, privately owned utilities cannot use this method of raising capital.) Many major public works projects, such as the hydro projects on the Columbia River in the northwestern United States, have been funded through the use of government appropriations.

The appropriations process has several advantages for public utility companies.

1. The capital does not need to be repaid.
2. No interest charges are assessed.
3. Funds come directly from the governmental body that oversees the utility's operation.

Despite these advantages, raising capital through the appropriations process is a time-consuming and uncertain business. Governments work slowly, so the capital may not be available as quickly as the utility might desire. Also, if the project is a long-term project requiring continuing appropriations, the project may not be funded fully throughout its construction lifetime. For example, the Clinch River Breeder Reactor has been subjected to a long series of funding difficulties over the years. As its political popularity has waned, legislators have been increasingly reluctant to provide the necessary funding. Finally, in October 1983, funds were cut off completely. The same problems can arise with other public power projects. For this reason, even public power systems often prefer to raise capital through the sale of bonds rather than the political appropriations process. Although the bonds must be approved by the taxpayers, they give the utility greater flexibility in managing its financial operations.

6.6.3 Sales of Equities

Unlike public utilities, private corporations cannot raise money through the appropriations process. But they are not limited to borrowing in order to raise money. Corporations can raise money by selling *equities,* commonly called *stocks.* Raising money in this way offers several advantages over borrowing:

1. Because shareholders are owners rather than creditors, the company does not need to guarantee repayment of the principal.
2. The proceeds gained from the sale of equities belong to the corporation, and there are no restrictions on the use of the money. However, if enough shareholders object to the ways in which the money is being used, they can force a change in corporate policies.
3. Although utility companies generally pay dividends to their shareholders, the company can stop paying dividends in the event of difficult economic conditions. In 1979, General Public Utilities (GPU) stopped paying dividends on its stock because of the accident at the Three Mile Island nuclear power plant. This allowed the company to reduce expenses at a time when the cleanup of the accident placed a very serious strain on corporate resources.

However, raising capital by equity sales has serious limitations.

1. The prices of equities fluctuate with market conditions. This means that a corporation cannot be certain of raising a certain amount of money through the sale of equities.
2. The value of equities is related to the value of a company's *assets.* Assets are the property that the company owns, along with certain types of financial obligations. A company with few assets will have difficulty selling securities at a favorable price.
3. Sale of additional equities is often opposed by existing shareholders because it can dilute the value of existing shares.
4. Issuing common or preferred stock is a time-consuming, difficult process. When money must be raised quickly, sale of stocks can take too long.

Although sales of common stock can be an important source of capital for energy projects, they are not as widely used as bond sales for financing large projects.

6.6.4 Rate Increases

One of the newest methods of raising capital is by means of rate increases. In the past, utility companies frequently saved a portion of their profits for reinvestment. They did not need to request special rate increases to cover new projects. But today utilities rarely use this method to raise money for large-scale energy projects for the following reasons:

1. In the past, utility costs declined steadily. This made it possible for utility companies to set aside a portion of their profits without having to raise rates. But today, with utility costs rising steadily, utility companies cannot enjoy these "invisible" profit increases. In fact, most utilities must request frequent rate increases simply to keep pace with increasing costs.
2. Because of the enormous scale of new energy projects, utilities cannot set aside sufficient funds out of current revenues.
3. Public utility commissions review rate increase requests very carefully. If a company's profits exceed a certain level, the utility commissions will not grant the rate increases.

In response to these conditions, a number of utilities have begun to request special rate increases for *construction work in progress* (CWIP). Under CWIP regulations, current ratepayers can be asked to pay a portion of the construction costs of future energy plants. Because the ratepayers will benefit from the power produced by the plants, they are asked to help defray the costs of the plants.

From the utilities' point of view, CWIP plans have several advantages over alternative methods of raising capital:

1. The principal does not need to be repaid.
2. No interest payments are required.
3. New capital is available continuously.
4. Capital formation is closely tied to the level of customer demand for electricity; that is, as more customers purchase electricity, more capital will be available for future energy projects.

Despite their obvious advantages as a method of capital formation, CWIP plans have been attacked by environmentalists and consumer advocates. The objections fall into three basic categories:

1. Although it is the ratepayers who provide the capital for energy projects, stockholders receive the profits generated by those energy projects.

2. Current ratepayers, the ones who pay for the project, are not necessarily the ones who will benefit from the future power to be produced by the projects.
3. Under CWIP plans, the ratepayers who are providing the capital for the project do not have the right to determine whether the project should be carried out in the first place.

Recently, CWIP plans have been rejected by public utility commissions in several states, including New Hampshire and Pennsylvania. But the debate is far from over. It remains to be seen whether CWIP will be seen as a legitimate method of financing energy projects.

6.7 RETURN ON INVESTMENT

Energy projects are not designed to lose money. If everything goes according to plan, the project will enable the developers to recover all the capital costs, including interest, and all the operating costs. In the case of projects undertaken by privately owned energy companies, the project should also make a profit.

6.7.1 Repayment of Capital

Whether the original capital for the project is borrowed or paid out of past earnings, it is necessary to recover the capital; otherwise, the project has been an economic failure. The repayment does not have to occur immediately, but it must be completed before the energy project has lost its value. Usually, the capital costs are recovered gradually, through *amortization* or the use of a *sinking fund.* When a debt is *amortized* the lender is paid back in periodic installments. In the case of a *sinking fund,* the developer makes contributions to a special fund earmarked for recovering the capital.

The following formula will determine the payment required to recover the capital.

$$\text{payment} = P \times \frac{r\,(1 + r)^n}{(1 + r)^n - 1} \tag{6.7.1}$$

where P = amount of capital (principal)

r = interest charged per payment period

n = number of payments

A simpler if less precise method is to consult the table of repayment factors (Table 6.7.1), and multiply the amount of principal by the value given in the table. These values assume that interest is assessed once a year and payments are made annually.

TABLE 6.7.1 Repayment Factors for Capital Investments

Years to Pay Back Loan	Interest Rate (%)														
	6	7	8	9	10	11	12	13	14	15	16	17	18	19	20
3	0.374	0.381	0.388	0.395	0.402	0.409	0.416	0.424	0.431	0.438	0.445	0.453	0.460	0.467	0.475
5	0.237	0.244	0.250	0.257	0.264	0.271	0.277	0.284	0.291	0.298	0.305	0.313	0.320	0.327	0.334
10	0.136	0.142	0.149	0.156	0.163	0.170	0.177	0.184	0.192	0.199	0.207	0.215	0.223	0.230	0.239
15	0.103	0.110	0.117	0.124	0.131	0.139	0.147	0.155	0.163	0.171	0.179	0.188	0.196	0.205	0.214
20	0.087	0.094	0.102	0.110	0.117	0.126	0.134	0.142	0.151	0.160	0.169	0.178	0.187	0.196	0.205
25	0.078	0.086	0.094	0.102	0.110	0.119	0.127	0.136	0.145	0.155	0.164	0.173	0.183	0.192	0.202
30	0.073	0.081	0.089	0.097	0.106	0.115	0.124	0.133	0.143	0.152	0.162	0.172	0.181	0.191	0.201
40	0.066	0.075	0.084	0.093	0.102	0.112	0.121	0.131	0.141	0.151	0.160	0.170	0.180	0.190	0.200
50	0.063	0.072	0.082	0.091	0.101	0.111	0.120	0.130	0.140	0.150	0.160	0.170	0.180	0.190	0.200

BOX 6.7.1 ILLUSTRATIVE EXAMPLE

Suppose that a group of farmers has decided to invest in a small (100-kW) hydro plant. Their plant will cost $250,000, and the capital will be obtained through a bank loan at 12% per year. Payments and interest calculations are made monthly, and the loan is to be paid off in 15 years. Substituting for the values in Eq. (6.7.1), we compute the following:

$$P = \$250{,}000$$

$$r = 0.01 \quad (12\% \text{ per year} = 1\% \text{ per month})$$

$$n = 180 \quad (15 \text{ years} \times 12 \text{ payments per year})$$

$$\text{payment} = \$250{,}000 \times \frac{0.01 \times 1.01^{180}}{1.01^{180} - 1} = \$3000$$

Looking at the values in Table 6.7.1, we arrive at an approximate annual payment figure of 0.147 × $250,000, or $36,750. For the purpose of estimating costs, this is close enough to the actual figure of $36,000 per year.

6.8 OPERATING COSTS

It is difficult to generalize about operating costs because each resource has different costs. For oil-fired electric power plants, fuel is the main operating cost. But hydro systems have no fuel costs at all. Maintenance and repair costs are high for wind energy systems, low for passive solar heating systems. A project developer usually estimates the probable operating costs by studying other projects with similar characteristics.

6.8.1 Fixed and Variable Costs

Some operating costs are *fixed*: that is, they do not change with the amount of energy being produced. Taxes, insurance, easements, and building maintenance are examples of fixed operating costs. Other costs are *variable*: they will increase as more energy is produced. Fuel, equipment maintenance, and labor costs are examples of variable costs.

The total cost of a project is the sum of the capital costs and the operating costs, both fixed and variable. As Figure 6.8.1 indicates, this cost will tend to increase as energy production increases. But that is not the whole story. The cost *per unit of energy* will actually decrease as energy production increases. Clearly, the lowest unit cost can be achieved if the plant is producing at its designed capacity.

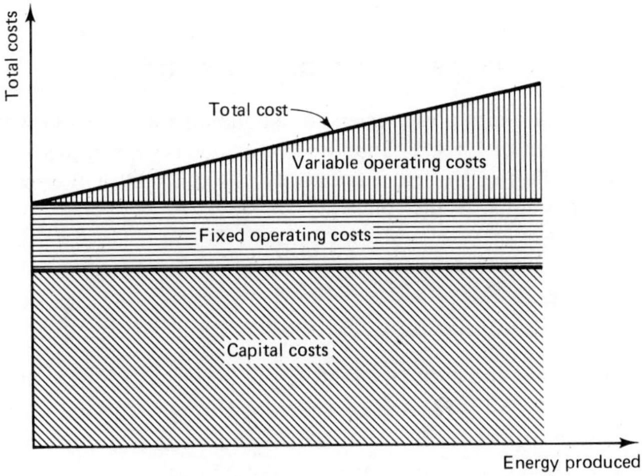

Figure 6.8.1 Energy costs versus energy produced.

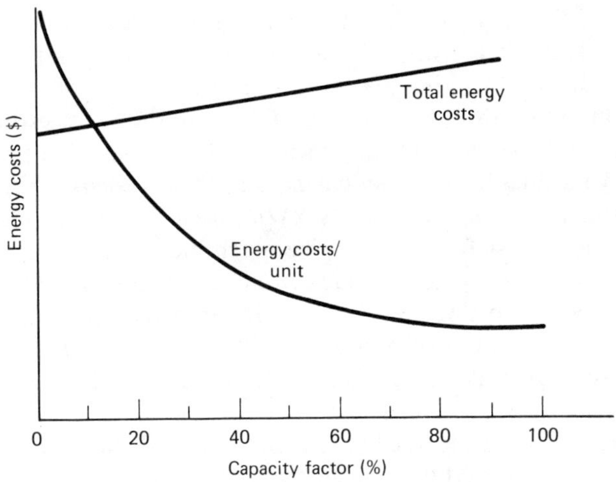

Figure 6.8.2 Energy costs as a function of the plant capacity factor.

Capacity factor is an important measure of how fully an energy facility is being used. It is figured according to the following formula:

$$\text{capacity factor (CF)} = \frac{E}{P \times T} \tag{6.8.1}$$

where

E = energy produced, kWh

P = maximum power output, kW

T = hours of operation

BOX 6.8.1 ILLUSTRATIVE EXAMPLE

In Box 6.7.1 a group of farmers were building a hydro plant with a maximum power output of 100 kW. Suppose that the facility is completed according to plan, and that during the first year it produces 475,000 kWh of electricity. The capacity factor is equal to the following:

$$CF = \frac{475{,}000}{100 \times 365 \times 24} = 0.54$$

In general, the higher the capacity factor, the better. A higher capacity factor means a lower unit cost. A word of caution, though: capacity factor should not be used as the sole criterion for choosing between two design alternatives. The only fair method is a direct comparison of energy costs (or energy savings).

6.8.2 Marginal Costs and Average Costs

One final distinction needs to be made before we leave the subject of cost comparisons. This is the distinction between average costs and marginal costs of energy-producing facilities.

When utilities establish their rates, they base their cost estimates on the average cost of all energy-producing facilities. This means that the higher cost of new energy facilities is somewhat balanced by the lower cost of older facilities. So while a utility may charge 10 cents/kWh for electricity, the electricity produced from the newest facilities may be much more expensive.

When a new power plant is being considered, its cost should be compared, not with the average cost of power production (the *average* cost), but with the cost of alternative methods of producing additional power (*marginal* cost). In many cases, this can make a significant difference in evaluating the feasibility of a project.

Consider, for example, the hydro plant described above. The capital costs total $36,000 per year, and taxes, insurance, operation, and maintenance cost an additional $10,000 per year. The plant produces 475,000 kWh per year, for an average total cost of 9.5 cents/kWh. This is actually more than the utility's cost of 7.8 cents/kWh. So the project does not appear to be cost-effective.

But the figure of 7.8 cents/kWh is an average figure for the utility. If the utility were to build a new thermal power plant, the electricity from the plant would cost 13.5 cents/kWh. This is substantially higher than the cost of power from the hydro plant. For this reason, the utility agrees to buy power from the hydro plant.

These are only representative figures. But they illustrate an important point: cost-effectiveness depends on the perspective of the person doing the analysis. If the cost of the hydro project is compared to the cost of electricity *to the farmers,* the project is not cost-effective. But from the point of view of the utility, the hydro project is less costly than the alternative of building a new thermal power plant. Whenever cost comparisons are being made, it is essential to use the same basis for comparison. Average costs must be compared with average costs, and marginal costs must be compared with marginal costs.

6.9 ENERGY AND INFLATION

The late 1970s were a time of unprecedented price increases for many countries in the industrialized world. In the United States, the consumer price index increased by 12.2% in 1978, 13.3% in 1979, and 12.4% in 1980. These price increases were considered to be a major factor in the presidential election of 1980. In other countries, price increases have been even more serious: Israel and Argentina routinely experience inflation rates over 100%.

Economists have never fully understood why prices increase. Some blame government deficits, others point to labor unions and rising labor costs, and still others claim that corporate control of the marketplace is the underlying cause. But there is good reason to believe that during the late 1970s, increases in the price of energy were one of the main forces that drove all prices higher.

If we compare the prices of energy with prices in the economy at large, we find that the periods of the greatest overall inflation were also the periods with the highest increases in energy prices (see Figure 6.9.1). The correlation is certainly striking. Not only do the periods of price increases correspond, but the periods of stabilization in energy prices correspond to the periods of reduced overall inflation.

This in itself is hardly enough to prove that energy prices were responsible for the overall inflation. In fact, it could very well work the other way: in a time of generally rising prices, we would certainly expect the prices of any given commodity to rise as well. But although economists are unsure of the causes of inflation in general, they have a very good idea of the reasons behind the increase in energy prices.

The first period of sharp energy price increases began in 1973, after the Arab oil embargo. The supply of petroleum in the United States was less than the demand. Long lines appeared, and gasoline prices shot up from less than 30 cents to more than 50 cents in most parts of the country. Overall, energy prices increased by more than 50% in 1974. 1974 also marked the beginning of double-digit inflation.

Between 1974 and 1979, energy prices rose more slowly, and the inflation rate also fell to a more manageable level. But in 1979, Iran cut off oil shipments to the United States because of hostilities between the two countries, putting a major dent in the U.S. oil supply. Gas lines reappeared, and gasoline prices began moving upward again, reaching $1.50 per gallon in some areas. Overall, energy prices increased by 26.5% in 1979 and 40.7% in 1980. At the same time, the inflation rate returned to double figures.

The second round of energy increases finally began to have an effect on energy demand. Fuel-efficient cars dominated the market, home insulation be-

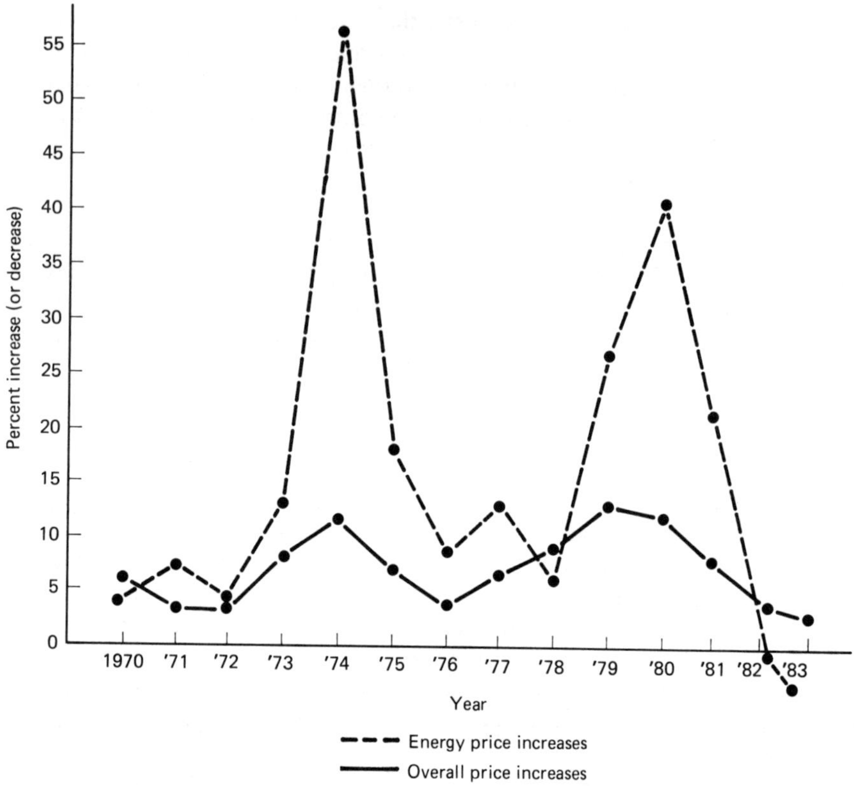

Figure 6.9.1 Inflation and energy price increases. The solid line represents the consumer price index; the dashed line represents the index of energy/power costs. (*Source:* U.S. Bureau of Labor Statistics.)

came a best-seller, and sales of solar water heaters began to take off. Soon, energy consumption began to drop from the high point reached in 1979. As a result of this decline in demand, an "energy glut" developed, and the rise in oil prices came to an abrupt halt. Gasoline prices began to fall, dropping below $1.00 per gallon in early 1983. Overall, energy prices actually declined in 1982 and early 1983. At the same time, the annual rate of inflation dropped to less than 4%.

Energy costs contribute to the costs of almost all products. Transport of raw materials uses diesel fuel. Production of metals requires high-temperature smelters. Many industrial processes use process steam. Petroleum is an essential feedstock for the plastics industry. The list goes on and on, but the point is clear: when energy prices increase, costs increase for the manufacturers of most products. Some economists believe that this is what happened during the 1970s. In their view, energy supply and demand, not fiscal and monetary policy, brought about the serious price increases of the 1970s and the reduced inflation of the early 1980s.

6.10 ENERGY AND EMPLOYMENT

In Chapter 2 we established that energy can be converted into useful *work*. In a less technical sense, *work* is also what most people do to earn money. The relationship between these two concepts is very close. In fact, work in the form of mechanical or electrical energy can often be substituted for human labor. In the lumber industry, for example, one logger with a chain saw can do the work that used to require four or five. In the textile industry, power looms have displaced large numbers of workers. Today, electronic office machinery is taking over many functions that used to be performed by clerical workers. When economists speak of increases in productivity, they are usually referring to the substitution of inanimate energy for human labor.

There can be no question that many of the jobs replaced by energy have been dreary and repetitious. But the people whose jobs were replaced have often been unable to find other work. For most people, even a repetitious job is better than no job at all. During the early nineteenth century, British textile workers destroyed new machinery which was eliminating large numbers of jobs. More recently, many thousands of automotive workers have lost their jobs due to the introduction of robotized assembly lines. But perhaps the most striking example of this job replacement is in American agriculture. Between 1920 and 1970, the number of agricultural workers declined by 70% while agricultural production increased steadily. Clearly, the introduction of tractors and other powered machinery was a highly significant factor in this process.

But if energy replaces jobs, it also creates new areas of employment. Along the Gulf coast in the southern United States, the oil industry is a major employer. In the Appalachian region, many families have four or five coal miners living in the same household. Indirectly, energy production permits the development of industries such as radio and television, automobile manufacturing, railroads, and airlines. So at the same time that energy replaces jobs in some segments of the economy, it creates whole new sectors employing many thousands of workers.

Most of the new jobs created by energy production are skilled jobs in areas such as engineering, heavy equipment, and finance. These jobs are usually not available to the people who have been displaced. So even if the total number of jobs remains constant, the substitution of energy for labor causes a shift in the nature of the nation's work force.

Some economists have suggested that it is not energy production, but the overall expansion of the economy, that creates the new jobs. If the economy stops expanding, or even expands too slowly, unemployment increases. Since no system can go on expanding forever, these economists believe that in the long run, large-scale energy production inevitably leads to greater unemployment. But the unemployment problem cannot be solved by simply cutting back on energy production. If we look at the energy-poor regions of the world, especially the Third World, we see enormous unemployment rates: 30 to 40%, or perhaps even more. Clearly, we cannot solve the unemployment problem by reducing energy production and use.

For another perspective on the relationship between energy and employment, some economists have looked at the numbers of jobs provided in different sectors of the economy. In general, they have found that in the nonrenewable sector of the energy industry, very few jobs are created compared to the rest of the economy. For example, a 1977 study found that over $100,000 of investment was required to create one job in the oil and electric power industries, compared to an average of less than $20,000 for an average job in manufacturing. In fact, the energy industry required more investment per job than any other sector of the economy.

This means that when capital is invested in the production of petroleum, natural gas, or electricity, there is a net *loss* of jobs because the money is unavailable for investment elsewhere. One estimate suggests that for each quad of primary energy that is converted to electricity, there is a net loss of 75,000 jobs in the economy at large.

Few people have suggested that the United States should curtail production of oil and electricity in order to increase employment. But some economists have suggested that by investing in conservation and renewable resources, we can create more jobs while meeting the nation's energy needs. They point out that conservation measures, such as installing insulation or more energy-efficient furnaces, require very low capital investment per job created, and that saving energy is far less expensive than producing new energy from any source. Solar energy advocates offer similar statistics in support of their technologies. For the most promising solar applications, such as water and space heating, large numbers of jobs would be created in manufacturing and installing the systems—many times more than producing equivalent amounts of energy in the form of coal, oil, natural gas, or nuclear-fueled electricity.

One difficulty with these arguments is that they compare two different kinds of enterprise: installation of end-use equipment versus production of primary energy resources. Manufacture and installation of furnaces provides roughly the same number of jobs as manufacture and installation of solar heating systems. Once the solar heating systems are installed, the number of jobs required for repairs and maintenance is certainly no greater than for gas or oil furnaces. Since no external fuel is needed in the case of solar energy systems, it can be argued that a new house with a solar heating system will create *fewer* jobs in the long run than a house with a gas or oil furnace. But if the production of oil and natural gas prevents the investment of capital in other sectors where more jobs might be created, as suggested earlier, then conservation and solar energy can indeed be said to create more jobs than can conventional, nonrenewable energy resources.

The argument concerning energy and employment is only part of a deeper philosophical dispute. The issues include not only energy and employment, but centralization, labor relations, and concepts of economic growth. The arguments are complex, and both sides can offer statistics in support of their case. For readers who are interested in pursuing these questions further, we recommend a course in economics or a careful exploration of the Further Readings section.

STUDY QUESTIONS

1. Describe a situation where you could see how supply and demand affect the price of an item. Describe the consequences of changes in the supply and demand.
2. Do you think that the prices of energy should be left completely to the free marketplace, or do you think the government should try to influence prices? Explain your reasoning.
3. Investigate how your electric utility company sets its rates. Are there any features of the rate structure that you would like to see changed? Explain what should be changed and why.
4. Suppose that you are a member of a public utility board and the utility board that you regulate has just asked for a rate increase. What questions would you want to ask the utility before making your decision on the rate hike?
5. What are the federal government expenditures for research and development of different energy resources in the current fiscal year? How is this likely to affect the supply and price of energy resources in the United States?
6. Interview a bank officer concerning the bank's policy on making loans for energy projects. In particular, try to find out the following information:
   (a) What kinds of energy projects would be eligible?
   (b) Would the interest rates for the project be above, at, or below the prime interest rate?
   (c) What would be the length of the loan?
   (d) What percentage of the project could be financed?
7. Visit an energy facility. Find out when it was built, how long it took to build, and how much it cost. How does this compare with the cost of building a new facility?
8. If no money needs to be borrowed for an energy project, no interest charges need to be paid. Does this mean that no interest needs to be included in the capital costs of the project? Explain your reasoning.
9. Interview a local business to determine how their energy costs have affected the prices they charge. Does it seem that increases in energy prices are a cause of inflation?
10. Do you agree that more jobs would be created by a greater reliance on solar energy and conservation, or do central power plants provide more jobs? How would you answer someone on the other side of the issue?

PROBLEMS

1. Which represents a better bargain?

   Fuel oil at $1.50 per gallon
   Electricity at 15 cents per kilowatthour
   Natural gas at $11.00 per thousand cubic feet

2. Suppose that you are building a wind energy project. The 5-kW generator costs $12,500 installed. You need to borrow $10,000 for 20 years at 2% above the current prime interest rate.
   (a) How much will you have to pay back each year? (Assume one payment per year.)
   (b) What will be the total cost of the wind system, including interest and the $2500 down payment?
3. A public utility needs to raise $100 million through the sale of bonds. Look up the rates of interest being paid on new utility bonds. (The *Wall Street Journal* is a good source of this kind of information.)
   (a) What will be the annual interest payments to the bondholders?
   (b) What will be the annual capital costs to the utility, assuming that the bonds are being paid off by a 30-year sinking fund?
4. Suppose that a passive solar house costs $5000 more than a comparable house without passive solar heating, and assume that the house is financed by a 30-year mortgage at 13%.
   (a) What is the annual cost of the passive solar heating system? (Disregard operating costs.)
   (b) How many gallons of fuel oil would have to be saved each year to make the house cost-effective?
5. Assume that the capital costs of a nuclear power plant are $236 million per year, the fixed operating costs are $65 million per year, and the variable operating costs are 1.2 cents per kilowatthour. The plant has a capacity of 1000 MW and operates at a capacity factor of 61%. What is the cost per kilowatthour of electricity produced by the plant?
6. If the utility that owns the nuclear plant in Problem 5 is permitted a return on investment of 16%, what is the cost to the customer of electricity produced by the nuclear plant? (Disregard the effects of transmission and distribution costs.)

FURTHER READINGS

FARRIS, MARTIN T., and ROY J. SAMPSON. *Public Utilities: Regulation, Management, and Ownership.* Boston: Houghton Mifflin, 1973.

GARVEY, GERALD. *Energy, Ecology, Economy.* New York: W. W. Norton, 1972.

GROSSMAN, RICHARD, and GAIL DANEKER. *Energy, Jobs, and the Economy.* Boston: Alyson, 1979.

HEALY, TIMOTHY J. *Energy and Society.* San Francisco: Boyd & Fraser, 1976.

LEBEL, PHILLIP G. *Energy Economics and Technology.* Baltimore: Johns Hopkins University Press, 1982.

LOVINS, AMORY B. *Soft Energy Paths: Toward a Durable Peace.* New York: Harper & Row, 1977.

SAMUELSON, PAUL W. *Economics,* 11th ed. New York: McGraw-Hill, 1980.

7
Environmental Analysis

7.1 THE CHANGING ENVIRONMENT

Over the centuries, humanity has changed the shape of the world. For example, when the first European explorers reached the North American continent, nearly all the land east of the Mississippi was covered by dense forests. Clear-running streams flowed across the landscape. Innumerable game animals populated the forests, together with enormous flocks of birds. When the passenger pigeons migrated, the flocks were so dense that they darkened the skies, and hunters could put dinner on the table by shooting at random into the air.

Today, large areas of the eastern United States have been cleared for cities and farmland. In many areas of the Midwest, it is impossible to find areas that have not been logged over. Many rivers are seriously polluted, some so badly that they can no longer support aquatic life. The passenger pigeon is extinct, and many formerly abundant game animals have become endangered. Even the bald eagle, the national symbol, was recently classified as an endangered species in the lower 48 states.

Energy production and consumption have played a large role in this transformation of North America. Forests have been cut for firewood, the land has been mined for coal, and farmlands have been paved over to make roads and parking lots for our automobiles. These are only the most visible environmental changes. The air, the water, and most important, natural ecological systems, have all been altered by the development of an energy-intensive civilization in North America.

Some people believe that environmentalists are simply opposed to progress. They fear that the technological progress of the past century will be rolled back in the name of wilderness preservation. But for the most part, environmentalists do not seek a return to the "good old days" before the invention of the automobile. They are concerned that environmental collapse threatens not only wilderness and wildlife, but also the very civilization we live in.

In this chapter we look at some of the ways in which energy production and use have affected the environment. In doing so, we hope to establish a few principles of environmental analysis, so that future energy development can be accomplished with a minimum of damage to the environment. We cannot return to the days when North America was a wilderness—indeed, we would be foolish to reject the advantages of modern technology—but we can help to protect our environment from further degradation. Nowhere is this awareness of our environment more essential than in the development and use of our energy resources.

7.2 WHAT IS AN ECOSYSTEM?

Any system of life supports, together with the organisms it supports, is called an *ecosystem*. All ecosystems, taken together, constitute the *biosphere*.

An ecosystem functions according to three essential principles:

1. *Nutrient cycling*. This is the continuous exchange of chemicals between organic matter and its inorganic surroundings. Organic matter grows, lives, dies, and decomposes. Organisms grow on the available supply of nutrients in their environment, while that environment receives waste matter and the decomposed bodies of the organisms.

2. *Energy supply*. An ecosystem will not work without an external energy supply. On earth, the primary source of this energy is the sun. Recently, an ecosystem was discovered which depended on geothermal energy from the center of the earth for its energy supply, but the vast majority of ecosystems are based on solar energy.

In solar-powered ecosystems, sunlight is used by plants to combine simple inorganic molecules into more complex organic molecules. The chemical energy stored in these organic molecules can then be used by the plants or by other organisms which feed on them.

Although nutrient exchange in an ecosystem goes both ways, energy exchange is a one-way street. The solar energy absorbed by plants is transferred to animals which feed on the plants. From these animals, it is transferred in turn to other animals and bacteria. But eventually, the energy is given off into the environment as waste heat, rather than being returned to the plants that originally captured the energy. Additional energy must enter the ecosystem in order for the cycle to continue.

3. *Interrelationships among the components*. Delicate relationships exist among the many parts of the ecosystem to meet the needs and requirements of each component of the system. If these needs are not fully met for all components, a swift and radical transformation will occur, changing the fundamental nature of the ecosystem until stability can be regained or until the entire ecosystem collapses. For example, insects feed on plants, and they in turn are kept in check by birds. If a farmer sprays plants with large amounts of insecticide, some of the insects will be killed. But many birds will also be killed, because the pesticide becomes more concentrated as it moves higher in the food chain. When the bird population declines, this can lead to an explosive increase in the number of insects, so that crop losses are greater than they were before the spraying. This can lead to more spraying, further loss of birds, and ultimately a complete collapse of the ecosystem. In a stable ecosystem, however, the many components are balanced. This ensures the future of the ecosystem itself, and thus of all the individual components.

One consequence of these three principles is that every ecosystem has a finite *carrying capacity*. The carrying capacity is the highest level of biological activity that can be sustained by an ecosystem. Once the level of activity exceeds the carrying capacity, the ecosystem will begin to break down. This breakdown can be caused by the improper functioning of any of the three principles just discussed.

An ecosystem can be small or large. A tiny pond, the great Arctic tundra, or an enormous desert—each can be considered as a single ecosystem. In fact, the definition of an ecosystem is not always a clear-cut matter. Ecosystems do not have distinct or fine boundaries. A transitional zone always exists: a shared or common region between two nearby ecosystems. Forest animals often interact with a water ecosystem—a bear catching salmon on their way upstream, for example. And the boundaries themselves may not always be stable. Ecosystems expand and contract depending on environmental conditions and pressures from neighboring ecosystems. The Sahara Desert, for instance, has expanded drastically since the beginning of recorded history.

7.2.1 Ecosystems and the Biosphere

Despite the tremendous diversity and range of ecosystems, all living things must exist within certain limits. Life cannot exist without water—at least, not in an active state. Similarly, life as we know it cannot function in very high temperatures. Some people used to believe that life might exist on the planet Venus. However, these speculations ended when it was discovered that the surface of Venus maintains a temperature of 400 °C (700 °F). Thus there is a narrowly circumscribed arena where biological processes can take place. Human and other life forms are found almost exclusively in a very thin layer above, at, and below the earth's surface. The air that surrounds the earth, the water that flows and fills the streams and reservoirs of the earth, and the upper crust of the planet are all likely places to find life forms. While that arena may seem vast indeed from the perspective of an inhabitant, it is rather small and limited when seen in the context of the universe at large.

The term *biosphere* is used to describe the sum total of all spaces, occupied or unoccupied, where ecosystems can exist on this planet. Accordingly, we shall include the continental shelves, the oceans, the land, underground water, the lower atmosphere, and the ozone layer of the atmosphere in our discussion of the biosphere. We have included the ozone layer because, although it is located well above the region where life forms exist, it plays a very important role in the development and maintenance of life forms in the biosphere; indeed, most researchers believe that the ozone layer is largely created by living organisms.

BOX 7.2.1 ENVIRONMENT AND BIOSPHERE

The terms *environment* and *biosphere* are often used interchangeably. In principle, *environment* is a much broader term, because it includes everything around us, animate and inanimate, and by extension, the entire universe. But because we are living creatures, the environment that matters most to us is the biosphere. After all, that is the environment from which we draw our very existence. So the term *environment* has become essentially synonymous with *biosphere,* and we shall use it in that way.

As the preceding discussion should make clear, the *energy* required to maintain life must come from outside the biosphere. However, the biosphere contains all the *materials,* or nutrients, necessary for the maintenance of life. The air provides carbon dioxide, which plants need in order to carry out photosynthesis, and also contains oxygen, which both plants and animals need to carry out *respiration,* or the oxidation of organic matter. Nitrogen, an essential component of plant cells, is also provided by the atmosphere. Water, which is essential to all life, is provided by the ocean, groundwater, and lakes and streams. Nearly all living organisms are composed primarily of water, and plants require water together with carbon dioxide and sunlight as the raw materials for photosynthesis. The soil provides a rich variety of nutrients for plants and animals, as well as a protective covering for sensitive plant roots. The earth also provides living space for innumerable microscopic organisms, insects, and burrowing animals.

Within the biosphere, plants and animals inhabit more narrowly circumscribed areas called *habitats.* For example, trout can only live in cool, fresh water, and many snakes can only live on land in relatively warm climates. An organism's habitat is defined as the portion of the biosphere that meets the minimum physical conditions for the organism to survive. These conditions may be very strict; for example, nearly all freshwater fish require water with a fairly high pH (not too acidic). No matter how clean a lake may be, no matter how rich in nutrients, the fish will not be able to survive if the pH falls below a certain level.

The simple presence of a suitable habitat does not necessarily mean that a certain plant or animal species will be able to survive in that habitat. Other conditions, sometimes very subtle, also influence whether a species can inhabit a certain region. Competition from other species may make a certain region unsuitable, even though all the physical conditions are right. For example, when rabbits were first introduced into Australia, several species of marsupials were threatened because they could not compete successfully with the rabbits. In other cases, cooperative relationships with other species are essential. Apple trees cannot survive without bees to pollinate them. If the entire bee population of a region should be exterminated, apple trees would soon die out as well.

In describing these complex ecological factors which permit a species to inhabit a certain area, biologists speak of a biological *niche*. A niche is the particular set of ecological characteristics which permit an organism to exist—characteristics that go far beyond what would be included in a description of a suitable habitat. Since one type of organism can depend so heavily on the presence of other species, it is easy to see how environmental change can influence the survival not only of the organisms that are directly affected by the change, but of many other organisms as well.

7.2.2 Unity of the Biosphere

Although we can perceive distinct boundaries between the air, earth, and water, these boundaries are by no means absolute. Pollutants from the air can easily dissolve in the water and permeate the land. For instance, sulfur dioxide emissions from fossil-fueled power plants create acidic compounds in the atmosphere (called *acid air*). These chemicals are readily dissolved in rain and snow, which falls to the ground and acidifies ground and surface water, sometimes so much that fish can no longer live in the affected lakes and streams. Similarly, in the late 1950s, radioactive air pollutants from nuclear weapons tests were deposited as fallout, polluting both land and water.

In general, it can be said that because of nutrient recycling in ecosystems, pollution in one part of the system will appear, sooner or later, in the remaining parts. Some people are inclined to say, "We don't live near the waste depositories, so we shouldn't worry about toxic wastes" or "Air pollution isn't really a problem; we need to be working on the problem of water quality." In fact, these distinctions can be somewhat misleading. We cannot sacrifice the quality of the air in order to protect the water, because both the air and the water are important parts of the global ecosystem. To make intelligent decisions on energy and the environment, we need to recognize the interrelatedness of the entire biosphere. What affects one part affects the whole.

The environmental effects of energy production and use can be grouped into five different categories: resource depletion, environmental pollution, climate change, land use, and loss of wilderness. We will look at these five categories of environmental effects and then examine the steps that are being taken to reduce the adverse environmental effects of energy technology.

7.3 RESOURCE DEPLETION

Resource depletion occurs when resources are being used more rapidly than they are being replenished. Obviously this is the case with nonrenewable resources. For example, when we use 20 trillion cubic feet of natural gas per year (our current rate of consumption), we are definitely "using up" or depleting our natural gas reserves. There is no possible way for natural gas to be regenerated beneath the earth's surface at that rate. As soon as any nonrenewable resource is mined and put into our energy distribution and consumption system, it becomes unavailable for future generations.

In Chapter 2 we discussed the principle that energy can be neither created nor destroyed. When we mine coal or drill for oil, we are not truly "producing" energy; rather, we are drawing on a finite resource of available energy and making a portion of it unavailable for future use. In other words, we cannot find any more energy, because energy cannot be created. Once we have used a certain store of energy, we cannot use it again. This means that the available energy base always decreases.

This physical principle has implications for the environment. The environment consists not only of air, land, and water, but also the resource base on which all species, including the human species, feed and thrive. Humanity is especially dependent on nonrenewable natural resources: fossil fuels, metals, and other minerals we depend on for our daily survival. We must make use of these natural resources in order to survive. But as we use those resources, we are making them unavailable for future generations.

It may appear at first glance that renewable resources cannot be depleted. But, in fact, they will only continue to be available if we use them no faster than they are being replenished. For example, woodlands can supply us with firewood indefinitely if we are careful how trees are removed and replaced. But this same forest can be entirely destroyed, and irreversible soil erosion processes can be given a chance to establish themselves, if we practice careless and indiscriminate logging of forested areas.

An increased level of energy consumption generally implies an increased use of other resources, such as iron, aluminum, copper, synthetics, and wood products. Some of these resources are renewable and some are not. If we depend heavily on the nonrenewables and do not recycle or reuse them, clearly future generations will have fewer of these resources at their disposal. The questions that must be answered, then, are these: How many resources must we consume to ensure an adequate standard of living, and how many resources may we consume without depriving our descendants of an adequate standard of living?

There are no easy answers to these questions. Clearly, we cannot go back to the Stone Age, when humanity depended almost entirely on renewable resources. But we must not squander our resources so lavishly that we leave nothing for future generations. All resources, even renewable ones, should be used prudently and with an eye toward the future. Some resource depletion is certainly inevitable, but we can minimize its effects by exercising a bit of foresight.

7.4 POLLUTION

Whereas resource depletion is a problem caused by diminishing supplies of desirable materials, pollution is an excessive supply of undesirable materials. For most people, pollution consists of highly visible substances fouling the air, land, or water. We are all disgusted by the sight of huge smokestacks belching black smoke into the air, or of drainage pipes discharging putrid wastes into a stream. But there is another type of pollution, less visible but perhaps more insidious than large-scale dumping of waste products. Some artificial chemicals can be toxic in concentrations of only a few parts per million or billion (ppm or ppb). Polychlorinated biphenyls (PCBs), DDT, and dioxins are some of the most notorious examples. These chemicals may be invisible, tasteless, and odorless, but the damage they cause can be far-reaching and long-lasting. In fact, some synthetic chemicals, such as PCBs and dioxins, are essentially immune to the normal processes of decay. Once they appear in the environment, we may reasonably say that they are there forever.

Excessive amounts of heat and noise are often considered pollution, as are unsightly objects such as junk cars and empty oil drums (visual pollution). Most people are strongly aware of noise and visual pollution, but thermal pollution is much more difficult to detect. Even a temperature change of a few degrees Fahrenheit can be sufficient to change the ecological characteristics of a body of water. But because such a small temperature difference can be almost impossible to detect, thermal pollution can go unnoticed for long periods of time.

7.4.1 Air Pollution

If we take our cue from nature, it seems perfectly natural to exhaust combustion waste products to the atmosphere. Table 7.4.1 presents a clear view of the composition of the lower atmosphere. When we exhale, we are essentially disposing of our bodily waste products, water vapor and carbon dioxide, into the atmosphere. These waste products are created by the slow oxidation processes that go on inside our bodies, and the biosphere has no difficulty handling those waste products. When forest fires occur, or when natural decomposition takes place, the gaseous by-products are also released to the air. But nutrient recycling rapidly turns these by-products into plant mass, so the air's carrying capacity is not permanently reduced by the release of these combustion products.

For thousands of years, nutrient recycling also dealt very effectively with the combustion products generated by human industry. The waste products being

TABLE 7.4.1 Composition of the Lower Atmosphere

Gas	Approximate Percentage by Volume
Nitrogen	78.0
Oxygen	20.9
Argon	0.9
Water vapor	0.07
Carbon dioxide	0.03
Neon	Trace
Helium	Trace
Krypton	Trace
Hydrogen	Trace
Nitrogen oxides	Trace
Methane	Trace
Ozone	Trace

generated were simple in nature—gases released from wood fires, and dust and dirt from plowing. But when industry began to use fossil fuels for combustion purposes, rather than biofuels such as wood, a fundamental change occurred in the process of atmospheric disposal of wastes. Now the amount of combustion products released into the air exceeded the natural rate at which those combustion products could be converted into plant matter.

At first, even this increase was not significant due to the great mass of the air that surrounds us—on the order of 5.5 quadrillion (5.5×10^{15}) tons, and the quantity of fossil-fuel combustion products was small by comparison. But with explosive growth in human populations, worldwide industrialization, widespread use of automobiles, the chemical revolution of the mid-twentieth century, and the splitting of the atom, things have changed considerably.

The most deadly effects of air pollution have been felt in major population centers. In 1952, 4000 deaths in London were attributed to air pollution; and in 1965, another 400 deaths from air pollution were reported in New York City. Today, the burden of waste products on the atmosphere is beginning to be felt not only in major cities such as New York, London, and Los Angeles, but also in rural areas and even over the oceans. Acid rain has devastated large areas of the Adirondacks in New York State as well as Scandinavian countries, and radioactive fallout from atomic explosions has been detected thousands of miles from its source. In fact, it is fair to say that no area of the world has been unaffected by air pollution caused by human activities.

Air pollution comes in a variety of forms. In this chapter we consider only those forms that are caused directly or indirectly by artificial processes, especially energy production and consumption (see Table 7.4.2). We shall consider these pollutants one at a time. However, it should be kept in mind that air pollution is a very complex natural phenomenon. Several kinds of pollutants may be in the atmosphere at once, and their combined effects may be much different from the effect of any single pollutant.

Smog. When automobiles burn gasoline, not all the hydrocarbons are completely oxidized. Automobile exhausts emit quite a variety of hydrocarbons, together with carbon monoxide and nitrogen oxides. The nitrogen oxides and hydrocarbons interact in more than 100 ways in the presence of solar radiation, especially ultraviolet radiation. The products of these reactions typically consist of fine droplets which give a bluish haze to the atmosphere. Many of these compounds, such as aldehydes, peroxyacetylene nitrates (PAN), ketones, and ozone, are irritating, and some are even toxic to various species. In human beings they cause irritation and disorders of the eyes and the respiratory system, including the lungs and throat, together with resulting mental disturbance. The compounds are quite toxic to plants, especially pines and other coniferous trees.

TABLE 7.4.2 Air Pollution from Energy Resources

Energy Resource[a]	Air Pollutants of Concern
Oil (transportation)	Hydrocarbons, CO, nitrogen oxides, CO_2, lead oxides, particulates
Oil (power generation)	CO_2, nitrogen oxides, particulates, hydrocarbons
Natural gas	CO_2
Coal	Sulfur oxides, particulates, CO_2, H_2S, radiation, CO, hydrocarbons
Wood	Hydrocarbons, CO, particulates, CO_2
Nuclear	Radioactive gases or steam
Geothermal	Hydrogen sulfide, radioactive gases

[a]Other energy resources, such as hydro power, wind, solar, wave, tidal, and OTEC, do not cause air pollution in particular. They have other environmental impacts.

Not all smog is caused by human activities. The Smoky–Blue Ridge Mountain area of the southeastern United States is a typical natural smog area. The hydrocarbons released by the plants react with naturally occurring nitrogen oxides to produce the smog effect. But by far the most common cause of smog (also called *photochemical smog*) is automobile emissions. Scientists believe that at least 90% of all photochemical smog is caused by automobiles.

New cars are required to meet strict emission control requirements, which has helped to reduce the levels of smog in some uban centers. But older cars have less effective emission control systems, and even on new cars the systems do not always work perfectly. Use of leaded gasoline can damage catalytic converters, which are the main emission control devices in nearly all new U.S. cars. Many owners do not bother to repair the emissions control systems on their cars, and some have intentionally disconnected emissions control equipment in the belief that this will improve their fuel economy. Finally, emission control standards have not been implemented as rapidly as expected. The federal government has been reluctant to impose added burdens on the financially troubled automobile industry. Indeed, there is some question whether the strict standards for 1985 model cars will ever be imposed.

Figure 7.4.1 This wet scrubber system cleans the dust-laden gas collected during coke production. (*Source:* Bethlehem Steel Corporation.)

Acid pollution. When sulfurous fuels are burned, such as bituminous coal and low-grade petroleum, they give off sulfur dioxide. Coal-burning electric power plants are the main producers of sulfur dioxide, but it is also produced in cities where coal is burned for residential heating. Unless precipitators or other pollution-control devices are used, the sulfur dioxide is released into the atmosphere, where it combines with water to form sulfuric acid. Besides sulfur dioxide, high-temperature combustion processes produce several different oxides of nitrogen. These oxides, represented by the general symbol NO_x, combine with atmospheric water to produce nitric acid.

Through means which are still poorly understood, acidic particles in the atmosphere combine with water vapor to produce acidic precipitation. Some meteorologists believe that the small acidic particles actually cause the precipitation. Although the term *acid rain* is usually used, the acidic water may fall in the form of rain, snow, or fog. Sometimes the acidic particles fall directly without being incorporated into any form of precipitation. But however they fall, these acidic particles cause serious damage. Perhaps the easiest to observe are the effects on inanimate materials. Sulfuric acid in the atmosphere has been blamed for discoloring paints, dissolving white marble and limestone, and damaging other materials. The Lincoln memorial in Washington, D.C., the ancient Greek temples in Athens, and the Taj Mahal in northern India have all been damaged by these sulfur emissions.

The environmental effects of acid precipitation have been especially serious in Scandinavian countries, eastern Canada, and the Adirondack region of New

York State. In many areas, acidification has made freshwater lakes incapable of supporting aquatic life. In 1982, the Stockholm Conference on the Acidification of the Environment reported that 20,000 of Sweden's 90,000 lakes have been severely acidified. In Norway, 13,000 km^2 of lakes are "dead," with another 20,000 km^2 on the critical list. In Canada, 50,000 lakes are considered acidified.

Land species are also affected by acid precipitation. West German biochemist Bernhard Ulrich has found that some 1 million hectares of evergreen forests in central Europe are adversely affected by acidic precipitation and deposition of acidic particles. In the United States, large numbers of red spruce trees have been killed in the mountains of northern New England, almost certainly from the effects of acid precipitation. Although no agricultural effects from acid rain have been proven, it is well known that pH levels (acidity or alkalinity) in the soil have a significant impact on crop development and growth.

Other problems have also been associated with acid precipitation. For example, it is believed that acidic groundwater leaches metals into groundwater deposits, poisoning aquatic animals and contributing to human health problems. Current research suggests that the leaching of aluminum may be the cause of the tree kills that have been observed in the northern Appalachians.

Acidic precipitation may also affect human health directly. Tiny acidic particles are believed to lodge in the lungs, causing respiratory problems, and ingestion of aluminum has been associated with many serious nervous disorders, including Alzheimer's disease. In general, it is safe to say that acid pollution is one of the most troublesome environmental effects of energy production.

So far it has been very difficult to establish a direct cause-and-effect relationship between air pollution in one area and acid precipitation in another. For example, it is widely believed that emissions from electric power plants in the American Midwest have caused acidification of lakes in Quebec and the Adirondack region of New York State. But before the responsibility can be assigned with any degree of certainty, it must be established that the pollutants in question definitely originated from the suspected power plants. Since the power plants are in compliance with national air quality codes, they cannot be forced to change their emission patterns without such direct proof. But funds for research into acid precipitation, like most environmental programs, have been severely curtailed in recent years.

One especially troublesome problem with acidic air pollution is that the pollutants often travel across international boundaries. Particularly when tall smokestacks are used to minimize the effects in the immediate region of the power plants, the pollutants may be carried hundreds of miles before doing their damage. In response to this problem, the United Nations established a Long-Range Trans-Boundary Air Pollution Convention. The United States signed it in November 1979. The treaty was accepted in 1981.

Researchers continue to search for ways to deal with acid precipitation. Affected lakes have been treated with lime to reduce their acidity, but this does not solve the problem of heavy metals leached into the water. New coal-fired power plants are equipped with "scrubbers" to remove sulfur dioxide, but older plants continue to emit large amounts of sulfur and nitrogen oxides. As the effort to solve the acid rain problem continues, we should realize that we are only beginning to understand the environmental effects of energy production technology.

Particulate pollution. Another important constituent of air pollution is dust and other light particles which remain suspended in the air for prolonged periods of time. These particles come from various combustion sources, such as power plants and automobiles, as well as other industrial, commercial, and agricultural processes. Not all particulate pollution originates from power plants. There are natural sources of particulate pollution, especially volcanoes. Over a short period of time, natural sources of particulate pollution can be more significant than artificial sources, and some people are fond of saying that nature is the "real" polluter. But over the long term, the continued effects of artificial pollution greatly outweigh the effects of natural events.

Particulate pollutants can act as carriers of harmful chemicals. They contribute significantly to the production of photochemical smog (see above), and they are detrimental to human health when they enter the lungs. Lung cancer,

(a)

(b)

Figure 7.4.2 Do environmental controls work? Emissions from power plants and industrial smokestacks release hundreds of tons of pollutants to the atmosphere every day. (a) How fumes would pour from the stack at a steel plant if an electrostatic precipitator control system had not been installed. (b) The difference made by the emission control system. (*Source:* Bethlehem Steel Corporation.)

emphysema, and other respiratory ailments can be caused or aggravated by the presence of particulate pollutants in the atmosphere.

Atmospheric processes and the dispersal of pollutants. Although weather forecasting has improved greatly over the past decade, atmospheric processes are still imperfectly understood. But most researchers agree that the atmosphere handles pollutants by four methods: *diffusion, dilution, dispersal,* and *assimilation.* (See Box 7.4.1 for an explanation of how these processes work.)

BOX 7.4.1 DIFFUSION, DILUTION, DISPERSAL, AND ASSIMILATION

Over time, any materials will tend to become less concentrated, or more *diffuse.* This is a basic physical principle, related to the second law of thermodynamics (see Section 2.2.1). Thus, if a car emits concentrated carbon monoxide into a garage, that carbon monoxide will tend to spread out from the exhaust pipe. This process of *diffusion* will take place regardless of any air currents or other forces that might tend to disperse the pollutants. In the long run, through the process of diffusion, the concentration of carbon monoxide will tend to be the same throughout the garage.

Most automotive mechanics are aware of the danger of carbon monoxide, so they provide a hose to carry the exhaust outside the garage. The exhaust gases mix with the outdoor air and their concentration is greatly decreased, thus reducing the danger to human health. Another approach is to bring air into the garage with a fan. In both these cases, we are disposing of the exhaust gases through *dilution.* The total amount of exhaust has not decreased, but we are mixing that exhaust with a large amount of fresh air, thereby reducing its concentration to a safe level. This process of *dilution* is related to the process of *diffusion:* since the exhaust gas will disperse equally throughout a volume of air, it stands to reason that a larger volume of air will result in a lower concentration of the gases.

Once the exhaust gases have been diluted, they are carried over a large area by the atmospheric winds, a tremendous system of air movements fueled by solar energy. The winds carry the pollutants or waste products over a range of hundreds of miles, distributing them so that they are even less concentrated. This process is called *dispersal.*

Finally, the forces of gravity separate the heavier particles from the lighter ones. Larger particles such as ash and dust settle to the ground or other surfaces. Lighter particles remain in the air, where they form the core of raindrops or snowflakes. Other soluble and insoluble pollutants are swept down by the raindrops, washing the pollutants into the ground and surface water. Trees and other green plants absorb many of the pollutants from the water and use some of them as nutrients. Their leaves also draw pollutants directly out of the air. This final process is called *assimilation.*

When these processes of diffusion, dilution, dispersion, and assimilation work properly, the level of atmospheric pollutants will not increase—as long as the production of pollutants does not exceed the rate at which they can be disposed of. But what if the winds do not blow? What if the cleansing agents in the ecosystem cannot function properly?

In still air, the rate of dilution is limited by the rate of natural diffusion—the tendency of molecules to distribute themselves equally. This is a slow process, usually too slow to dispose of pollutants as rapidly as they are produced. Heavy particles, such as soot and ash, are especially slow to diffuse. So when the winds are slow, the rate of waste production in heavily populated or industrialized areas can be greater than the carrying capacity of the atmosphere.

Under normal atmospheric conditions, the air is warmer near the ground and cooler in higher levels of the atmosphere. Since air rises, this guarantees a certain amount of air movement even when winds are weak. But occasionally the normal stratification of the atmosphere is reversed, so that the cooler air is near the ground with warmer air above it. This is called a *thermal inversion.* When this condition occurs, the dispersal of atmospheric pollution is greatly decreased, and the concentration of waste products can rise very quickly. Most "pollution alerts" in large cities occur during periods of thermal inversion. Los Angeles's air pollution problems can be traced in part to the fact that thermal inversions occur there more than 75% of the time. This aggravates the already serious smog problem. Other areas with frequent temperature inversions include Fairbanks (Alaska), Chicago, and the southern Appalachian region.

Of the many types of pollution, air pollution may be the most difficult to control. Governmental efforts to protect air quality have been only partially successful (see Section 7.9). Because of the unpredictable nature of the weather, measurements of air pollution vary significantly from day to day and even from

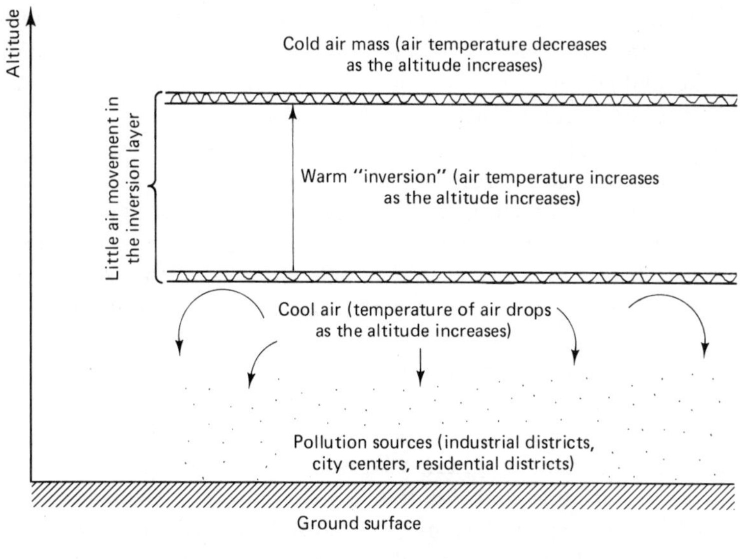

Figure 7.4.3 Thermal inversion: how polluted air gets trapped.

hour to hour. More significantly, the large-scale movement of the atmosphere makes air pollution a very widespread problem. As we pointed out before, if a corporation in the midwestern United States uses tall smokestacks, the air quality in the immediate vicinity of the plant may not be significantly reduced, but the pollutants released from the high stack may cause acid rain in New England—or was the acid rain caused by industrial plants in the northeast? It is this kind of problem that makes air pollution so difficult to deal with.

7.4.2 Water Pollution

In 1969, the Cuyahoga River in Cleveland, Ohio, caught on fire. This was not the first time the Cuyahoga had burned, nor was the Cuyahoga the first river to catch on fire. But the blaze emphasized the poor condition of U.S. rivers and streams at a time when pressure for environmental protection laws was beginning to have some effect. Three years after this incident, the Clean Water Act was enacted.

Water pollution can be divided into three basic categories: pollution of streams and rivers, pollution of underground water, and pollution of the oceans. Each kind of water pollution has its own special characteristics, but all are related to our increasing industrialization and our use of nonrenewable energy resources.

River and stream pollution. Rivers and streams provide us with a wealth of benefits. For many people, rivers are a source of drinking water. Fish from freshwater streams are an important food source for many people. Rivers also provide recreation, wildlife, and a host of other benefits.

Industries also use rivers as a source of water for many purposes. The water can be used to generate low-temperature process steam, although for electrical generation purer water is required. River water is used in the condensers of many thermal power plants. Major rivers also serve as transportation routes for coal, iron, and many other industrial materials.

Unfortunately, many cities and industries use rivers and streams primarily for the disposal of sewage. As the wastes from industries and cities have increased in volume and toxicity, this has placed an intolerable burden on rivers. In the Hudson River, contamination from PCBs, which are used in the manufacture of electrical transformers, has made fish unsafe to eat. The Niagara River in Buffalo is heavily polluted with dozens of toxic materials. The problem of river pollution is not confined to the United States. The Rhine, the Danube, and many other important European rivers have been heavily contaminated by discharges from oil refineries and other industrial plants.

The reason for much of this pollution is economic: it takes money and energy to process sewage and industrial wastes for proper disposal. Dumping them in rivers is cheaper and requires less energy. Since the water in a river is continually being replenished by new rainfall, it is often assumed that the river or stream will clean itself. Over long periods of time, this is probably true.

BOX 7.4.2 SOURCES OF WATER POLLUTION

Here are some of the important sources of water pollution, directly related to energy production, transportation, and consumption.

1. *Coal mining.* Surface and subsurface mining activities cause surface and groundwater pollution, respectively.
2. *Uranium mining and tailings.* Chemical contamination and radioactive pollution of groundwater may result.
3. *Oil wells.* Groundwater may be contaminated by leaks in casings that are used in oil wells. Offshore production may cause ocean water pollution.
4. *Oil transport via pipelines or tankers.* Spills and leaks, accidental or routine, cause major harm to sea organisms and shore life.
5. *Storage tanks.* Leaks and spills from petroleum product storage sites pollute both surface and groundwater.
6. *Power plants.* Power plants routinely discharge an assortment of chemicals during maintenance and cleanup operations, resulting in groundwater contamination and acid rain. Coal-plant waste lagoons leak into groundwater, and radioactive water may be released by nuclear power plant malfunctions.
7. *Agriculture.* Nitrogen fertilizers, pesticides, herbicides, insecticides, and manure contaminate the water table, as well as nearby lakes and streams.
8. *Industry.* Industrial energy supply systems, processes, and operations routinely discharge pollutants to bodies of water. Industrial waste lagoons may leak chemical contaminants into the ground, leading to groundwater contamination.
9. *Landfills.* A variety of pollutants can leach out from landfills into groundwater.
10. *Municipal sewage.* Sewage contains a host of harmful chemicals and microorganisms.
11. *Highway salt.* Salting operations during winter cause surface and groundwater pollution.
12. *Air pollution.* Air pollution resulting from a host of human activities eventually ends up causing water pollution.

But toxic materials can settle out into the streambeds, where they will continue to pollute the water for many years.

Another type of river pollution comes from power plants that use river water in their cooling cycles. When the heated water is discharged into the river, the water temperature increases, causing changes in the vegetation and fish populations. Although this thermal pollution cannot be seen (or smelled) as easily as municipal and industrial wastes, it adds significantly to the heavy burden of river and stream pollutants. During normal operation, nuclear power plants also pollute rivers with tiny amounts of radioactivity. These releases can increase during accidents; for example, during the cleanup at Three Mile Island, radioactive tritium was dumped into the Susquehanna River.

As people have become more aware of the problem of polluted streams and rivers, sewage treatment facilities have been upgraded. Approximately 30 billion dollars has been spent on sewage treatment plants across the nation. In many cases, this has led to significant improvements. The Cuyahoga, for instance, now supports some kinds of fish. But efforts are currently under way to weaken the Clean Water Act. In October 1982, the U.S. Environmental Protection Agency (EPA) introduced a proposal to make it easier for states to "downgrade" rivers and streams, thus increasing the amounts of pollutants that can be legally permitted in the streams. Fewer polluters have been prosecuted in recent years, and environmentalists fear that conditions will deteriorate still further in the next few years. It remains to be seen whether the nation's rivers and streams will ever reach the "fishable/swimmable" condition called for in the Clean Water Act.

Pollution of underground water. In many regions, underground *aquifers* supply water for most uses. For example, on Long Island in New York State, groundwater is the main source of drinking water for millions of residents. But that water is becoming increasingly contaminated by toxic chemi-

Figure 7.4.4 Treated water flows out of a settling basin. As it passes through the effluent sump, the water is automatically tested for pH, temperature, turbidity, and dissolved oxygen. (*Source:* Bethlehem Steel Corporation.)

cals and industrial by-products. Dozens of public wells on Long Island have been found to be contaminated, and many more wells remain untested. Similar problems can be found in many other industrialized regions of the country. According to an EPA/Cornell University study, 63% of all private wells in the United States are contaminated.

Groundwater pollutants come from a variety of sources, including oil wells, refineries, chemical plants, and landfills. By far the largest source of pollutants is landfills, including some that are operating legally with appropriate discharge permits. In the past, land disposal of toxic materials was not closely regulated because it was assumed that natural filtration by soil, clay, and gravel would purify the water before it reached underground aquifers. As many communities are discovering, this is not always the case.

Groundwater does not flow freely and rapidly, like rivers and surface streams. This means that pollutants cannot simply be washed downstream. Typically, the pollutants form a "plume" that gradually spreads outward through the groundwater deposit. Even after the source of the contaminants is discovered and eliminated, the pollutants can continue to spread, endangering water supplies in an ever-larger area.

Ocean pollution. Even the immense oceans are beginning to feel the effects of water pollution. In this case, oil production and transport is the major culprit. An oil spill during ocean transport of crude oil can cause serious ecological damage. In the notorious *Torrey Canyon* disaster of 1967, an oil spill of approximately 100,000 gallons caused very serious damage to wildfowl and other aquatic species. Although 100,000 gallons seems like a huge amount, in global terms it is only a tiny fraction of the petroleum and petroleum products which are annually discharged into the oceans. Routine discharge of petroleum from oil tankers and other vessels amounts to 3.5 million tons annually. Additional oil finds its way into the oceans through accidents in port and in the open sea, during oil exploration and production, in storage and pipeline transportation, and through discharge of industrial sewage containing lubricants and other petroleum products. Altogether, these sources contribute 6 million tons of oil pollutants to the ocean each year. We do not know what the impact of these pollutants will be in the long run, but navigators report an alarming increase in the occurrence of "pelagic tar"—large globules of congealed petroleum—in the open oceans.

Perhaps the most notorious case of ocean pollution occurred in early 1983, when an Iranian oil well blew out and began dumping enormous quantities of crude oil into the Persian Gulf. Because the well was in a war zone, technicians were unable to reach the well and make repairs. After nine months, the well was still discharging crude oil into the ocean, and no end was in sight.

Protection of the oceans is an especially difficult problem because the oceans are not part of any country's territory; therefore, no country is clearly responsible for protecting them. Since most oil exploration is now taking place in offshore waters, pollution of the oceans is likely to continue.

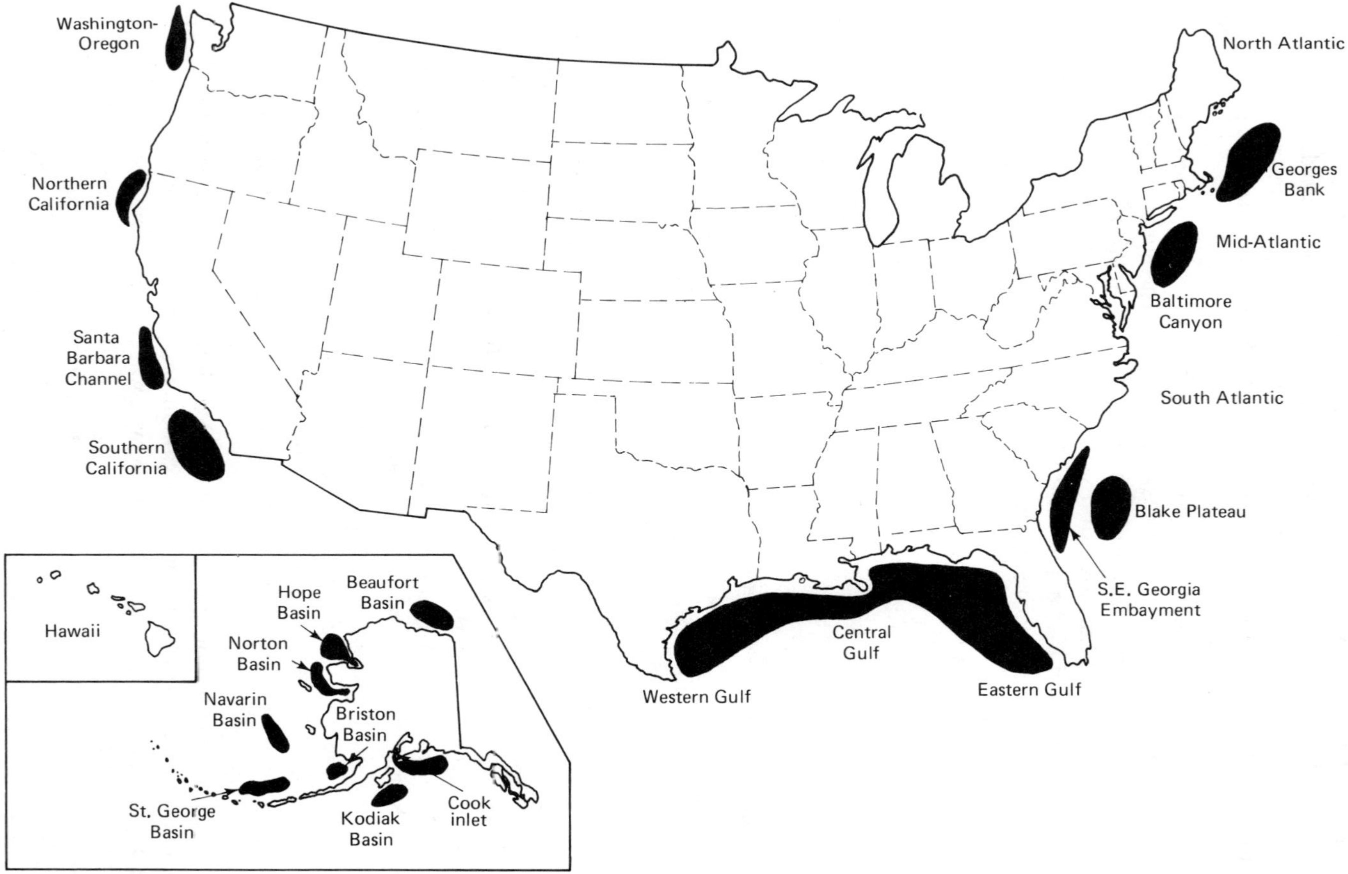

Figure 7.4.5 Outer continental shelf areas under oil leasing consideration. (*Source:* U.S. Geological Survey, Department of the Interior.)

7.4.3 Toxic Wastes

During the past few years, we have all heard of problems with toxic chemicals in "Love Canal," kepone in Hopewell, Virginia, PBBs in Michigan, and dioxins in Times Beach, Missouri. These four episodes share a common bond. In each case, the problems were caused by the unintended release of highly toxic organic substances to the environment. Although the mechanisms of release differed widely, many of the effects were the same: serious health problems, enormously expensive cleanup efforts, and long, complex court battles over the question of responsibility for the contamination. Indeed, some environmentalists believe that the last two decades of the twentieth century will be characterized by ever more serious cases of toxic waste pollution.

So many new synthetic organic chemicals have been produced that the American Chemical Society has developed a computerized filing system to keep track of trade names, generic names, chemical formulas, and structures. Approximately 1000 new chemicals are developed every year. Of the many thousands of new chemicals that have been developed during the past few decades, some are highly toxic to various species and warrant safe disposal. Unfortunately, the disposal of these materials has not always been carried out with the necessary degree of care.

A substance is defined as hazardous if it can be corrosive, reactive, or toxic, or if it can ignite, cause an explosion, release undesirable fumes, or pose a threat to the general health. At present, more than 85 industrial process wastes and more than 410 commercial products and chemical substances have been classified as hazardous. In the United States alone, more than 60 million metric tons of hazardous wastes are generated every year.

The problem of hazardous wastes is quite serious and complex for four basic reasons:

1. These substances are often very persistent. Because they do not occur naturally, nature has devised no effective way of dealing with them.
2. They may have subtle or delayed effects on us or our environment which often go unnoticed in the early stages.
3. The commercial and economic realities of industrial development do not provide ample time and incentives to study these materials in depth.
4. When we do analyze toxic substances, we often fail to account for the complexity of ecosystems and for the *synergistic* action of some chemicals.

Toxic materials and the energy industry. Although toxic substances are associated primarily with the chemical industry, they are closely connected with the energy industry and energy-using products. For example, PCBs are a by-product of the electric power industry, where they are used in transformers for high-voltage power lines. Lead is a direct component of automotive fuel, the main product of the U.S. oil industry. Asbestos, a cancer-causing toxic substance, was used for many years in brake linings, electric heaters, and flue pipes for high-temperature furnaces and kilns. Over the years, these energy-related products have been a major source of toxic contamination of the environment.

BOX 7.4.3 TOXIC WASTES: THE WORST OFFENDERS

Lead. For years, tetraethyl lead has been added to gasoline in order to improve its combustion properties. This allowed refiners to use a lower grade of petroleum distillate as automotive fuel. However, when automobile engines burn the gasoline, they release large quantities of lead into the air. This lead ultimately winds up contaminating soil and groundwater, especially in urban areas. Because the body cannot dispose of lead, many residents of urban areas have suffered serious lead poisoning as a consequence of automotive emissions. Unleaded gasoline is now required in most new motor vehicles.

PCBs. PCBs are complex organic chemicals called *polychlorinated biphenyls.* PCBs used to be used widely in electrical capacitors and transformers because of their heat-resistance and electrically insulating properties. During the manufacture of these electronic products, large amounts of PCBs have been generated. These waste products have not always been disposed of wisely. For example, in Bloomington, Indiana, PCBs were dumped into the city sewage system for decades. The chemicals eventually made their way into the waste sludge, which was given to the public for use as crop fertilizer. Before the practice was discontinued, many acres of prime farmland had been severely contaminated. PCBs have also been found in the Hudson River, in such concentrations that communities which drew their drinking water from the Hudson were forced to search for alternative supplies. PCBs are so widespread that every human on the face of the earth probably has some PCBs in his or her tissues. PCBs have even been found in the tissues of Antarctic penguins.

Dioxins. Dioxins are among the most toxic substances known to science. They cause serious birth defects, health effects, and possibly genetic damage. Dioxins were introduced into the environment of southeast Asia during the war in Viet Nam, when millions of acres were sprayed with *Agent Orange,* a herbicide that contained dioxin-producing chemicals. The chemical is believed to have caused serious effects on residents of Viet Nam and on U.S. soldiers stationed in the area. In late 1982, dioxin at levels 100 times higher than the legal limit were found in a suburb of St. Louis, requiring the EPA to evacuate the residents of the area. Tests are being performed to find the extent of the contamination, which was worsened by serious flooding in the region. In 1983, discovery of a high level of dioxin contamination in New Jersey led to evacuation of residents in the affected area.

7.4.4 Radioactive Pollution

Radioactive wastes are properly classified as a form of toxic wastes, but they deserve special attention because of their uniqueness. No one can see, hear, feel, smell, or taste radiation. Yet the effects can include cancer, birth defects, and genetic changes. In very large doses, radiation can cause immediate death.

Even though radiation and radioactive pollution are commonly linked with nuclear weapons and nuclear power reactors, these are not the only sources of radioactive pollution (see Table 7.4.3). Radiation enters the environment from a number of natural sources:

1. The sun is a huge fusion reactor. Radiation from the sun can cause skin cancer, especially for avid sunbathers and people living in very sunny climates.
2. Many natural minerals, including granite and other building materials, give off some radiation.
3. Our own bodies are somewhat radioactive. Some of the minerals contained in our tissues include radioactive isotopes.

These natural sources of radiation cannot easily be controlled by human activities. However, radiation also enters the environment from artificial sources, including the following:

1. Atomic and hydrogen bomb tests
2. Nuclear power plants
3. X-rays and other medical procedures
4. Television sets
5. Miscellaneous radioactive products such as luminescent watches and smoke detectors

TABLE 7.4.3 Radiation Exposure and the General Public

Source of Radiation	Exposure (mrem)
Average natural background radiation	38–150
Average additional levels due to human activities (TV sets, microwave ovens, nuclear power plants, X-rays, etc.)	80–125
Average total exposure	210–255
Legal limit of exposure for minors and the general public	500
Recommended maximum exposure beyond background and medical radiation sources (International Commission on Radiological Protection)	170

Radiation and radioactivity. For many people, the terms "radiation" and "radioactivity" are both mysterious and indistinct. But the terms have very specific meanings, and the differences between them are worth knowing.

Radiation is a particle or set of particles emitted by a material as its atoms decay. As Box 7.4.4 shows, radiation falls into three categories: alpha, beta, and gamma. The radiation can cause health problems by affecting chromosomes and other parts of living cells.

Radioactivity is the capacity to produce radiation. If a material is radioactive, it will emit radiation over a precisely predictable period of time. Radioactive materials do not injure cells directly, but they emit the radiation that does the damage.

If a person is exposed to a radioactive material, two dangers are possible. First, the radiation produced by the material during the time of exposure may cause health problems. Second, some of the radioactive material may incorporate itself into the person's body, continuing to expose the person to radiation over a long period of time.

BOX 7.4.4 RADIATION TYPES

When a nucleus decays, it emits one of the three following kinds of radiation:

1. *Alpha radiation.* This is a high-energy helium nucleus with a positive charge of 2. It cannot travel far and will easily be stopped by a piece of paper. But if an alpha-emitting particle is absorbed into the body, the particles can cause serious cellular damage.
2. *Beta radiation.* This is a high-energy electron with a negative charge. Beta radiation penetrates somewhat farther than alpha radiation: up to 20 mm through human tissues.
3. *Gamma radiation.* This is a high-energy photon, similar to an X-ray. It can penetrate large distances through most substances. Lead or earth shielding can stop gamma rays.

Radioactive half-lives. Each radioactive material has a characteristic *half-life.* This means that half the nuclei in a given quantity of the material will decay during the period of time designated as a half-life. Suppose, for example, that we have 1 million atoms of a material with a half-life of 10 minutes. After 10 minutes, half of the atoms will have decayed, so that now we have only 500,000 atoms. After another 10 minutes, only 250,000 atoms will be left. After 4 hours, or 24 *half-lives,* only about one-millionth of the original material will still remain; in other words, the material will be essentially gone.

Some materials have very short half-lives: as short as a few thousandths of a second. Other materials have very long half-lives: many thousands of years. Plutonium, a radioactive element produced in nuclear reactors, has a half-life of 24,000 years.

BOX 7.4.5 DECAY OF 1,000,000 ATOMS OF STRONTIUM

In order to understand how the processes of decay work, let us follow the decay processes of two fairly common radioactive substances, strontium-89 and strontium-90. Radioactive strontium is produced in atomic bombs and in nuclear power reactors. It is found in two isotopes: strontium-89, with a half-life of 59 days, and strontium-90, with a half-life of 29 years. Let us compare the length of time required for decay of these two elements.

Comparison of Decay Time for Strontium-89 and Strontium-90

	Number of Radioactive Atoms	
Elapsed Time	Strontium-89	Strontium-90
Start	1,000,000	1,000,000
59 days	500,000	996,000
120 days	250,000	976,000
1 year	125,000	909,000
4 years	3,700	682,000
50 years	0	303,000
100 years	0	92,000

Clearly, an element with a half-life of 29 years must be contained much longer than one with a half-life of 59 days.

Radiation decay products. Although a radioactive material will decay very substantially over a period of several half-lives, this does not mean that all danger is gone. The element may decay into other radioactive materials, called *radiation daughters*. These daughter products may be as radioactive as the original radioactive materials, or even more so. The daughter products may decay into other radioactive materials, and so on through several steps. Eventually all radioactive materials will decay into stable isotopes, but in the meantime they must be isolated from the environment for very long periods of time.

Disposal of radioactive materials. A fundamental difference between radioactive wastes and chemical wastes is that chemical pollutants can often be neutralized and rendered harmless, whereas radioactive products cannot. No process has yet been discovered for reducing the radioactivity of any element. Chemical processes are useful for combining the radioactive materials with other substances to facilitate disposal, as we will see. But radioactivity can only be reduced by natural decay processes, sometimes requiring many thousands of years.

A number of processes have been proposed for the disposal of radioactive waste materials. In the most seriously proposed plan, high-level wastes are to be chemically combined with vitreous materials to form glassy capsules of radioactive wastes. The capsules have low solubility and can easily be contained in large containers for long-term disposal. Proponents of nuclear power believe that vitrefaction of wastes provides a complete solution to the radioactive waste problem, but critics point to the extremely long periods of time for which wastes must be contained, claiming that no tests have been done that can accurately predict the behavior of materials over hundreds of centuries.

Releases of radiation to the environment. Radiation has been released into the environment on a number of occasions. Atomic and hydrogen bombs (including the bombs dropped on Hiroshima and Nagasaki) have been the main source of radioactive pollution. Tests conducted in southern Nevada are believed to have contaminated huge areas of southern Utah, and the fallout from recent Chinese nuclear tests was found to have circled the entire globe several times.

During the production of fuel for nuclear power plants and weapons, large volumes of low-level wastes are generated. In the early years of nuclear science, tailings were used as a building material for roads and even for the foundations of houses and schools, most notoriously in Grand Junction, Colorado. Other tailings were simply heaped up in piles, which have given off radioactive gases, or dumped into artificially created ponds.

Radiation is also produced at nuclear weapons production plants. This can be another source of radiation pollution. Recently, several kilograms of plutonium were found near a weapons plant in Rocky Flats, Colorado. Oddly enough, these radioactive materials were found not by the operators of the plant, but by environmentalists working independently. The operators of the plant were not aware that any radioactive materials were missing from the plant.

Radiation has also been released to the environment from nuclear power plants, although the aggregate amounts released have been very small to date. The most serious releases from a nuclear power plant occurred in 1979, when a nuclear accident at the Three Mile Island plant in Pennsylvania led to the release of radioactive materials, including iodine-131 and other biologically active isotopes. During the process of cleanup from the accident, additional radioactive materials have been released into the environment. Accidents at other nuclear power plants have released smaller amounts of radioactivity.

Perhaps the most worrisome releases of radioactive material have been the escape of radioactive wastes which were thought to have been disposed of safely. At Hanford, Washington, radioactive wastes have migrated from storage tanks and trenches. Barrels of wastes buried under the ocean have leaked onto the ocean floor. It is believed that in the USSR, buried nuclear wastes exploded, contaminating an enormous area of land in the Urals regions. Although the Soviet government has not officially acknowledged this accident, both Western and Soviet scientists are quite convinced that such an accident has taken place (see Medvedev, *Nuclear Disaster in the Urals*).

Biological effects of radioactivity. It is difficult to make general statements concerning the biological effects of radioactive pollution because the effects are so varied. Depending on the level of radiation exposure, the type of radiation (alpha, beta, or gamma), and the type of cells exposed, the effect can be death, severe impairment of organ function, or a delayed effect causing cancer, leukemia, or genetic change.

At high enough levels of radiation exposure, cells will die. If a sufficient number of cells die, the affected organ will die. If the dead organ is a vital one, the body's functioning can be fatally disrupted. A radiation level that kills 50% of the exposed cells is called a *lethal dose 50* or LD_{50}. For the human species, LD_{50} is approximately 400 rem (see Box 7.5.6). Generally speaking, this is a very high dose and it is unlikely that human beings will be exposed to this high a level of radiation from radioactive pollution except in the event of a nuclear war or a catastrophic accident at a nuclear power plant.

It should be emphasized that scientists have not been able to find any level at which a person can be exposed to radiation with no danger whatsoever; there is always a finite possibility of biological damage. Over the years, the permissible exposure levels have been reduced to reflect our concern with this problem (see Table 7.4.4). But it has also been clearly demonstrated that the higher the exposure to radiation, the greater the possibility of radiation-induced injury. For these reasons, containment of radioactive materials is one of the most important environmental issues of our time.

TABLE 7.4.4 Maximum Permissible Exposures for Workers in the Atomic Energy Industry

Year	Maximum Annual Exposure (mrem)
1934	30,000
1948	15,000
1958	5,000[a] (not to exceed 1,250 in any 3 months)

[a]Specially-classified workers may be exposed to a maximum of 3000 mrem in 3 months; however, their lifetime exposure since age 18 must not exceed 5000 mrem per year. New standards are said to be under consideration.

Source: International Commission on Radiological Protection.

BOX 7.4.6 MEASURING RADIOACTIVITY

Because radiation is found in a variety of forms and acts through a large number of biological pathways, several different methods exist for measuring radiation. The first unit to be developed was the *roentgen*, named after the discoverer of X-rays. One roentgen of X-rays or gamma radiation will produce 2.08×10^9 ion pairs in 1 cm^3 of free air at O °C.

The roentgen has two serious deficiencies as a measure of radiation: it only measures gamma radiation, and it only measures the ionization produced in the atmosphere. Different forms of radiation work in different ways. For example, external gamma radiation can be very damaging because of its capacity to penetrate tissues. External alpha radiation is relatively harmless, especially for people wearing clothing over most of their bodies. But if alpha-producing isotopes are ingested or inhaled, they can cause serious damage: *internal* alpha radiation is quite dangerous indeed. To measure environmental and health effects, it is necessary to have units that account for the type of radiation and the nature of the exposure. In response to this need, two other units have been developed: the rad and the rem.

One *rad* is the amount of radiation that will cause 1 gram of material to absorb 10 mJ of energy. This unit accounts for different types of radiation and the different densities of various materials. But it does not account for the fact that some tissues are highly sensitive to radiation whereas other tissues are not affected so severely. In order to account for this variable, a third unit has been developed: the *rem* or *roentgen equivalent man*. One rem of radiation is defined as the amount of radiation that will produce the same biological effects in a human being as 1 roentgen of gamma radiation. Thus a relatively small amount of radiation can produce 1 rem of exposure if it affects a sensitive organ such as the bone marrow, whereas the same amount of radiation will produce less than 1 rem if it affects a less sensitive organ such as the foot.

Each of these units can be subdivided into smaller subunits; thus, 1 rem = 1000 *millirems*. Radiation exposure for human beings is usually measured in *millirems*, although very high exposures may be measured in rem.

7.5 CLIMATIC EFFECTS

It is not precisely understood how changes occur in the climate. We know very well how individual atmospheric events take place, but the aggregate effect of several influences is very poorly understood. Similarly, we can identify isolated climatic effects of energy production and use, but it is more difficult to understand the cumulative impacts of many energy projects all over the world. Let us start by looking at local climatic changes, as they are the easiest to observe and the most immediate in their effects.

7.5.1 Local Climatic Change

Energy projects can have an immediate effect on the climate of a region. For example, a hydroelectric power site will tend to moderate temperature extremes in the nearby area. The lake will increase the humidity, creating foggy conditions in some localities. In the winter it will create increased precipitation in the form of snow. The fog, mist, and icy conditions may cause hazardous driving conditions.

A thermal power plant will increase the local air temperature to a slight degree, either by releasing heat through the cooling towers or by raising the temperature of nearby bodies of water, which in turn raises the air temperature. In addition, if the plant uses wet cooling towers to reduce thermal pollution of nearby bodies of water, the water vapors from the towers will cause an increase in fog, mist, and icy road conditions.

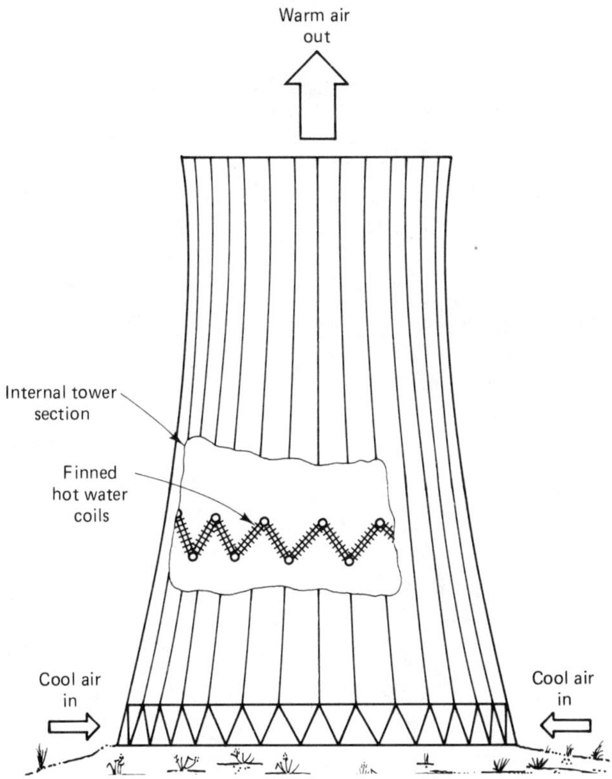

Figure 7.5.1 Dry cooling tower. Cooling towers are used to reduce thermal pollution from coal- and nuclear-fueled power plants. Dry cooling towers require less water than do wet cooling towers (see Figure 3.8.2). This may permit the siting of power plants away from large bodies of water. (*Source:* U.S. Department of Energy.)

Deforestation often accompanies energy production, especially in strip-mined regions or areas where forests are clear-cut to produce firewood. Where forests are removed, the summers tend to be hotter and the winters colder. The average wind speed in the area will increase because the natural windbreak of the forest has been removed. Precipitation will generally decrease because the trees no longer provide water to the atmosphere through transpiration. Air pollution will also increase, producing additional climatic effects.

Cities and urban areas are usually warmer than nearby rural areas because of the large heat output from industries, apartment and office buildings, and automobiles. In New York City, for example, during the winter, human sources produce 2.5 times as much heat as the solar radiation received by the city. It is easy to see why some meteorologists refer to urban areas as "heat islands." As a result of these higher temperatures, a special climatic relationship develops

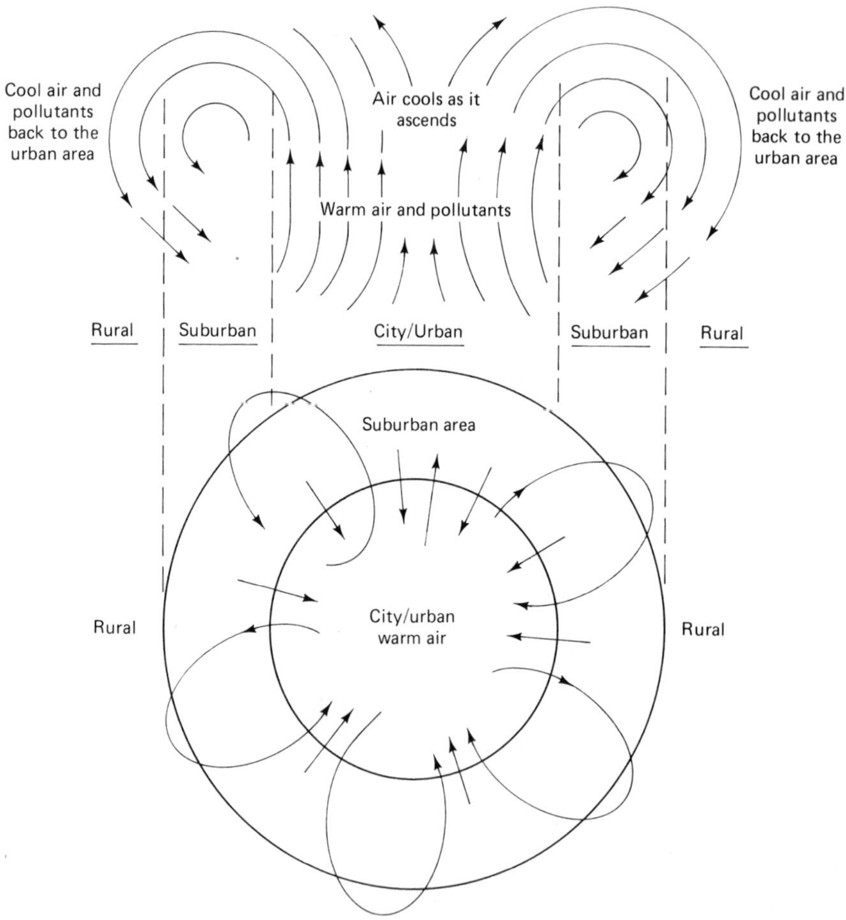

Figure 7.5.2 Air circulation caused by "heat island" effect.

between cities and the rural communities surrounding them. The warm air over the urban areas tends to rise, being replaced by cooler air from the surrounding areas. As the heated, polluted air rises, it spreads outward from the city center and cools, falling to the earth. This creates a circulating pattern which draws the polluted air back toward the urban center. This pattern of circular stagnation produces a variety of unpleasant climatic effects in urban areas. In general, cities have higher temperatures, fewer sunny periods, more rain, and lower visibility than surrounding areas. This is true even if we discount the decreased visibility caused by air pollution over urban areas.

The use of automobiles, which accounts for a large percentage of world energy consumption, also has significant climatic effects. First, the building of roads and shopping centers causes a substantial amount of deforestation, with its attendant climatic effects. Second, automobile exhaust causes an increase in local temperatures, especially in urban areas. Finally, automobile emissions change the composition of the air. Particulate emissions create photochemical smog, which decreases the amount of sunlight reaching the earth. Particulate matter from automobiles increases the amount of precipitation locally. The carbon dioxide content of the air is increased whenever fossil fuels are burned. This can have serious climatic effects at the global level, as we shall see in the next section.

It should be emphasized that these local climatic effects are less dramatic than the normal fluctuations characteristic of climate systems. In the short term, changes caused by human activities are negligible compared to the powerful effects of worldwide climatic systems. But over the long term, the slight changes caused by human activities may lead to more serious consequences, even on the global level.

7.5.2 Global Climatic Effects

The twentieth century has seen a tremendous increase in human alterations of the world's ecosystems. Many researchers believe that equally profound alterations of the world's climatic system may be in the offing unless steps are taken to reduce human impact on the environment. Even though the precise nature of these possible global alterations is not known, it is possible to make a few educated guesses. Here is a list of the major factors influencing global climate.

Solar radiation
Reflective properties of the earth's atmosphere
Heat-retention properties of the atmosphere
Temperature
The extent of the ocean's surface
The extent of forestation/vegetation on the land surface
The size of the polar ice caps

These factors can be affected significantly by the following products of an energy-intensive world economy:

1. *Increases in atmospheric concentrations of carbon dioxide* (CO_2). It has been estimated that in the last 100 years the level of carbon dioxide in the atmosphere has increased by approximately 12 to 15% as a result of human activities, including burning of fossil fuels and large-scale deforestation. In absolute terms, this may not seem to be a particularly significant increase: from approximately 290 parts per million to 335 parts per million. But if current trends continue, by the year 2000, the total increase could be on the order of 25 to 30%: double the increase that has been observed through all of recorded history. Although the total concentration of atmospheric CO_2 will still total less than 1 part per thousand, the anticipated climatic effects of this increase can be very serious.

Like glass, CO_2 allows incoming solar radiation to pass but blocks outgoing long-wave (infrared) thermal radiation from the earth (see the discussion of the greenhouse effect in Box 4.1.2, Section 4.1). This means that solar energy is allowed to reach the earth, where it raises the temperature of the surface, but the heat created is not allowed to escape back into outer space. As a result of this effect, the temperature of the earth's surface is increased.

With the increase in atmospheric concentration of CO_2, the average temperature of the earth could increase by 1 or 2 °C. People living in far northern reaches might appreciate the temperature increase. But the rise in temperature could melt a portion of the polar ice caps and thus cause the level of the oceans to rise. This could seriously affect coastal cities—and many of the world's most important cities are located on seacoasts. An increase in the world's air temperature would certainly result in significant changes in precipitation patterns—perhaps the creation of large deserts in some of the world's prime agricultural regions, and drastic increases in precipitation in other areas of the world, possibly causing erosion damage. Polar regions would get more snowfall as a result of the increased global temperatures.

2. *Particulate pollution.* Dust and other suspended particles in the atmosphere can create a variety of effects. They tend to have the exact opposite effect from carbon dioxide. The dust particles reflect incoming light, thereby reducing the amount of solar radiation reaching the earth and producing a cooling effect. This effect varies depending on whether the particulate pollution is in the upper or lower atmosphere, over land or over water. In a few cases, particles in the atmosphere absorb heat energy, causing a general warming of the environment.

It has been suggested that since the effect of carbon dioxide pollution is to raise the earth's temperature and the effect of particulates is to reduce the earth's temperature, the two effects will cancel each other out, resulting in no net climatic change. There may be some mitigation of climatic effects, but there may also be an uneven distribution of the two contrary effects, with the result that the earth's weather system may be destabilized. At present, few responsible researchers are willing to assume that the effects of particulate pollutants will cancel out the effects of increased carbon dioxide concentrations in the atmosphere.

3. *Deforestation.* We have discussed the local effects of deforestation. Globally, extensive deforestation reduces the capacity of the world's ecosystems to assimilate CO_2. It is estimated that between 1880 and 1940, deforestation accounted for almost two-thirds of the total increase in the atmospheric concentration of CO_2. Thus the global effects of carbon dioxide concentration are significantly increased by widespread deforestation.

4. *Thermal pollution.* Whenever energy is consumed, heat is added to the atmosphere as waste heat from manufacturing processes, friction, exhaust, and so on. If human heat output reaches or exceeds 1% of the energy represented by the solar radiation reaching the earth, experts expect the global climate to be affected by the additional thermal load. Fortunately, present human energy conversion is substantially smaller than this 1% figure, so the global climatic effects of thermal pollution are not yet highly significant.

7.6 LAND USE

Energy production can have a major impact on local and regional land-use patterns. These impacts should be carefully assessed before any energy project is carried out.

For example, when coal is strip-mined in agricultural areas, the topsoil that was formerly used for growing crops is stripped from the surface and put

Figure 7.6.1 Devastation created by the surface mining of coal. (*Source:* Office of Surface Mining, U.S. Department of the Interior.)

to one side. The process of land reclamation is long and uncertain; so far, strip-mined areas have not been returned to agricultural production very rapidly or successfully. Since the mineral rights to a plot of land may be owned by a commercial interest, landowners are sometimes powerless to stop the mining of their land, even though they may suffer economically from the change.

Hydro projects require very large areas of land which, once they are submerged, cannot be used for housing, farming, manufacturing, or many other purposes. Often farms and homes in the region must be condemned and vacated before the energy project can be carried out. Hydro projects also change the ecology of a region. If a free-flowing river is to be dammed to produce electric power, the project may cause destruction of wilderness and the loss of endangered species. For these reasons, local residents and environmental groups often raise objections to a project.

Nuclear power plants take up much less room than hydro plants with comparable power capability, but the land occupied by a nuclear power plant may remain contaminated for centuries after the plant has ceased to operate. This is also true of radioactive waste sites.

Figure 7.6.2 Reservoir and dam site at Glen Canyon Power Plant Project. (*Source:* Bureau of Reclamation, U.S. Department of Energy.)

7.6.1 Distribution of Land-Use Impacts

In many cases, the consumers who benefit from an energy project are not aware of the land-use impacts of the project. Energy users tend to be concentrated in urban areas, while energy projects are located in rural areas. This is true because of three basic factors:

1. Energy production often precludes other use of the land.
2. Land in and around urban centers is often costly and heavily populated.
3. Energy facilities must be located near the areas where the energy will be used.

Since urban land is already occupied for other purposes, energy facilities are rarely located within urban centers. Land acquisition would be very expensive, and it would be necessary to displace large numbers of people through condemnation proceedings. (Table 7.6.1 indicates the acreage requirements of a variety of energy facilities.) Such action would not be welcomed by urban residents, who might exert political pressure to stop the projects. But because the energy is required in the cities, the plants must not be located too far away from the cities.

Figure 7.6.3 High-voltage power lines. The transmission of electricity is often carried out at high voltages in order to reduce transmission losses over hundreds of miles. This calls for huge towers, thick wires, and a right-of-way for the power lines. Utility companies usually maintain access to the power lines by spraying herbicides to kill off the vegetation—a cause of controversy in some regions. (*Source:* U.S. Department of Energy; photographer, Jack Schneider.)

TABLE 7.6.1 Land Areas Converted to Energy Use per Effective Plant-Year[a]

Form of Energy	Acres
Electricity	
Coal	
Mines	300–600
Power plants[b]	
Station site	3
Cooling lakes (optional)	40
Scrubber wastes	20–30
Nuclear	
Mines	60
Power plants[b]	
Station site	1–7
Total site (including exclusion areas)	7–70
Cooling lakes (optional)	70
Mill tailings	8
Solar electric[b]	70–300
Biomass	10,000
Other energy	
Oil shale (conventional)	
Mining	2000–3000
Additional shale-disposal areas	10–25
Oil shale (in situ)[b]	7

[a]Based on 19.6 trillion Btu delivered to home-heating end uses, taken for the electricity technologies as equivalent to 1 GW-plant-year of 6.57 billion kWh.

Note: The relative land requirements for power plants depend on many factors not listed here, and therefore, this table cannot be taken at face value. Security, future expansion, transport services, transmission stations, and other factors contribute to this uncertainty.

[b]Land use for sites occupying a fixed number of acres for a long period are normalized to a 30-year lifetime and then prorated annually; that is, a typical cooling lake for a coal-fired plant will take up an actual area of $40 \times 30 = 1200$ acres.

Sources: From Sam H. Schurr et al., *Energy in America's Future* (Baltimore: Johns Hopkins University Press for Resources for the Future, 1979). Data originally from William Ramsay, *Unpaid Costs of Electrical Energy: Health and Environmental Impacts from Coal and Nuclear Power* (Baltimore: John Hopkins University Press for Resources for the future, 1978), Chap. 8 (for electricity technologies); Harry Perry, "Production of Liquids and Gases from Other Resources," draft prepared for the National Energy Strategies Project (Washington, DC: Resources for the Future, 1978) pp. 17-100, 17-105, and 17-106 (for oil shale); William Ramsay, "Electricity from New Technologies," draft prepared for the National Energy Strategies Project (Washington, DC: Resources for the Future, 1978) pp. 16-26 and 16-61 (for solar and biomass). The last two reports are available from RFF's Center for Energy Policy Research.

If we look at the locations of nuclear power plants in the United States, we can see that the power plants are situated primarily near urban centers: the eastern seaboard, the industrialized Southeast, the Great Lakes, and the California coast. But the plants are generally located in rural areas a few dozen miles from the urban centers: close enough to provide the power but far enough away to avoid many of the political obstacles that would be encountered if the plants were sited directly in the urban centers.

Increasingly, however, there is a shortage of prime sites for power plants near large cities, and some of the newest power developments have occurred in remote regions. For example, the James Bay project of Hydro Quebec is located approximately 600 miles north of Montreal and even farther away from the New England cities that are expected to provide a market for the surplus power generated by the project. This enormous project, when completed, will flood hundreds of square miles of former wilderness—land that will not return to its original state for many decades, if at all. Power lines will require additional land use. The James Bay is the habitat of many arctic animals, including polar bears, whose lives may be profoundly altered by the project. These impacts on wilderness areas will probably increase as energy developers are forced to look further away from the cities to locate their energy projects.

7.7 LOSS OF WILDERNESS AND WILDLIFE

Over millions of years, many species have died out due to climatic change, disappearance of habitat, competition from other species, or other factors that may never be understood. However, during recent years the pace of extinctions has increased. Wildlife specialist Norman Myers suggests that every day on the average, one species disappears from the face of the earth. In a few years, we may be losing one each hour.

The cumulative impact of these disappearances is quite alarming. Thomas Lovejoy of the World Wildlife Fund believes that pressures of human resource development may wipe out half a million species by the year 2000—perhaps as much as one-sixth of all the species on the planet. In 1978, the Smithsonian Institution listed approximately 2000 plant species—nearly 10% of the known U.S. native plant species—as "endangered."

Many species have been driven into extinction or near-extinction by changes in their habitat. Urban development, construction of energy facilities, and introduction of competing species can threaten species with extinction. Often these threats have resulted from inadequate planning and research, or a simple lack of concern for wildlife and wilderness.

Protecting wildlife habitat is not a simple matter. Although there may be substantial agreement that an energy project will change habitat, there may be substantial disagreement concerning the desirability of the change. For example, a hydroelectric power plant causes the flooding of substantial amounts of land upstream from the dam. This may destroy wildlife habitats. But the reservoir itself can provide habitats for a variety of wildlife, although usually not the same species which inhabited the area before. Analysts often disagree on the desirability of these changes, which can make wildlife protection very difficult.

Another serious obstacle to habitat protection is the claim that economic progress is more important than wilderness and endangered species. For example, a few years ago it was thought that the Tellicoe Dam project in Tennessee threatened a species of snail darter. Environmentalists opposed the dam project on the grounds that a species, once extinct, can never be recovered, while proponents of the dam claimed that the snail darter was an insignificant fish and much less important than the anticipated benefits of the project. The project was ultimately approved by a special act of Congress despite expectations that the snail darter would become extinct. Fortunately, other populations of snail darters were established. But it is worth noting that, in the final analysis, protection of an endangered species was judged to be less important than completion of an energy project.

Energy development need not endanger wildlife, although protection of wildlife may raise the costs of a project or cause a slight drop in efficiency. For example, when the Alaskan oil pipeline was built, special pipeline crossings were constructed to avoid blocking the migration routes of wild animals such as moose and caribou. In the case of hydroelectric dams, fish ladders can be installed to avoid interference with fish migrations. These modifications do not prevent completion of the energy projects; in fact, by rendering the projects less objectionable to environmentalists, they may actually facilitate their completion.

Why Protect Wilderness and Wildlife?

Wildlife and wilderness protection measures, such as fish ladders or the special crossings on the Alaskan pipeline, add to the cost of a project and may reduce its energy output somewhat. But these measures can also provide economic benefits. The fish ladder may provide economic benefits to fishermen upstream, and increased tourist traffic can contribute to the prosperity of scenic wilderness areas. Thus it is often possible to argue that the measures are economically justified, and environmentalists use such arguments whenever possible.

In an important sense, though, the main benefits of wilderness and wildlife protection are noneconomic. Wilderness areas have a special meaning to us in a world where technology and civilization are developing at an inconceivably rapid pace. Wilderness can be an oasis in the midst of a technological society: a place to replenish our spiritual energy. Like all other animals, human beings developed in a natural environment. By preserving wilderness and wildlife, we can maintain our connection to the natural environment from which we emerged.

7.8 WHY WORRY ABOUT ECOLOGICAL CHANGE?

Many people argue that ecological change is part of nature's grand design: a way of stimulating evolution and biological progress. They argue that if a species becomes extinct as a result of ecological change, this proves that the species was not sufficiently adaptable. They also point out that humanity has always altered ecosystems: there are areas of China where agriculture has so thoroughly changed the face of the land that biologists have been unable to identify the types of vegetation that originally inhabited the region. Some have even claimed that people can do a better job of controlling the environment than nature.

This argument overlooks several important facts about our relationship to our environment:

1. We cannot foresee all the consequences of our actions. For example, it was recently discovered that the propellant in most aerosol containers was causing a depletion of the ozone layer of the atmosphere. If this depletion had been allowed to continue, the result would have been widespread cancer, changes in vegetation, and possible climatic changes. The developers of aerosol containers had never imagined that their simple product could cause such far-reaching consequences. All too often, changes in ecosystems cause results that could not have been foreseen.
2. All parts of the ecosystem are interrelated. When we affect one part of the system, we are likely to affect other parts at the same time. For example, when we spray to kill undesirable insects, we are also likely to kill birds and desirable insects, such as bees. In some cases, when spraying has been carried out in areas where fruits are cultivated, it has been necessary to import bees from unaffected regions in order to save the fruit crops.
3. Even if humanity is unusually adaptable, other parts of the system may not be. We may be able to adapt our behavior to changed conditions. But other species cannot. As we alter habitats and pollute the air and water, we are creating conditions that many species have never faced before. If they fail to survive, we may also fail to survive, for we depend on other species for our food, and for the very air we breathe.

Many environmentalists are most concerned by what they see as a decrease in the *resilience* of ecological systems. Most ecosystems can absorb a fair amount of abuse without collapsing, largely because there is large variety of organisms in the system. If one organism is eliminated, another will fill the *niche* that was vacated. But as the diversity of ecosystems decreases, much of this resilience is lost, and we are pushed closer to the brink of ecological collapse. Some biologists fear that if major ecosystems are pushed over the brink, the survival of humanity may be jeopardized.

How much change can an ecosystem sustain before a breakdown occurs? Are all changes necessarily bad? These are difficult questions. It can be argued that some changes caused by human endeavor are beneficial. Examples in this category might include the "greening" of desert land due to reforestation, or the creation of a new freshwater lake by a hydroelectric project. Thus while we need to recognize the ecological effects of energy development, we should not assume that all those effects are necessarily bad. But neither should we assume that the environment has an unlimited capacity to adapt.

Tables 7.8.1 and 7.8.2 provide information on the impact of various energy sources on people and the environment. This information is not intended to be definitive. Many of the conclusions will need to be reassessed as the effects of new energy technologies become better known. But the important point here is that we should make a continuing effort to understand the environmental consequences of our actions.

TABLE 7.8.1 Environmental Impact of 1000-MWe Energy Systems[a]

System	Air Emissions (10^3 units)	Water Discharges		Solid Waste (10^3 units)	Land Use (10^3 acres)
		tons	10^{13} Btu		
Coal					
Deep-mined	383.0 tons	7,330	3.05	602.0 tons	29.4
Surface-mined	383.0 tons	40,500	3.05	3,267 tons	34.3
Oil					
Onshore	158.4 tons	5,990	3.05	—	20.7
Offshore	158.4 tons	6,070	3.05	—	17.8
Imports	70.6 tons	2,520	3.05	—	17.4
Natural gas	24.1 tons	810	3.05	—	20.8
Nuclear[b]	—	21,300	5.29	2,620 tons	19.1
radiation	489,000 curies	2,680 curies		1,400 curies	

[a]Operating at a 0.75 load factor with common prevailing practices. MWe, megawatts of electric power.

[b]Nuclear systems release radiation in various stages of the fuel cycle.

Sources: Council on Environmental Quality, *Energy and the Environment* (Washington, DC, 1974). Adapted with permission from Phillip G. LeBel, *Energy Economics and Technology* (Baltimore: Johns Hopkins University Press, 1982).

7.9 GOVERNMENTAL ACTION

Environmental protection is generally regarded as a function of government. This makes sense for four basic reasons:

1. The environment is common property. That is, it is not owned by any one person or group of people.
2. Past abuses have damaged the environment to the point where protective action is needed.

TABLE 7.8.2 Environmental Analysis of Various Energy Resources[a]

Type of Energy	Energy Significance	Nonrenewable	Renewable	Air Pollution	Water Pollution	Land Impact	Visual Pollution	Thermal Pollution	Wilderness Impact	Noise Pollution
Oil	4	×		3	2	4	3	2	3	2
Natural gas	3	×		1	1	1	1	2	1	1
Coal	4	×		4	3	3	2	3	3	1
Synfuels	Exp.	×		1	3	3	2	2	2	2
Nuclear fission (normal operation)	2	×		1	1	1	1	2	1	1
Nuclear fission (accident)		×		4	4	4	?	3	4	?
Nuclear fusion	Exp.	?	?	?	?	?	?	?	?	?
Geothermal	1	?	?	1	1	1	2	1	2	1
Solar	2		×	0	0	1	1	0	1	0
Hydro	2		×	0	1	3	1	0	2	0
Wood	2		×	4	1	1	2	1	1	1
Wind	1		×	0	0	1	2	0	1	2
Biomass	1		×	1	1	2	1	1	1	0
Tidal	1	?	?	0	1	0	1	0	1	0
OTEC	Exp.		×	0	1	0	1	1	?	0
Wave power	Exp.		×	0	1	0	2	0	2	1
Conservation	3		×	1 (indoor)	0	0	1	0	0	0

[a]Exp., experimental; 0, negligible; 1, some; 2, considerable; 3, very much; 4, extreme; ?, unknown.

Source: Based on a chart developed by Jane Albee, Vermont Technical College, Randolph Center, VT.

3. The welfare of humanity depends on a sound and healthy environment.
4. Economic factors are not enough in themselves to guarantee adequate protection of the environment.

To protect the environment, governments enact laws and regulations. Other governmental activities include incentives, directives, guidelines, and penalties. All these strategies can be used to reduce damage to the environment.

The history of environmental laws and regulations goes back almost a century in the history of the United States (see Table 7.9.1). The Rivers and Harbors Act of 1899 marked the beginning of environmental protection. Other laws have been passed over the years. But the watershed in modern-day environmental protection occurred with the 1970 signing of the National Environmental Policy Act (NEPA).

TABLE 7.9.1 Noteworthy Environmental Laws

Law Number	Year	Name
PL 55-425	1899	Rivers and Harbors Act
PL 68-238	1924	Oil Pollution Act
PL 88-206	1963	Clean Air Act; amended 1970
PL 89-272	1965	Solid Waste Disposal Act
PL 91-190	1969	National Environmental Policy Act
PL 91-596	1970	Occupational Safety and Health Act
PL 92-500	1972	Water Pollution Control Act Amendments
PL 94-469	1976	Toxic Substances Control Act
PL 96-510	1980	Superfund Law (Response, Compensation, and Liability)

7.9.1 Salient Features of NEPA

NEPA recognizes the need for assessing and evaluating the impact of our activities on natural ecosystems. To deal with these environmental impacts, the NEPA calls on the federal government to act in cooperation with state and local governments and groups. The government's responsibilities fall into six basic categories:

1. To provide all practical means, including financial and technical, to create and maintain conditions for environmental protection and preservation.
2. To promote the use of renewable resources and recycling of depletable resources.
3. To preserve important historic, cultural, and natural aspects of our national heritage.
4. To achieve a healthy balance between population, resource utilization, and individual choice and diversity.

Figure 7.9.1 Reclaimed surface mine. The 1977 Surface Mining and Reclamation Act regulations requires mining companies to recontour and revegetate the mined area to its original state. (*Source:* Office of Surface Mining, Department of the Interior.)

5. To recognize the global and long-range character of environmental problems and therefore to initiate and take active part in international cooperative efforts to curb the decline in our environmental quality.
6. To function and plan so as to fulfill the responsibilities of each generation as trustees of the environment for future generations.

The Act established its objectives through two principal means: the Council on Environmental Quality and the Environmental Impact Statement (EIS). Together with the Environmental Protection Agency (EPA), established in 1970, the NEPA provides the basis for federal environmental policymaking.

7.9.2 The Council on Environmental Quality

The Council on Environmental Quality is not a regulatory agency or commission, but simply a research and advisory council. The Council's main function is to assist and advise the president in the preparation of the nation's annual Environmental Quality Report. The Council reviews all Environmental Impact Statements, although it does not approve or disapprove an EIS. It can only make recommendations to the appropriate agencies. In addition, the Council interprets the NEPA for governmental agencies.

In conducting its studies, the Council on Environmental Quality consults with individuals as well as public and private organizations. It uses this input to work with the president in establishing environmental policy.

7.9.3 The Environmental Protection Agency

The Environmental Protection Agency (EPA) was created by a governmental reorganization in 1970. Nine functions, which had previously been assigned to five different agencies, were combined under one "umbrella" organization. These functions were related to pollution control and environmental conservation.

Since the establishment of the EPA in 1970, environmental protection has become a major concern of government. The Clean Air Act requires the EPA to comment in writing on the environmental impacts (air and water pollution, water supplies and quality, solid-waste disposal, radiation pollution, noise pollution, toxic materials, etc.) of any regulation, project, or legislation related to any government agency. Depending on its evaluation of the Environmental Impact Statement, the EPA approves the project, disapproves it, or requires additional information for further evaluation. Another major function of the EPA is the administration of the "Superfund," a fund established by corporate assessments which is used for cleaning up hazardous waste sites.

Although the EPA has the responsibility for administering the nation's environmental laws, the EPA is not authorized as a law enforcement agency. This means that when violations of the law are found, the EPA must refer matters to law enforcement agencies via the Justice Department. The EPA maintains an office of enforcement which acts not only to recommend prosecution of lawbreakers but also to arrange for voluntary compliance with environmental laws. Voluntary measures are generally less expensive than lawsuits, and in many cases they bring about more rapid compliance with environmental standards.

In recent years, the role of the EPA has become highly controversial. Funding for the agency has been cut repeatedly, with some of the largest cuts coming in the area of energy programs. This has happened at the same time that the agency has accepted major new responsibilities, including the management of the Superfund. Environmental organizations have accused the government of wanting to undermine environmental protection through budgetary means, but the EPA has responded by claiming that the agency functioned inefficiently in the past, and that state governments are assuming many environmental protection functions.

7.9.4 The Environmental Impact Statement

The Environmental Impact Statement (EIS) is the heart of governmental environmental protection. Under NEPA, all governmental agencies are required to take a close look at the environmental side effects of their programs and actions. An EIS is required for any major action that is expected to affect the quality of the human environment. Only projects that receive financial assistance or an operating license from the federal government are required to file Environmental Impact Statements. However, because so many state and local projects depend on federal funding or licensing, most energy projects require Environmental Impact Statements.

Recently, a court has ruled that state and local governmental projects can also be required to prepare EISs. The case involved the city of Davis, California, where a highway was proposed but no EIS was filed. The judge determined that it was essential to consider the environmental impacts of the project.

There have been a few instances where projects have been given the go-ahead despite adverse environmental impacts that were identified in the EIS. The Alaskan oil pipeline was one such project. But in response to the EIS, several design changes were instituted which reduced the environmental impacts of the project. On the other hand, many projects have been scrapped because of the information brought out by the environmental studies required to produce an EIS. For example, a proposed coal gasification plant in New Mexico was canceled after the EIS identified serious environmental impacts.

The range of issues addressed by an EIS includes three basic areas of the environment:

1. *The natural environment.* This area includes questions of air quality, water supply and quality, soil conservation and hydrology, fish and wildlife populations, fish and wildlife habitats, noise levels, radiation, and disposal of hazardous substances.
2. *Land-use planning and management.* This includes energy supply and natural resources development, protection of environmentally sensitive regions such as floodplains, wetlands, beaches, dunes, watersheds, steep slopes, and unstable soils.
3. *Socioeconomic issues.* This heading includes matters such as population density, changes in neighborhoods, outdoor recreation, and historic, architectural, and archaeological preservation.

The EIS must be prepared using a systematic interdisciplinary approach and should incorporate all relevant analytical tools, yet it must be written in a rather nontechnical, easy-to-understand style and language. The EIS is also required to describe the effects of nonimplementation of the proposed action.

As the preceding discussion makes clear, preparation of an EIS is a rather complex, time-consuming affair. The EIS and the whole concept of NEPA have been criticized as costly, time consuming, and inefficient. In many cases, preparation and approval of an EIS can take two years or longer. Some project managers may disagree with federal agencies concerning the amount of detail required in an EIS. These cases occasionally wind up in the federal courts. Another criticism is that the EIS process only applies to government sponsored or licensed projects, while many private projects are not required to go through the procedure.

On the positive side, a study by the Council on Environmental Quality showed that the EIS helped planners in making logical and sound decisions.

It has also been pointed out that an EIS conducted during the early stages of a project can cut down on costly revisions later in the construction phases of the project. On balance there appears to be general agreement that the concept of the EIS is a good one, even though there is disagreement on the details of the process. As in most human endeavors, perfection is elusive and there is always room for improvement. What matters most is that we take the time and make the effort to look carefully at the environmental consequences of our actions in energy and other fields.

7.10 ENVIRONMENTAL ORGANIZATIONS

In most democratic countries, businesses and other organizations try to influence the government in order to advance their own interests. Businesses and other profit-making groups lobby for favorable tax and tariff policies, social organizations lobby for programs, and trade groups lobby for favorable regulations. In this climate, it should come as no surprise that environmental groups also engage in lobbying activities.

During the past two decades especially, many environmental groups have come into being on the local, regional, national, and even international level. These groups typically begin as small, grass-roots organizations and draw their support from large numbers of individuals rather than from corporate or governmental organizations. Although most of these organizations are financially shaky, they have managed to survive, and some have even had a considerable impact on national environmental policy. For example, many of the environmental protection features of the Alaskan oil pipeline would not have been in place if it were not for the struggle of a number of environmental organizations. In this section we will identify a few of the larger environmental organizations. Many others exist on regional and local levels, often working in concert with these larger organizations.

Sierra Club. This is certainly the best known of the many environmental organizations, although not the largest. But because of its well-known lobbying and legal actions, the Sierra Club has come to symbolize active environmentalism. The organization works in a number of different areas to influence public policy:

Testifying before governmental policymaking boards

Bringing lawsuits against violators and potential violators of environmental laws and regulations

Publishing periodicals and books to promote an awareness and appreciation of the environment and of environmental issues

Organizing local political activities through its many chapters.

At present, the Sierra Club has more than 300,000 members across the United States, with the largest concentration in California and the West coast.

Friends of the Earth. During the 1970s, some members of the Sierra Club, including its director, David Brower, decided to take a stand against the development of nuclear electric power generation. They were opposed by the board of directors of the Sierra Club, so they formed their own organization, Friends of the Earth (FOE). Today, the group has national offices in 14 countries and active lobbies in many capital cities, including Washington, D.C. FOE uses civil lawsuits, political actions, and publications, including full-page advertisements in national newspapers, to arouse public awareness and interest in environmental issues.

Besides its political activities, FOE has supported a number of important energy research projects. FOE energy-related publications include a number of significant books, including Amory Lovins, *Soft Energy Paths: Toward a Durable Peace;* Hugh Nash, ed., *Progress As If Survival Mattered;* and John S. Steinhart et al., *Pathways to Energy Self-Sufficiency: The 2050 Study* (see Section 1.6 for a discussion of the latter).

National Audubon Society. Best known as an organization for birdwatchers, the Audubon Society has recently branched out more widely into broader areas of wilderness preservation and environmental protection. Concern for the preservation of endangered species has led the Audubon Society to take positions in opposition to hydro projects, nuclear power plants, and other energy projects which threaten wilderness.

The Audubon Society works through local chapters and the national organization to influence public opinion. One particularly effective tactic used by the Audubon Society is the production of films showing wilderness areas that are threatened by development. Publications, lectures, and educational programs also help to get the Audubon message across.

National Wildlife Federation. This has always been considered the most conservative of the many environmental organizations, and it has also enjoyed the largest membership. Recently, the National Wildlife Federation, like the Audubon Society, has begun to take positions on energy issues as they relate to the preservation of species and wilderness areas. Through the publication of its magazines, the National Wildlife Federation reaches a very large membership.

This very limited sampling represents only a fraction of the many groups working on energy and environmental issues. Appendix C lists many such groups, together with their addresses and specific areas of interest. Readers are encouraged to write to the individual organizations for fuller statements of their organizational objectives and stands on energy issues.

STUDY QUESTIONS

1. What is an ecosystem? Give an example of an ecosystem and explain the principles of how an ecosystem functions.
2. Explain the importance of diversity in an ecosystem. How does energy development affect biological diversity?
3. In light of your understanding of how ecosystems function, what are the ecological implications of resource development?
4. What is the "carrying capacity" of an ecosystem? In what cases do you think the earth's carrying capacity is being reached or exceeded?
5. What are the natural processes that clean the environment? How effectively do these processes deal with the wastes from energy production and use?
6. What is smog? How is it caused, and why is it a problem? How would you propose to deal with the problem of smog?
7. What is acid precipitation? What are its causes? What measures have been proposed to deal with it?
8. In recent years, many industries have pressured the U.S. government to relax clean air and water standards. Do you think these standards should be weakened? Explain your reasoning.
9. Coal is being advanced as a solution to America's need for energy independence. What would be the environmental consequences of a large-scale shift to coal in the United States?
10. Compare the environmental impact of electric power production using the following resources:
    (a) Oil
    (b) Coal
    (c) Natural gas
    (d) Hydro
    (e) Nuclear
11. What is meant by the "nuclear fuel cycle?" Describe the environmental effects of various stages of the fuel cycle.
12. Should radioactive wastes be handled differently from other kinds of wastes? Why or why not?
13. Evaluate the following statement: "Since we have not proved conclusively that anyone has died as a result of acid rain, there is no need to enforce air quality standards."
14. What is the greenhouse effect? What environmental problems are associated with the effect?
15. What toxic or hazardous substances are used by industries in your area? How do the industries dispose of their wastes? Are these waste disposal methods adequate in your opinion?
16. Discuss the environmental impacts of the following renewable resources: wind, wood, solar, and hydro.
17. What is the importance of wilderness and wildlife? How are they affected by energy production and consumption?

18. Is it important to have international cooperation in solving the energy problems created by today's energy consumption? Why or why not?
19. The earth has often been compared to a spaceship. Explain this comparison as it relates to the environment.

FURTHER READINGS

CARSON, RACHEL. *Silent Spring.* Greenwich, CT: Fawcett, 1962.

CHEREMISINOFF, PAUL N., and ANGELO C. MORRIS. *Environmental Assessment and Impact Statement Handbook.* Ann Arbor: Ann Arbor Science Publishers, 1972.

CONGRESSIONAL RESEARCH SERVICE. *Environmental Protection: An Historical Review.* Washington, DC: U.S. Government Printing Office, 1981.

COUNCIL ON ENVIRONMENTAL QUALITY and U.S. DEPARTMENT OF STATE. *The Global 2000 Report to the President.* Washington, DC: U.S. Government Printing Office, 1980.

DETWEILER, ROBERT, ET AL. *Environmental Decay in Its Historical Context.* Glenview, IL: Scott, Foresman, 1973.

Energy and Environment (articles reprinted from *Scientific American*). San Francisco: W. H. Freeman, 1980.

FOWLER, JOHN. *Energy and Environment.* New York: McGraw-Hill, 1975.

JACKSON, WES. *Man and the Environment.* Dubuque, IA: Wm. C. Brown, 1973.

MCKENZIE, GARRY D., and RUSSEL O. UTGARD. *Man and His Physical Environment.* Minneapolis: Burgess, 1975.

MEDVEDEV, ZHORES A. *Nuclear Disaster in the Urals.* New York: W. W. Norton, 1979.

MILLER, G. TYLER, JR. *Energy and Environment.* Belmont, CA: Wadsworth, 1980.

NEBEL, BERNARD. *Environmental Science.* Englewood Cliffs, NJ: Prentice-Hall, 1975.

ROBERTSON, J. CRAIG. *A Guide to Radiation Protection.* London: Macmillan Press, 1976.

SCHURR, SAM H., ET AL. *Energy in America's Future.* Baltimore: Johns Hopkins University Press for Resources for the Future, 1980.

UDALL, STEWART. *The Quiet Crisis.* New York: Holt, Rinehart and Winston, 1970.

WINTER, JOHN V., and DAVID A. CONNER. *Power Plant Siting.* New York: Van Nostrand Reinhold, 1978.

8

Social Analysis

In the preceding chapters we discussed energy issues "from the inside." Physical, economic, and environmental analysis can all be done in relatively "pure" theoretical ways. But when it comes to energy in the real world, social and political considerations are at least equally important.

In making energy decisions, people are certainly concerned with efficiency, economy, and environmental issues. But they are also greatly concerned with the ways in which their use of energy affects their everyday lives. If an energy resource is incompatible with the needs of a society, that energy resource should be rejected in favor of one better suited. In this chapter we identify some of the important social issues related to energy development and use.

8.1 ENERGY AND POPULATION

Although it is difficult to generalize about humanity, we can say one thing without hesitation: humanity's numbers are growing. Not only is the human population at the highest level in history, but the rate of growth is also higher than ever before. As Table 8.1.1 indicates, the population of the earth has doubled four times in the past 2,000 years and is expected to double again by the year 2010. What is more, the population doubling time keeps shrinking, as is shown clearly by Table 8.1.1.

Clearly, this is explosive, exponential growth. How long can it last? What

TABLE 8.1.1 World Population

Population	Year	Doubling Time (yr)
250 million	0	
500 million	1650	1600 (approx.)
1 billion	1850	200
2 billion	1930	80
4 billion	1975	45
(4.58 billion)	(1982)	
8 billion	2010	35 (est.)

Source: George C. Beakley and H. W. Leach, *Engineering: An Introduction to a Creative Profession* (New York: Macmillan, 1977).

will cause a slowdown or halt in this human proliferation? How will it happen? What will be the consequences? These questions are vital to the survival of our species, but the answers remain elusive. At best, we can make only educated guesses. But we can be more certain of the relationship between population growth and energy consumption.

Every human being requires a food equivalent of between 2000 and 3000 kcal per day. Currently, world food production is sufficient to provide that much nourishment for every person on the earth. However, not everyone gets that much food energy. Underdeveloped countries see a large portion of their citizens go hungry, while in more developed countries, a large portion of the population is overweight. Even in highly developed countries, hunger is a serious problem for much of the population. These problems are clearly caused by unequal food distribution rather than a general shortage.

But while world food production is now sufficient to feed the earth's population (assuming more equitable distribution), the situation is likely to change as the world's population increases further. Another doubling in the world's population would require a substantial increase in food production. Assuming that humanity currently produces one-third more food than it requires (a very generous assumption), a 50% increase in food production would be required to meet the needs of a doubled world population in the year 2010. This is a tall order. Can we realistically expect to increase food production to that level in 25 years?

In the past, world food production has been able to increase dramatically. Yields per acre of cropland have been rising significantly, largely through the use of inorganic fertilizers made from petroleum products and natural gas. But as petroleum and natural gas production begin to decline, there may not be enough energy resources available to sustain a 50% increase in agricultural production. Thus the population problem is also an energy problem. Any decisions we make about energy resources will directly affect the capacity of the earth to feed its human inhabitants.

Increased world population will put tremendous pressures on other energy-related areas as well. The additional people will require housing, transportation, communications, and industrial products. As we have seen, all these things require energy in large quantities. Unless there are significant technological breakthroughs in many of these areas, we can expect serious energy supply problems as the world's population increases.

As we established in Section 7.2, every finite ecosystem has a certain *carrying capacity*, and the earth is no exception. As we look at the population problem, we must ask ourselves, what is the earth's carrying capacity for human beings? Estimates differ significantly. Some experts put the figure as low as 500 million, while others set it as high as 12 to 14 billion. Obviously, the current world population of 4.6 billion is not too far away from the limits of the earth's capacity to sustain human life. Two more doublings (requiring less than a century at current rates) will put the earth's population above even the most generous estimates. Clearly, the population problem needs to be addressed immediately.

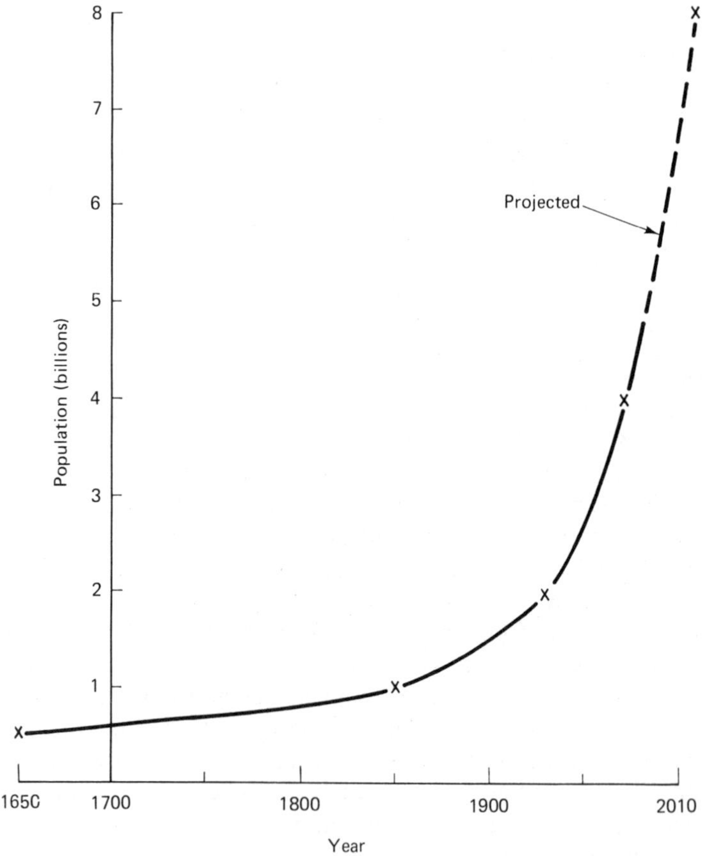

Figure 8.1.1 World population (1982: 4.68 billion).

8.2 STANDARDS OF LIVING AND ENERGY USE

The enormous increase in the world's population is a major cause of the increasing consumption of energy resources. But the rate of world energy consumption has increased more rapidly than the world's population. This is because energy use per capita has increased very rapidly during the past century. This increase has occurred primarily in the more industrialized areas of the world.

It is generally accepted that energy use and standard of living are closely related. But the relationship is a complex one. In countries where natural resources are in plentiful supply, each person can use a large share because there is enough to go around. Thus, availability of energy resources leads to an increase in standards of living. But at the same time, a high standard of living leads to increased use of energy resources. A person who can afford a personal automobile uses much more energy than a person who can only afford a bicycle. Similarly, in countries with high standards of living and well-developed industrial economies, the high standard of living leads to a high consumption of energy resources.

These two processes reinforce each other. Higher energy use leads to a higher standard of living, and a high standard of living leads to higher energy use. These related processes have led to a serious imbalance in energy use

BOX 8.2.1 STANDARD OF LIVING AND LIFE-STYLE

The Smith and Chung families have very different standards of living. The Smiths drive a new Cadillac, while the Chungs have only an old bicycle. The Smiths' seven-room house has a color television set, a stereo set, a washing machine and dryer, a telephone, and running hot and cold water. The Chung house has no electricity or hot water, and the Chungs wash their clothes at the community washhouse. Clearly, there is an enormous difference in their standard of living.

Now consider the Smiths and their neighbors, the Joneses. The Jones family also has a car, but it is a fuel-efficient import. The Jones family has a washing machine, but they hang their clothes on a line to dry. The Smiths keep their house at 75 °F in the winter and 72° in the summer. The Jones house is much cooler in the winter and warmer in the summer. The family prefers it that way. The Smiths buy sodas in aluminum cans and smash the cans in their trash compactor. The Joneses buy sodas in returnable bottles, which they store in the garage. The Smiths rake their leaves into plastic bags and put them out for the garbage man. In spring they spread commercial fertilizer on their lawn. The Joneses rake their leaves into a bin at the edge of the yard, where they turn into leaf mold. In the spring, the Joneses spread leaf mold on their lawn and garden to help fertilize the soil.

Do the Smith and Jones families have different standards of living? Not at all! They both have the same conveniences. The difference lies in the way they choose to live. The Jones family uses *far* less in the way of natural resources while they enjoy a standard of living every bit as high as the people next door.

throughout the world. Each inhabitant of a highly developed country may use many times as much energy as high or her counterparts in the less developed countries. For example, in 1982 the per capita consumption of energy in the United States was 310 million Btu (MBtu). In Brazil, the figure is closer to 20 MBtu, and in China (1975), per capita energy use averages approximately 2 MBtu. Thus, per capita energy use in the United States is nearly *150 times* as high as in China.

People in all parts of the world hope that those in the less developed countries will soon be able to raise their standards of living. But this can only aggravate the problem of energy supply worldwide. As Third World countries, with their tremendous populations, try to emulate standards of living in the developed countries, the strain on the world's natural resources increases dramatically. Table 8.2.1 will give you some idea of the present contrasts in standards of living.

TABLE 8.2.1 Comparison of Indicators of Life in Sweden, United States, and India[a]

Indicator/Item	Sweden	United States	India
Population (millions)	8.3	226.5	700.0
Area (sq mi)	173,732	3,623,420	1,269,420
Population density (sq mi)	47.85	64.0	519.13
Urban population (%)	82.7	79.2	21.9
Per capita income ($)	9274	8612	150
Literacy rate (%)	99	99	36
Electricity (billion kWh)	91.8	2294	107
Rail transport (billion passenger—km)	6.8	18.1	192
Freight transport (billion ton—km)	15.6	1.4	154
Automobiles [millions (per capita)]	2.88 (0.347)	120 (0.5298)	1.03 (0.00147)
Television sets [millions (per capita)]	3.1 (0.373)	150 (0.664)	1.1 (0.00157)
Radios [millions (per capita)]	8.3 (1.0)	450 (1.98)	19.6 (0.028)
Telephones [millions (per capita)]	6.4 (0.77)	189 (0.834)	2.6 (0.0037)
Newspapers (per 1000 population)	584	282	15.25
Life expectancy (yr; male/female)	72.5/78.7	69.8/77.5	54/53[b]
Energy use per capita (MBtu)	166.3	310.0	4.9

[a]Latest available data, some of which are up to five years out of date.
[b]Approximation, exact data not available.
Sources: World Almanac—1983; Robert L. Loftness, *Energy Handbook* (New York: Van Nostrand Reinhold, 1978); U.S. Department of Energy, *Monthly Energy Review,* December 1982.

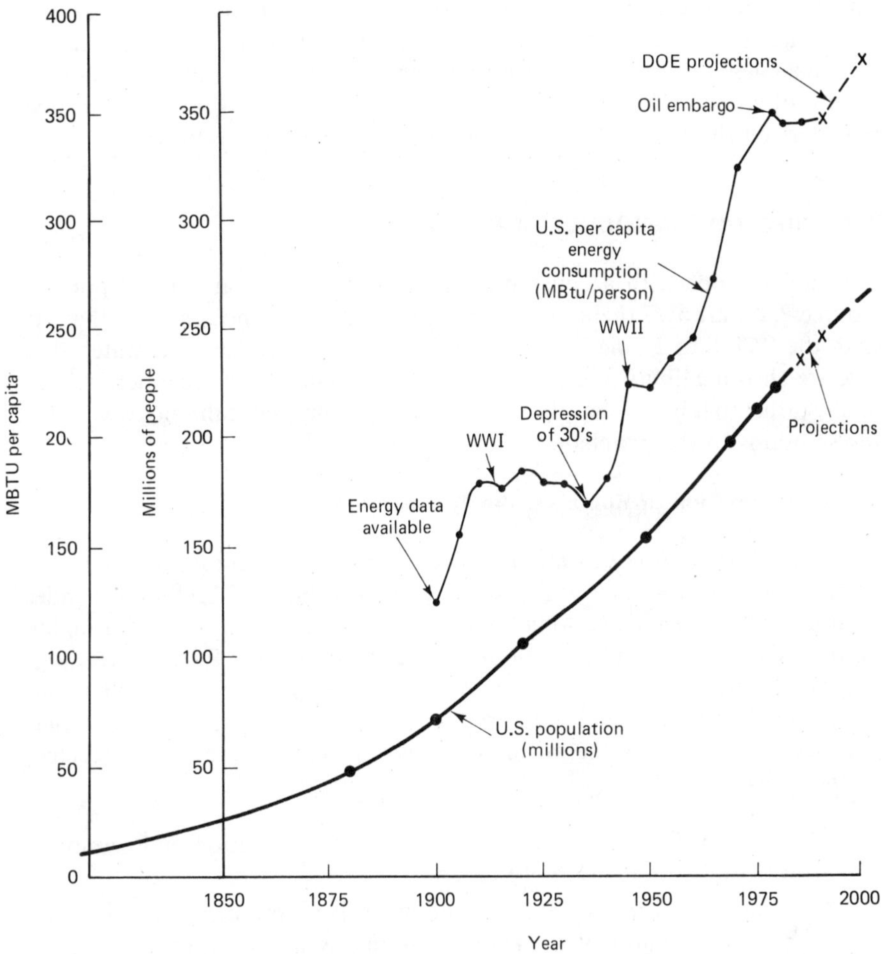

Figure 8.2.1 U.S. population and per capita energy consumption. (*Source:* U.S. Department of Energy.)

Not all differences in energy use can be traced to differences in standards of living. Even among developed countries, there are significant differences in energy use. For example, the United States, with approximately 5% of the world's population, accounts for approximately 30% of humanity's use of the world's energy resources. West Germany, with a standard of living somewhat higher than that of the United States, uses only slightly more than one-half as much energy per capita. Sweden, Switzerland, and Denmark are other countries with standards of living comparable to the United States, but which use much less energy per capita. In fact, the United States and Canada have the highest per capita energy consumption in the world. So although standards of living have much to do with levels of energy consumption, it is clear that other factors must be at work. In the case of the United States, traditionally low energy prices may help to explain the extremely high levels of consumption. Historical patterns of consumption have also been influenced by the "frontier spirit" that characterized the United States during the late nineteenth century.

8.3 COPING WITH RESOURCE DEPLETION

How can we cope with a constantly increasing demand on a finite pool of resources? As always, there are at least two schools of thought. On the one hand, the "Club of Rome," an influential body of scientists, has stated that there are absolute limits to growth. On the other hand, the "technocrats," an equally influential if less sharply defined group, claim that technology will provide solutions to the problem.

8.3.1 The Club of Rome vs. the Technocrats

In 1970, the Club of Rome sponsored a study on the future of humanity. The research work was done at the Massachusetts Institute of Technology under the leadership of Dennis L. Meadows. The research team published a highly controversial report entitled *The Limits to Growth* (see the Further Readings section). This was a study of the relationships between population growth, pollution, resource depletion, and food production. Based on computer models that simulated the behavior of the world's systems of production, the report reached the following conclusions:

1. The world will not be able to sustain present population and economic growth rates for many more decades.
2. Continued resource depletion would most likely produce confusion and ultimately a collapse of the world economic system within the next 30 to 50 years.
3. We have the physical and technical means to solve the world's population and resource problems.
4. Political and social factors provide the main obstacle to ensuring our future.
5. Because it takes many decades to change human and environmental systems, we must work now to prevent ecological crises and resource depletion. We should not wait to solve these problems until they have become pressing.
6. While it may not be possible to prevent resource depletion altogether, it is possible to reduce the *rate* of resource depletion.

Among the critics of the *Limits to Growth* concept is Burnham P. Beckwith, author of *The Next 500 Years* (New York: Exposition Press, 1968).

He claims that the *Limits to Growth* study overlooks several important factors that will help humanity avoid the kind of social collapse outlined in the *Limits to Growth* study.

1. Substitute materials can be used for scarce resources.
2. Food production will continue to grow through the use of new agricultural techniques, synthetic foods, and so on.
3. Pollution of land, air, and water will be reduced through technological advances and applications.
4. Population will be controlled through artificial birth control methods.

The technocrats further question the claim that capital investment increases pollution, which in turn reduces food production. They believe that continued, gradual economic and social progress is possible through technological advances. These advances will enable us to use poor grade raw materials and reduce demand for nonrenewable energy resources, through redesign of consumer goods and changes in social habits.

If this does not work, the technocrats believe that a redesign of the whole system or part of the system will be carried out in order to reduce or eliminate our dependence on the resource in question. Such examples as the substitution of aluminum for copper, and the development of synthetic fibers to replace scarce natural fibers, appear to back their theory.

The solutions proposed by the technocrats are very appealing, but what will be the consequences if their optimism proves unjustified? The technocrats base their conclusions on essentially one kind of approach—a technological approach—whereas the Club of Rome integrates a broad range of analytical perspectives, including economics, biology, and political science. So who is right? Readers should not draw conclusions based on the summaries we have presented here. Those who are interested should read the books by Meadows and Beckwith, both of which are listed in the Further Readings section at the end of this chapter.

8.4 SOCIAL IMPACTS OF ENERGY TECHNOLOGIES

More often than not, people believe that "technology" will save us. In fact, technology's impact on society is highly significant. But that impact is not necessarily good or necessarily bad. Rather, the uses and effects of technology depend on the social conditions in existence, the availability of technological capability, and the ability of technology to solve the social problems.

Proposed solutions to energy shortages often consist of meeting the projected demand in any way possible. Proposals for increasing energy production often call for reduced environmental protection and elimination of safety regulations. It is generally believed that the benefits of increased energy production will outweigh the problems caused by reduced regulations, and that increased technological development will invariably bring about a rise in the standard of living. In fact, technological development usually results in a mixture of benefits and problems. For example, consider the case of a new power plant.

The benefits of a new power plant are well known. The plant will produce a large supply of electricity, which in turn will facilitate industrial development, provide electric power for households, and contribute to a general increase in the standard of living. But while these benefits are distributed fairly widely across the society, the costs of the project tend to be felt most severely in the region where the plant is being built.

During land acquisition for the power plant, a certain number of landholders will be required to relocate. Although they usually receive a fair price for their land, they do not have the choice of remaining on their property. For many people, the strain of dislocation is very troublesome. Sometimes, families that have occupied the same land for generations must move to make way for a new power plant or transmission lines.

While construction of the plant is under way (which can last for 5 to 10 years), a large, temporary work force moves into the vicinity of the site. In many cases, these workers come from outside the region, causing a large population increase. This, in turn, leads to an increased demand for community services such as schools, fire and police protection, hospitals, road construction, water and sewage treatment, and housing. Community institutions frequently become overloaded, and the need for new construction can put a strain on the local tax base. Although the increased construction activity in the region provides additional tax revenues, the sudden increase in construction needs, coming all at once, can still require a large increase in tax rates.

Besides the need for public facilities, the influx of construction workers may cause a housing shortage, competition for available land, and a general increase in prices. Construction workers typically have higher incomes than permanent local residents, so they can frequently buy up the more desirable properties. For those who cannot purchase existing houses, a typical approach is to bring in mobile homes and develop densely populated "mobile home parks." Zoning problems can arise as these developments spring up in formerly rural regions, and large areas of farmland may be lost to haphazard development.

Together with changes in housing patterns and public facilities, the construction of a large energy facility is likely to cause changes in population profile, especially in the areas of age, race, sex, religion, and educational background. Such changes historically have been resented by local residents, for they have a far-reaching impact. The increased density of population, together with the changes in the population profile, can cause significant increases in crime, drug abuse, public health problems, and need for welfare services.

Once the power plant has been built, the need for construction personnel decreases substantially. Only a few plant operators are needed, and the construction personnel move on to another project. This sometimes leaves the community with excess facilities that must be maintained. A general decline in land and housing values often results as demand for housing falls. The "boom times" created by the influx of construction revenues and workers disappear, often turning into a "bust." Severe unemployment, added mental health problems, and a deterioration of the tax base can leave the community worse off than it was before the construction of the power plant.

8.4.1 Providing for Impacts of Energy Projects

Although power-plant construction frequently results in local problems, it is possible to anticipate many of these problems and minimize their impact. According to a 1977 study by the U.S. Nuclear Regulatory Commission, the best results can be achieved if the following steps are carried out beforehand:

1. The community should assess its capability of handling the stresses caused by the power plant.
2. The community should develop its plans to meet the impacts of the power plant before the construction starts.
3. The state government and the utility should cooperate to help the community by providing needed financial and other support.
4. A joint planning commission should be set up to respond to unexpected impacts unique to the local situation.

In other words, it is essential to investigate, prepare for, and prevent possible problems before they get out of hand. Social impacts, like environmental impacts, are subtle in their initial stages. By the time they are felt, quite a lot of damage may already have occurred.

8.4.2 Fairbanks, Alaska: A Case Study

Community life can be changed both directly and indirectly by the introduction of energy production or conversion technology. Especially in rural areas, the changes can be rapid and significant.

For a fairly dramatic example of what energy development can do to a community, let us look at Fairbanks, Alaska, a small city whose character was changed significantly by the building of the Alaskan oil pipeline.

In the early 1970s, the enormous project of building a "trans-Alaska pipeline" was proposed. The project roused opposition from environmental organizations, who feared that irreversible ecological changes would be brought about by the project. However, after the project was completed, it soon became apparent that the social effects of the project were the most serious problems.

Perhaps the most immediate effect of pipeline construction was the infusion of enormous amounts of money into the Alaskan economy. The result was serious inflation of prices, which were already high compared to those in the "lower 48." Along with the money came thousands of workers, most from outside Alaska. They were flown in from as far away as Texas, attracted by large salaries and benefits. Soon Fairbanks became a boom town, with all the attendant problems.

Housing shortages developed as thousands of workers flocked to the tiny town, many bringing their families with them. "Outsiders," as they are called in Alaska, bought land and drove up real estate prices. Trailer camps sprang up across the landscape, and rents skyrocketed, reaching $1000 a month for tiny apartments or trailers. Some older residents, unable to pay the high rents, were forced to live on the streets, and others left the area.

Historically, boom towns are troubled by alcoholism and prostitution, and Fairbanks was no exception. Dozens of new taverns opened to cater to the pipeline workers. Alcoholism became widespread, and many native women turned to prostitution in order to pay the inflated housing prices.

With the tremendous influx of "outsiders," old-time residents found their simple way of life disappearing. The high-paying pipeline jobs often were not available to long-term residents. Many left the area, and others changed their way of life to cater to the well-paid newcomers. The frontier had disappeared.

Economically, there can be no question that Alaska is prosperous. The state government receives enormous annual royalties from oil companies—so much so that instead of paying state income taxes, Alaskans receive a share of the oil wealth from the state government. But this economic prosperity has its price. The population of the state increases every year as more "outsiders" move to Alaska, attracted by tales of readily available jobs. The state Chamber of Commerce sends notices to colleges and employment agencies warning that few jobs are available today, but thousands flock to Alaska despite the warning.

Perhaps the greatest social impact has been on the natives. Having tasted the fruits of modern technology and cheap energy, the native Alaskans are rapidly abandoning their self-sufficient way of life and their old social customs. They now live in mobile homes, drive snowmobiles, and listen to transistor radios. Under the Native Land Claims Settlement Act of 1971, the natives were given back the land they originally owned, together with millions of dollars in compensation. But along with the land and the settlement, they received an initiation into modern-day corporate economics. Some native corporations have abandoned their old ways of life and have signed lease agreements permitting the oil companies to carry out further oil exploration and development on tribal lands.

No two social situations are the same, and the Alaskan experience may not be characteristic of future energy resource areas. But similar social impacts can be expected whenever large-scale energy projects are undertaken in formerly undeveloped areas. Of the proposed new energy resources—shale oil, tar sands, uranium, wind, large-scale solar—the majority are most abundant in sparsely populated regions.

Today, when energy projects are undertaken, community needs are usually taken into account during the planning stages. But human societies are complex and unpredictable organizations. It seems likely that future energy projects, like the Alaskan oil pipeline, will offer communities a mixture of benefits and problems.

8.5 WORKER AND PUBLIC HEALTH AND SAFETY

Many forms of energy pose health and safety hazards to human beings and other life forms. These hazards may be very great or relatively insignificant, depending on the process or system, the amounts of energy and power converted, and the susceptibility of the individual. It is also important to consider the degree of exposure; the danger to the general public may be quite different from the danger to a worker. Let us look at the various stages in the development of an energy resource and identify the probable hazards, both to workers and to the public at large. Then we can look at a few specific energy technologies and see how worker and public health are affected.

8.5.1 Developing Energy Resources

Exploration. Exploration includes preliminary testing and evaluation of a resource deposit. In general, the stages of initial exploration do not present serious dangers to workers or the general public, although workers in some industries are threatened during these early stages.

Mining and extraction of raw materials. Hazards to workers are more serious during the process of resource mining or extraction than during the process of exploration. In most cases, the general public is relatively safe, except for those living in the immediate area.

Transportation. Electricity is sent through high-voltage lines, gasoline and heating oil are transported in trucks, and natural gas is transported in pipelines and LNG tankers (see Chapter 3). All these transportation technologies present hazards to transportation workers and the general public.

Construction of conversion facilities. The construction of power plants, or other facilities designed to convert the raw energy resource to a refined product, poses some risks to workers and the general public. The dangers are comparable to those associated with any large-scale construction projects.

Power plant operation. Once they are in operation, power plants and hydro projects present significant dangers, both to workers and to the general public. Workers are especially threatened by conditions inside the plant, including the following:

1. High-speed (and high-power) operation of machinery
2. High-temperature steam, fluids, etc.
3. Leaks of flammable materials
4. Equipment malfunctions (explosions, flying parts, etc.)
5. Presence of radiation or highly active chemicals at the plant site
6. In-plant air pollution caused by refining and combustion

Members of the general public face a somewhat different set of hazards, including the following:

1. Air pollution, especially from coal-fired power plants
2. Flooding downstream from hydro projects
3. Hazardous road conditions due to fog, smog, and icy conditions near a power plant
4. Releases of radiation and radioactive materials from nuclear power plants

Industrial and household energy systems. Not all energy resources are converted to electricity in large power plants. Machines, furnaces, boilers, and stoves convert significant amounts of energy and present a fairly serious hazard to the energy users.

Many house fires are caused by improper installation or operation of wood stoves, gas furnaces and stoves, kerosene heaters, and portable electric heaters. Toxic fumes from gas or kerosene appliances can also cause deaths or injuries. Steam leaks and exploding boilers can cause burns and other injuries. In fact, household energy systems are more dangerous, for the amount of energy converted, than any large-scale energy-conversion technologies.

8.5.2 Assessing the Risks

Looking at all the hazards described above, we can see that the production, distribution, and use of energy resources can present serious threats to workers and the general public. The effects vary widely, from relatively minor annoyances to major catastrophes such as serious coal-mining accidents. It makes sense to take these hazards into account as we assess the benefits and the costs of various energy resources. But it is difficult to assign a precise cost figure to an accident or injury. What is the financial value of a lost limb, or a premature death? Yet it is essential to assign *some* value to the losses suffered by victims of accidents and misuses of energy; otherwise, there is no clear evidence that such a value exists. In this spirit, we will look at a few examples and try to assess the risks, both qualitatively and quantitatively.

When we look at specific energy resources, we find that all present some dangers to workers and the general public. But each energy resource presents its own particular set of hazards. In this section we will look at the major energy resources and identify the most significant dangers represented by each energy choice.

Coal. Perhaps the greatest dangers occur during underground mining of coal. Black lung disease, mine roof cave-ins, and mine fires are all dangerous, potentially fatal occupational hazards. In the year 1981, coal-mining accidents caused 153 worker deaths nationwide. In 1947, 119 miners were killed in a single mine accident in Centralia, Illinois.

BOX 8.5.1 BASE POWER PLANT SIZE

For comparison purposes, researchers, together with other students of the health and safety impacts of energy technology, analyze energy-conversion systems using a base size of 1000 MW. This is a rather large power output, but not uncommon. Many coal and nuclear facilities, and a few large hydro facilities, have output power capacities in this range. Some energy uses do not involve conversion to electrical energy. In such cases, for comparison purposes, the delivered energy output is expressed in terms of equivalent electrical power.

Assuming a 75% capacity factor and 87.5% system efficiency (from generation to end use), a 1000-MW electrical power plant produces the equivalent of 19.6 trillion Btu per year. This, then, is considered the energy equivalent of one 1000-MW electric power plant operating for one year.

Using these ratings, it is possible to arrive at a rough estimate of the comparative danger to the workers from a number of different energy supply technologies. These estimates are presented in the following table.

Occupational Health Impacts of Energy Technologies[a]

System	Deaths	Workdays Lost
Coal		
Deep-mined	4.00	8700
Surface-mined	2.64	3090
Oil		
Onshore	0.35	3610
Offshore	0.35	3610
Imported	0.06	690
Natural Gas	0.20	1990
Nuclear[b]	0.15	270

[a]Data for a 1000-MWe power plant per year. MWe, megawatts of electric power.

[b]Because the health effects of radiation do not appear for many years, these figures may understate the hazard.

Source: Council on Environmental Quality, Energy and the Environment, 1974; Phillip LeBel, *Energy Economics and Technology* (Baltimore: Johns Hopkins University Press, 1982).

Black lung disease, a respiratory disease caused by the inhalation of coal dust, is a less dramatic but perhaps more serious cause of deaths among coal miners. According to the U.S. Occupational Safety and Health Administration (OSHA), three out of four coal miners have some form of black lung disease. An estimated 4000 coal miners per year die of this disease. By 1979, more than 480,000 miners and their survivors had received compensation for black lung disease. Total claims so far have exceeded 1 million.

Further indirect health effects may result from the air pollution (airborne particulates) caused by mining, surface water pollution from surface degradation, and groundwater pollution from underground mining operations. These hazards have the greatest effect on people living in the immediate vicinity of the resource deposit.

While workers in the coal industry are seriously affected by mining accidents and mining-related diseases, the general public is more seriously threatened by the effects of coal combustion. In coal-fired electric power plants, the major concern is sulfur oxide pollution. Among the health risks caused by this pollutant are respiratory and heart troubles, lung abnormalities, and higher general mortality rates. Trace elements from coal combustion also appear in the stack gases, and these (lead, cadmium, nickel, radon, mercury, arsenic, etc.), as well as combustion products such as benzopyrene and nitrogen oxides, can be hazardous due to their carcinogenic properties. Fly ash, soot, and other particulate matter, coupled with sulfur oxides and other smog components, have detrimental effects on human health. Clearly, the toll is large.

One study that has attempted to assess the damage from coal burning power plants is *The Ramsay Report: Unpaid Costs of Electrical Energy: Health and Environmental Impacts from Coal and Nuclear Power* (Baltimore: Johns Hopkins University Press, 1978). This study claims that a single 1000-MW coal-fired power plant causes the following injuries each year due to sulfur emissions:

7000 chronic respiratory-tract illnesses

60,000 cases of aggravation of heart–lung symptoms, mostly among elderly people

10,000 asthma attacks

8 deaths

These health hazards can be reduced significantly through the use of electrostatic precipitators, or "scrubbers." New coal plants are required to install pollution control equipment, and many of the installations are highly effective. This technology requires the disposal of large amounts of "sludge," but the sludge is easily contained. However, older power plants are not required to install scrubbers, and they will continue to emit large quantities of sulfur dioxide for many years.

Figure 8.5.1 The longwall system of underground mining is used to carve out large blocks of coal in a continuous process. Hydraulic jacks stabilize the roof and the conveyor system as coal is sheared from a 600-ft-long working face and transported from the mine. (*Source:* Bethlehem Steel Corporation.)

Oil and natural gas. Oil exploration presents some dangers to workers during the process of test drilling. Unexpected blowouts can occur due to unusually high pressures in the tested deposits. In general, though, these hazards can be minimized by proper precautions and preventive procedures. Workers on exploration rigs are also subject to danger from storms and hurricanes.

Blowouts are not common in producing oil wells. However, the threat to workers from hurricanes and other maritime storms continues throughout the production process. In February 1982, an accident on an oil platform off the coast of Newfoundland caused 85 deaths. Fires on producing gas and oil rigs are also a source of occupational danger to workers in the industry.

Transport of oil and gas can be hazardous both to workers and to the general public. Tank trucks burn and explode, causing injuries and deaths to drivers and to members of the general public. LNG tankers present the possibility of a major accident such as the one that occurred in Cleveland, Ohio, in 1944. In that accident, 135 people were killed and several hundred injured when a 150,000-ft^3 cylinder of LNG failed. Industry representatives claim that this accident set the LNG industry back 20 years. Less dramatically, pipeline breaks can cause breaks and hazardous interruptions of gas service, while oil spills cause land and water pollution, with a variety of resulting health problems.

The greatest danger from oil and gas comes when they are used by the general public. Gas appliances cause many house fires and deaths from fumes, and gasoline fires cause many injuries every year. But the automobile is clearly the most dangerous to the general public. In the United States alone, over 50,000 lives are lost every year due to automobile accidents. Millions are injured every year, sometimes permanently. This single cause of death and injury overshadows all other energy-related safety hazards, both in the number of casualties and in the degree to which the general public is affected.

Wood energy. Wood energy presents problems during the gathering of the resource. Many loggers are killed or seriously injured in logging operations. Householders suffer an even higher rate of injury from accidents involving chain saws, axes, splitters, and other woodcutting tools. Many muscle and back injuries are caused by individuals carrying heavy loads of firewood. These injuries are especially prevalent among the inexperienced.

Wood stove operation causes thousands of home fires every year. In most cases, the stoves are installed or operated incorrectly, although defective stoves cause a few fires every year. Chimney fires are quite common in areas where wood is burned heavily, and chimney fires can turn into more serious house fires. In other cases, owners leave inadequate clearance between wood stoves and combustible surfaces. Properly operated wood stoves are not appreciably more dangerous than other energy-conversion devices; however, the probability of improper installation or operation is very great.

In addition to the danger from fires, combustion of wood is believed to cause respiratory problems for the general public, especially in urban areas. In Oregon, for instance, it was determined that wood burning caused more air pollution than all other forms of domestic energy use combined. Some of the materials in wood smoke are suspected of causing cancer, while others may lead to asthma, emphysema, and other respiratory problems. More efficient stoves, combined with new technological devices such as catalytic converters, may help to alleviate this problem. However, most of the stoves currently on the market lack sufficient emission control devices, and homeowners have not shown any great enthusiasm for cleaner-burning stoves. This issue has emerged only during the past few years, so the total dimensions of the problem are still unknown. But the threat to public health and safety may be very great indeed.

Nuclear power. Because nuclear power is a new technology, and because the effects of radiation are not well established, nuclear power is by far the most controversial energy resource. The issues are complex, and public understanding of the issues is still incomplete. But feelings run high on both sides of the issue. For these reasons, we are spending more time on the safety aspects of nuclear power than we devote to the other energy resources. Yet we can do little more than suggest the general outlines of the nuclear safety debate. For more detailed information on the nuclear safety issue, readers should consult some of the books listed in the Further Readings section of Chapter 3.

The dangers of uranium mining are not as immediate as coal-mining hazards. But there is increasing evidence that uranium mining by-products, including radon gas, may cause cancers and other radiation-induced sickness. The tailings left behind by miners cause additional radiation hazards. In Grand Junction, Colorado, many house foundations were built of concrete with uranium mine tailings used as aggregate. The level of injury caused by these radioactive foundations has yet to be fully determined. On July 16, 1979, in a little-publicized but very serious accident, a dam broke in Church Rock, New Mexico, draining a tailings pond and contaminating the water supplies of several native American villages downstream. Again, the health effects may not be known for decades.

Construction of nuclear power plants does not pose a serious threat to the general public. Like construction of any power plant, construction of a nuclear plant causes some noise and visual pollution in the immediate vicinity of the plant. But during the past few years, especially since the accident at Three Mile Island, construction of a nuclear power plant has caused local protests and some psychological stress for the residents of the area. These psychological effects seem to be more serious for nuclear plants than for fossil-fuel plants, probably because nuclear power is a much more controversial issue than coal power.

Once nuclear plants are operating, the threats to health and safety are due primarily to accidental releases of radioactive materials. During normal power plant operation, radioactive emissions are held below federal guidelines, which are designed to ensure public safety. The health effects of such normal operation are unknown, partly because many radiation-induced diseases do not appear for decades after the exposure. But even when these factors are taken into account, the projected health risks from properly operating nuclear plants appear to be very low.

During maintenance and cleanup operations, workers in the nuclear industry receive higher-than-normal doses of radiation. In this case, there may be health effects. Several studies of workers in nuclear-related industries have shown a significant increase in mortality. These studies have been questioned by industry representatives, and some researchers claim that they have been harassed and lost their funding because of industry pressures. As a result, few reliable safety studies are currently available, and the few that have been done are highly controversial. However, there is no question that radiation and radioactivity, even at fairly low levels, can pose a danger to workers in the nuclear industry.

Perhaps the most dramatic threat from nuclear power plants is the danger of a catastrophic accident. The most publicized type of accident is a "meltdown," in which nuclear fuel escapes from its protective cladding and melts through the floor of the containment. Release of radiation is possible in such an accident; and under the worst possible conditions, a meltdown could have catastrophic consequences. Such a meltdown was avoided during the Three Mile Island accident, but some researchers claim that the plant was within one hour of a meltdown.

Although there is general agreement that an uncontrolled meltdown could cause terrible consequences, there is little consensus concerning the probability of such a meltdown. Large numbers of minor accidents have occurred at nuclear power plants, some involving fuel melting. But so far no accidents have gone completely out of control. Nuclear industry representatives claim that this shows the safety of nuclear plants; however, nuclear opponents claim that the industry has simply been lucky.

In 1975, to clear up the uncertainty surrounding the probability of nuclear accidents, the U.S. Nuclear Regulatory Commission issued a report that estimated the probability of a catastrophic nuclear accident. According to this report, commonly called the Rasmussen Report, the chances of being injured in a nuclear power plant accident were similar to the chances of being injured by a falling meteorite. However, the report was withdrawn in 1979 because of controversy over the methods used to estimate accident probabilities. To this date, there is no reliable estimate of the probability of injury from a reactor accident.

The final danger of nuclear power is the presence of long-lived and hazardous waste products. Waste elements such as radon, carbon-16, plutonium, and strontium-90 are produced by nuclear power plants. Failure to contain these wastes could cause release of radiation and radioactive materials to the general public. Possible health hazards include stillbirths, genetic mutations, cancer, and leukemia. Most researchers believe that the health effects of radiation are cumulative; this means that a high level of exposure during one period of time cannot be balanced by a lower exposure at another time. As in the case of worker health studies, public health studies are not yet sufficiently comprehensive to provide an exact estimate of the effects of low-level radiation releases. It is instructive to note, however, that the legally permitted levels of exposure to nuclear industry workers have been reduced from 15,000 mrem in 1948 to 5000 mrem in 1958.

The main difficulty with radioactive wastes is their longevity and the uncertainty concerning their toxicity. As in most areas of nuclear safety, studies on nuclear waste containment are contradictory and highly controversial. Some radioactive wastes have leaked from storage containers in the ocean and at the Hanford nuclear facility in Washington state. However, the leaking containers do not represent the state of the art in radioactive waste disposal. Future effects of waste disposal are unknown.

Hydro power plants. Although the power production facilities at a hydro installation present few hazards, the process is not entirely without dangers. The construction of a hydro dam involves a great deal of earth moving and requires the use of heavy equipment. Many dam construction projects have been marred by accidents, sometimes causing fatalities.

Once the project is completed, the danger to the public increases. Many dams have failed during periods of flooding, causing thousands of fatalities. Other hazards caused by flooding include significant losses of property. Dam construction technology has improved in recent years, and forecasting of dam failures is more accurate. But living downstream from a major dam is still a cause for concern, especially in times of high water.

In rare cases, as in Aswan, Egypt, the construction of an artificial lake behind the dam has increased the incidence of infectious diseases. Schistosomiasis, a disease carried by snails, has become much more serious since the construction of the Aswan High Dam. These kinds of effects are local and highly unpredictable, but they must also be included among the hazardous effects of hydro power projects.

Geothermal power plants. The health and safety hazards from a geothermal power plant include hydrogen sulfide, a foul-smelling and toxic gas. Irritation to the eyes and lungs is especially important. Another pollutant is the radioactive radon gas that may be released together with the exhaust steam from the plant. There is also a small probability of burns from ruptured steam pipes.

Geothermal power generation is a relatively new technology so no statistically reliable figures are available on the incidence of injuries and health hazards from geothermal power plants.

Other renewable energy resources. Renewable resources are not without their health and safety risks. Production of photovoltaic cells is still a relatively new technology, but many of the manufacturing processes require the handling of highly toxic materials. One type of photovoltaic cell uses arsenic as a component of the semiconductor material; and many types of cells use electrodes made of highly toxic metals. In manufacturing processes involving these materials, there is always a danger that water supplies can become contaminated through improper waste handling.

Installation and operation of renewable energy systems can be dangerous. Many renewable energy technologies are installed by do-it-yourselfers and other amateurs, often without sufficient safety factors. This can produce accidents and hazardous system failures. Wind turbine towers frequently collapse, sometimes causing property damage, personal injuries, and deaths. Homeowners are occasionally injured by falling from towers while servicing or installing wind-power equipment. Injuries are also common during the handling and installation of rooftop solar collectors, although the job fundamentally presents no more dangers than other rooftop work.

8.6 WHAT CAN BE DONE?

In the light of the numerous public and worker hazards presented by the production and use of energy resources, we may be tempted to dismiss the risks. After all, everything is dangerous, right? "When your time comes, your time comes." But awareness of the hazards need not lead to despair or indifference. It is possible to assess the risks involved in any energy enterprise and then determine whether the benefits of the technology are worth the risks. Only then can we be reasonably certain that we have acted neither timidly nor foolishly.

In assessing risks, we should not seek to avoid *all possibility* of risk. This accusation is frequently leveled at proponents of increased worker safety, usually without justification. Such claims only confuse the issue. No one has the right to demand a totally risk-free environment, and few responsible public-interest groups do so. But the public does have the right to ask that risks should be assessed and then minimized if possible. An example may help to clarify the issue.

During the late 1970s it was discovered that one U.S. auto manufacturer had produced automobiles with improperly mounted fuel tanks. During a court case, it was alleged that the manufacturer continued to produce automobiles with improperly mounted fuel tanks even after company executives were aware of the danger to the public. Apparently, the company decided that it would be too costly to correct the design defect. In this case, the public interest clearly demanded that the problem should be corrected after the company was aware of the problem. This is a far cry from insisting on a totally risk-free product.

Under pressure from labor unions, researchers, and other concerned parties, many laws and regulations have been developed to protect workers from hazardous products and practices. For example, the Mining Safety and Health Administration (MSHA) regulates safety conditions in the mining industry, and the Occupational Safety and Health Administration (OSHA) monitors safety standards on other industries. These agencies have their critics, but it is generally agreed that they have helped to reduce worker injuries and fatalities.

Similar agencies have been developed to monitor public health and safety, although their functioning has frequently been highly controversial. The Nuclear Regulatory Commission oversees the safety of all nuclear power plants built or operated in the United States. The Environmental Protection Agency was created in 1970 to reduce health hazards to the general public by monitoring and protecting the nation's environment. State and local agencies and zoning boards regulate local energy development for the general well-being of the community. Although these agencies do not always function perfectly, there can be little doubt that they provide an important public service by minimizing the dangers associated with energy projects and other technological enterprises.

8.7 ENERGY AND HUMAN BEHAVIOR

Does the availability of plentiful and cheap energy change our values or our behavior? If so, how much and in what direction? We have all observed that when gasoline is cheaper, we tend to be more careless about our consumption. But this is only the most obvious change in our behavior. In the final sections of this chapter, we will look at some of the subtler, and perhaps more far-reaching, social effects of energy affluence.

There can be no question that abundant energy is a tremendously valuable asset for any society. However, these advantages have their social price. We are all well aware of the many advantages, but it is equally important to be aware of the costs.

Adequate energy availability in a society usually brings about general conditions of material prosperity. Manufacturing increases, transportation is more widely available, and the general standard of living increases. Increased energy availability also provides important social benefits: better nutrition, greater availability of services, better health, and better education. When there are sufficient energy resources, more time can be spared for other pursuits such as music, art, entertainment, recreation, and hobbies.

However, availability of leisure time does not necessarily lead to a better use of that leisure time. When there is enough energy available for all purposes, people are less inclined to discriminate between valuable and less worthwhile pursuits. In the most-energy-affluent societies, people spend hours every day watching programs on television, even though the same people complain about the poor quality of television programs. Consumer goods are abundant, but shoddy workmanship is common. People run out of time because, with more resources at their disposal, they want to do more and more things. Because it is possible to travel long distances quickly, people join more organizations, attend more meetings, and find themselves with less time on their hands. Paradoxically, even as abundant cheap energy increases people's leisure time, it appears to reduce their capacity to use their leisure time effectively.

High energy availability can also cause a loosening of family structures. Inhabitants of energy-affluent societies often have weak family ties because of greater personal mobility. Young people leave the family home and move to other areas to find work. This creates smaller family units. When older people become infirm, their children have often moved far away. Since there is often no room for older people in their children's small household, the elderly are placed in institutions. In less-energy-rich societies, family ties are much closer, and the young feel a strong obligation to care for the elderly.

When people relocate every few years, there is little time to develop close friendships. At the same time, people's circle of acquaintances tends to widen. Visits to close friends and relatives are less necessary. Often an occasional telephone call is the only contact between distant friends or relatives.

Not all people in energy-rich societies find themselves isolated from friends and family. In many cases, increased mobility can be the means of bringing together families that are separated by vast geographical distances. Family relationships in technologically developed cultures are often less formal than in less developed societies. But in general, by looking at cultures across the world, we find that the effect of abundant energy can be a weakening of human relationships.

8.8 ENERGY AND HUMAN VALUES

In energy-affluent societies, energy and technology are often expected to solve every problem. For example, the United States and the USSR are engaged in a technological arms race. Rather than seeking to resolve their differences by negotiation and international diplomacy, both countries place their hopes in the possibility of developing technologically more advanced weapons. This suggests that leaders of both countries have greater confidence in technology than in nontechnological approaches to problem solving. On a more local level, when a company is having difficulties with production, the company may invest in new equipment or consider installing robotized machinery. In many cases, the new machinery does not solve the problem, which may have been caused by inefficient management. but again the tendency is to look for a technological solution.

Together with this dependence on technology comes a tendency to devalue humanistic concepts and nontechnological achievements. In most industrialized countries, scientists and engineers are paid more highly than comparably educated professionals in other fields. Teachers, artists, and ministers are notoriously underpaid, and their opinions are given less value in national policymaking than the opinions of technical experts. Thus it seems that energy affluence brings with it a shift in the values we assign to different human pursuits.

Energy is sometimes described as a "slave" that we can use to fulfill all our personal and social goals. Some energy experts speak of energy in terms of the number of human slaves it is capable of replacing. Similarly, an engine is rated in terms of "horsepower" (although 1 horsepower is much more than the power of an average horse). But this way of describing the value of energy overlooks the versatility of human and animal labor. A 50-hp engine cannot do the same range of tasks as 50 horses; for example, 50 horses can perform 50 different light tasks at once, but a 50-hp engine can generally do only one thing at a time. Similarly, the value of human labor does not lie in the ability to lift heavy burdens but in the ability to do a variety of complex tasks. In the face of this criticism, proponents of energy development point out that historically, human labor has been used in degrading applications. They claim that the harnessing of energy resources liberates humanity from the degrading, meaningless labor which characterized much of past civilization.

In fact, both perspectives offer insights into the social implications of energy technology. Certainly, it is desirable to eliminate repetitive, monotonous work. In principle, this should free human energy for more useful pursuits. But by replacing many human functions with energy-using machines, employers not only create unemployment; they may also unknowingly limit their manufacturing techniques to those processes that require only accurate mechanical repetition. If human workers are assigned to more imaginative, useful tasks, the introduction of energy-intensive machinery will indeed be a liberating application of technology. But if the net result is simply a substitution of mechanical for human labor, no real liberation has taken place: only a substitution of one kind of slavery for another.

8.9 PROSPECTS FOR THE FUTURE

We have come a long way since our time as cave dwellers. The energy forms of fire, animals, wind, wood, water, coal, oil, gas, and gravity have all been employed in our service. Energy has been good to human beings. Indeed, most of what we identify as civilization, most of our social activities, depend directly on the availability of energy-conversion technology. Today, when energy is more widely available in greater quantities than at any time in recorded history, the possibilities for social growth and development are also greater than ever before.

Given a few days and a few hundred dollars, one can visit a friend or relative all the way across a continent, or across an ocean. We can visit remote societies in a matter of hours or days—journeys that would have taken months or years a few centuries ago. Communications systems made easy and accessible with our modern energy-conversion technologies can bring information to our office or living room from every corner of the world. Computer systems and information storage devices give us the capacity to handle unprecedented amounts of information.

The development of energy-conversion technology has also contributed greatly to the liberation of humanity. In an important sense, we are all free to become whatever we wish, and humanity's control of energy is largely responsible for giving us that freedom. A vast, almost unlimited future lies ahead of us. As the Roman poet Marcus Manlius wrote 2000 years ago:

> Nothing can withstand the power of the mind. Barriers, enormous masses of matter, the remotest recesses are conquered; all things succumb. The very heaven itself is laid open.

Unfortunately, despite all these possibilities, we also face a variety of unsolved social problems. Technology has never been and never will be an answer to all our problems. Social inequalities, for example, cannot be solved by the simple application of energy technologies, however liberating those technologies may be in principle. In fact, the very potential of technology can multiply the inequalities that exist between peoples: the more fortunate reach heights of affluence never before imagined, while disadvantaged people continue to live under conditions of desperate poverty. The fault does not lie with technology alone, although some may blame the technology. Human errors, like human achievements, are multiplied many times over by technological systems. But let us remember that it is we, not the technologies, that are at fault.

If the problems lie within us, so do the solutions. As we focus on the social and cultural implications of energy technology, we can see the potential for great progress or for great harm, often both at once. If this chapter teaches only one thing, it should be that the social and cultural side effects of energy technology are integral to the technologies themselves, and in some sense implied within them. Technology itself cannot help us avoid these consequences; for that, we must depend on our own wisdom. Wisdom is the source of all technology and of the values that can teach us the proper use of our technology.

STUDY QUESTIONS

1. Describe the relationships between energy consumption and population growth. Which is the cause of the other? What other factors are connected to both energy consumption and population growth?
2. Distinguish between standard of living, quality of life, and life-style. How do they affect energy use?
3. Does higher standard of living necessarily imply higher energy consumption? Explain.
4. Do you agree with the Club of Rome's analysis of the limits to growth, or do you agree with the technocrats? Explain your choice.
5. Have you witnessed the construction of an energy project in your community? What social impacts did you notice?
6. How can a community avoid the adverse impacts of energy projects? Have you seen any of these measures in practice? How effectively did they work?
7. How effectively do such abstract measures as gross national product, standard of living index, and per capita income measure social and human well-being?
8. Does the availability of large quantities of energy tend to standardize human behavior, cultures, and values? Or does it lead to greater individual freedom and pursuit of personal interests?
9. Why do some energy analysts refer to energy as a "slave?" Do you agree with this way of speaking? If not, can you think of a better comparison?
10. In your judgment, where does the fault lie for the social problems and inequities that characterize the late twentieth century: in technology, human behavior, or social systems? What steps should be taken to remedy the situation?
11. If you are currently working or have worked in the recent past, analyze the energy-using devices at your workplace. What safety hazards do those devices present to workers or the public at large? What steps could be taken to reduce the hazard?

FURTHER READINGS

BECKWITH, BURNHAM P. *The Next 500 Years.* New York: Exposition Press, 1968.

COLLINGRIDGE, DAVID. *The Social Control of Technology.* New York: St. Martin's Press, 1982.

DORF, RICHARD. *Technology, Society, and Man.* San Francisco: Boyd & Fraser, 1974.

Environment and Health. Washington, DC.: Congressional Quarterly, 1981.

HALEVI, MARCUS, and KENNETH ANDRASKO. *Alaska Crude: Visions of the Last Frontier.* Boston: Little, Brown, 1977.

MEADOWS, DONELLA H., ET AL. *The Limits to Growth: A Report for the Club of Rome's Project on the Predicament of Mankind.* New York: Universe Books, 1972.

TAPINOS, GEORGE, and PHYLLIS PIOTROW. *Six Billion People: Demographic Dilemmas and World Politics.* New York: McGraw-Hill, 1975.

WILSON, RICHARD, ET AL. *Health Effects of Fossil Fuel Burning: Assessment and Mitigation.* Cambridge, MA.: Ballinger, 1980.

WINTER, JOHN V., and DAVID CONNER. *Power Plant Siting.* New York: Van Nostrand Reinhold, 1978.

9

Political Analysis

In modern industrial societies, the distinction between government and the private sector is not always a clear one. Through the powers of licensing, rulemaking, taxation, and granting of research funds, government has a profound influence on all areas of commerce.

This is especially true in the field of energy production and use. Often energy resources are found on public lands, and government has a major role in the development and use of those resources. Other resources have significant strategic importance. Finally, and perhaps most important, the government has an interest in regulating the supply of basic public utilities such as electricity and natural gas. In this chapter we examine these different areas of government involvement.

9.1 ORGANIZATION OF THE ENERGY INDUSTRY

The energy industry can be owned, operated, or controlled in a variety of ways. Public discussion tends to center around public and private ownership, as though those were the only available options. In fact, there is a wide range of possibilities for ownership and control of facilities for producing and converting energy resources (see Table 9.1.1). Let us look at the most common types of energy institutions.

TABLE 9.1.1 Public and Private Sectors in the Electric Industry

Category of Utility	1980 Energy Production (billion kWh)	Percentage of Total
Investor-owned utilities	1786	78.0
Municipal utilities	87	3.8
Federal administrations	235	10.2
Cooperatives and others	182	8.0
	2290	100.0

Source: Statistical Abstracts of the United States.

9.1.1 Unregulated, Privately Owned and Operated Energy Companies

These energy companies are owned by investors, stockholders, or in some cases by partnerships or private individuals. They are operated by a manager or a board of directors. There is a wide range of companies in this category, from huge multinational companies such as Exxon or Westinghouse to tiny one-person firewood businesses, and everything in between. These companies are not limited in the percentage of profits they make, the salaries they pay their employees, or the ways in which they invest their funds. Like any company, they must pay taxes and obey the laws regulating environmental pollution, worker safety, and so on. But they are free to buy other companies, merge with other companies, change their area of business, or even go out of the energy business altogether, with little or no government control. As long as they obey the basic laws of the land, these companies are free to pursue their profits however they choose.

9.1.2 Regulated Private Utilities

These utility companies are known as investor-owned utilities (IOUs). Like unregulated private companies, these are owned by stockholders and operated by managers or boards of directors. But the important difference between the two types of companies is that the IOUs are legal *monopolies*. They are assigned exclusive territories in which they are the sole distributors of electricity or natural gas.

The concept of a guaranteed monopoly is not entirely consistent with the principles of supply and demand, but it was developed to avoid duplication of facilities. For example, suppose that there were three different gas companies, each with its own system of natural gas pipelines. Construction and maintenance of these pipelines would create great disruptions in normal transportation and commerce, and the extra cost of maintaining the extra pipelines would draw off resources that could be used for other purposes. This is why companies such as electric utilities and phone companies are granted regulated monopoly status, and other companies, such as oil companies, are not.

One obvious drawback of a legal monopoly is that customers need to be protected from overcharges. If there were only a single unregulated electric company serving a city, it could charge extremely high prices. Customers would be unable to choose a different electric company, and the company could make enormous profits with no risk whatever. To prevent such practices, utilities are allowed only a limited rate of return, generally somewhat lower that the rates of return available to unregulated companies. At the same time, the utilities are *guaranteed* that "fair rate" of return, typically about 14%. Without these guarantees, utilities would not be able to attract investors.

Because of the complexity of regulating investor-owned utilities (IOUs), most states maintain public service boards or commissions to establish and enforce utility policies. Rate increases, capital expenditures, and profit margins are all subject to public hearings and complex decision-making processes. Many officials of IOU corporations object to the hearing process, which can be very time consuming. But this is part of the price that utility companies pay for the advantages of regulated monopoly status.

9.1.3 Public Utilities

Many municipalities or city districts are served by public utility companies rather than by investor-owned utilities. Unlike unregulated private companies or investor-owned utilities, these companies are operated on a strict *not-for-profit* basis. Production and distribution facilities (power plants, transmission lines, pipelines, etc.) are publicly owned and financed through bond issues underwritten by the local governmental authority. The utility is run by hired managers who are accountable to the elected officials of the appropriate governmental agency.

Although most public utilities are owned and operated by municipal governments, there are also a few regional publically owned energy organizations such as PASNY (Power Authority of the State of New York), APA (Alaska Power Administration), TVA (Tennessee Valley Authority), BPA (Bonneville Power Administration), SWPA (Southwest Power Administration), SEPA (Southeast Power Administration), and the Bureau of Reclamation. TVA and BPA were organized during the 1930s to develop the hydroelectric generating capacities of the Tennessee and Columbia Rivers. Since then, TVA has begun generating electric power using other types of power plants, both nuclear and fossil.

Today, there are over 2000 public utilities, although not all of them own generating capacity. About 700 of these public utilities generate at least some electric power; the rest buy and sell electric power from other companies. The federal energy agencies listed above (TVA, BPA, SWPA, and the Bureau of Reclamation) sell power to many of these public utilities at *preferential* rates. Federal law dictates that power must be sold to public utilities at lower prices than to investor-owned utilities.

9.1.4 Cooperatively Owned Utilities

Approximately 1000 cooperatives are involved in the electric power business. They are owned by their customers and operated by managers who are elected by the customers. Electric cooperatives do not make a profit. If any funds are left over after operating expenses have been met, they are returned to the customer. Because of their nonprofit nature, and thanks to low-interest loans, rural electrical coops have been able to supply electricity to remote rural areas at costs comparable to, and sometimes even lower than, the costs of electricity provided by investor-owned utilities.

The Rural Electrification Administration (REA) was created in 1935 under President Roosevelt. In 1936, the Norris-Rayburn Act extended its life by 10 years. The principal objective of the REA was to make low-interest loans available to rural cooperatives so that rural Americans could be served with reasonably priced electrical power. Before the creation of the REA, only 11% of U.S. farms were electrified. Private utilities were not interested in serving rural areas due to limited profitability. But with help from the REA, electric cooperatives brought electric power to virtually all farms in the United States.

Variations of these basic types of utility companies are possible to make the best of a given regional situation.

9.2 PUBLIC VERSUS PRIVATE OWNERSHIP

For a number of years, there has been a fierce debate between advocates of public and private energy companies. No definitive answers have been reached in this debate; rather, it seems that different forms of ownership work well for different situations.

9.2.1 The Case for Private Utilities

Advocates of private ownership claim that private utilities are a logical development of the American capitalist system. They argue that most sectors of the U.S. economy are in private ownership, and thus it makes sense for energy companies to be privately owned as well.

One important difference between publicly and privately owned energy companies is that privately owned companies make a profit. Proponents of private energy companies argue that profitability provides an incentive for efficient management and good service. They claim that government bureaucrats have no incentive to provide efficient service, and thus many publicly owned enterprises do not function well. The postal service is often cited as an example of a governmental service that has not always provided efficient service at a low cost.

9.2.2 The Case against Private Utilities

Opponents of privately owned utilities believe that public utilities provide better service at lower costs than do profit-making investor-owned utilities. They base their argument on three basic claims:

1. Because private utilities need to maintain a profit margin, they must charge more for power than public utilities, which do not earn a profit on the customers' payments. Thus, if all other costs are equal, a privately owned utility will charge higher rates than a public utility.
2. When utility rates are set for private utilities, the rates include a guaranteed profit on the company's capital investments. No profit is permitted on purchased power. This means that the larger a company's capital investments, the larger its profits. Critics claim that this makes private utilities tend to favor large capital projects regardless of the need for such projects.
3. To undertake major capital investments, regulated private utilities must convince public service boards of the need for the projects. If electrical consumption is increasing, a utility can present a convincing case. Critics argue that this leads private utilities to encourage excessive consumption rather than conservation. Bulk rates for large power users are one way that utility companies encourage increased consumption.

9.2.3 The Middle Ground

Although both sides make persuasive arguments, neither has been able to establish the clear superiority of its point of view. In some cases, public utilities have been significantly more efficient that privately owned companies. But in other cases, IOUs have been able to benefit from cost-cutting measures and provide good service to the public at a low cost. Private and public utilities alike have traditionally offered lower prices for large users, and both are now reforming their rate structures. Both public utilities, such as TVA, and privately owned utilities, such as Southern California Edison, are beginning to invest in conservation and renewable energy resources. So it seems that public or private ownership is not the real issue; rather, we should be concerned with the management and control of energy companies.

It seems indisputable that some form of public *control*, although not necessarily public ownership, is beneficial if utilities are to assist in developing a coherent and enlightened energy policy. In the final analysis, control is at least as important a concern as ownership. Investor-owned utilities are owned by shareholders, but they are controlled, at least in major policy areas, by public service commissions.

Without carefully considered policies on production, pricing, and distribution of energy resources, the erratic patterns of the past are likely to continue: alternating periods of shortage and surplus, with unstable prices, changing forecasts, and a lack of trust on all sides. Public control of energy resources is not especially popular, but it has evolved in response to very real needs. The current outlook suggests that energy production and distribution will increasingly become a matter of public policy in the United States as it is in other noncommunist countries, such as Great Britain, Japan, Israel, India, and Canada.

9.3 PUBLIC CONTROL OF ENERGY RESOURCES IN THE UNITED STATES

Even though the U.S. economy is based on free trade, dozens of government agencies, commissions, and rulemaking bodies have been set up to regulate various industries. In many cases, the industries themselves have promoted regulation.

The development of nuclear power was carried out with the active assistance of the Atomic Energy Commission (which was the forerunner of the present-day Nuclear Regulatory Commission). The U.S. Army Corps of Engineers has been instrumental in the design and construction of hundreds of power-producing dams all over the country. Public lands have been leased for gas and oil exploration since the nineteenth century. So there is certainly nothing new about governmental involvement in energy exploration and production.

In recent years, concerned citizens and energy enthusiasts have been bombarded by a series of initials: DOE, FEA, ERDA, AEC, NRC, TVA, BPA, EPA, NFS, FPC, and PUC are a few examples. All these intitials stand for various government agencies that oversee and control aspects of energy production and distribution. We will briefly describe the role these agencies play in the functioning of the U.S. energy industry.

1. The *Federal Energy Administration* (FEA) was created in mid-1974, after the first Arab oil embargo. Its purpose was to develop and direct a realistic and well-coordinated national energy policy. President Nixon created the FEA as part of his "Project Independence" program. The FEA was given the authority and responsibility to assure equitable distribution of energy resources to various sectors of the economy and all parts of the nation. In conjunction with FEA, President Nixon developed policies such as the 55-mph speed limit, bans on Sunday sales of gasoline, and standby gasoline rationing.

2. In 1975, the *Energy Research and Development Administration* (ERDA) was established to coordinate energy research in six major areas:

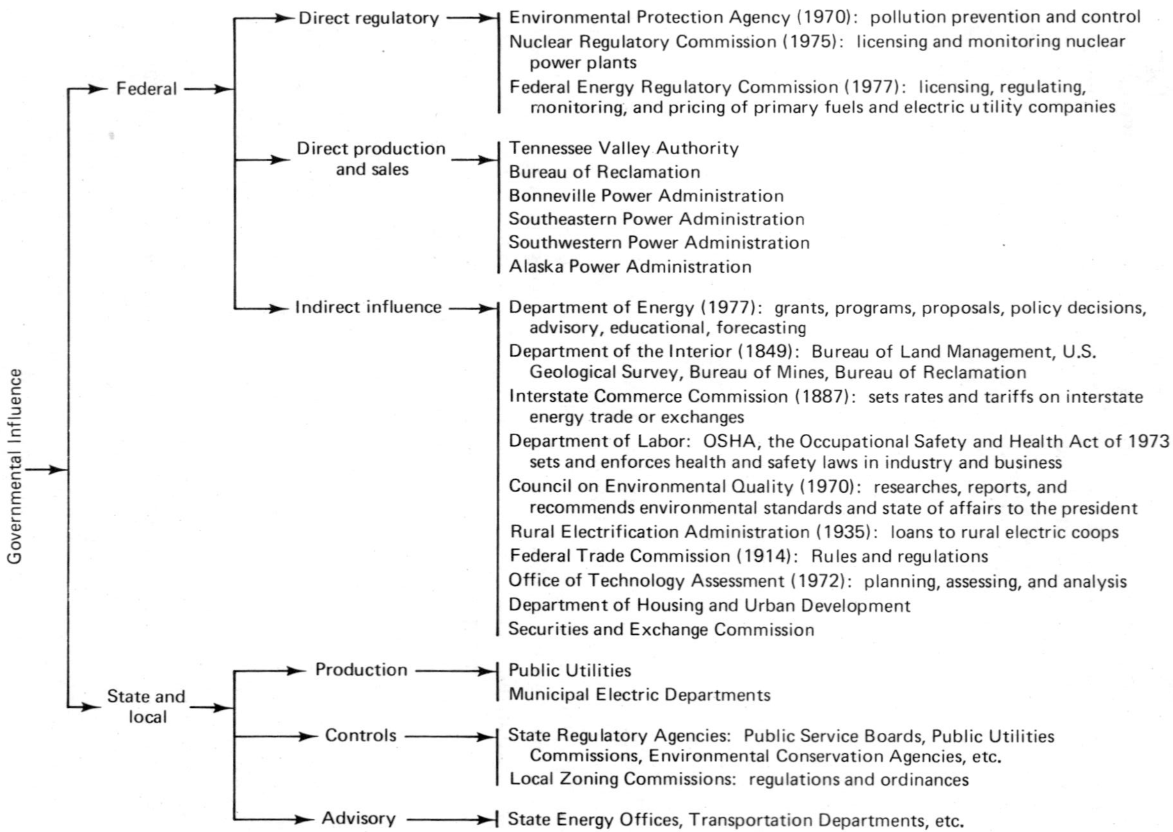

Figure 9.3.1 Governmental influence on the energy industry.

a. Nuclear power
b. Fossil fuels
c. Solar, geothermal, and advanced energy systems
d. Defense
e. Health and environmental protection
f. Conservation

From 1975 until the creation of the Department of Energy in 1977, ERDA remained the main agency for promoting and overseeing energy research in the United States.

3. Although the Federal Energy Administration and the Energy Research and Development Administration were fairly effective in dealing with the immediate problems caused by the Arab oil embargo, many people felt that long-term solutions required a more unified approach to energy policy. So, under President Carter, the main research and promotional agency (ERDA) was combined with the main regulatory agency (FEA) to produce a new cabinet-level agency, the *Department of Energy* (DOE).

The Department of Energy immediately assumed a significant role in determining U.S. energy policy. By coordinating regulations, federal expenditures, and promotional activities, the Department of Energy has been able to work in a more focused way to promote national energy goals.

When President Reagan took office in 1981, his original intent was to abolish the Department of Energy. However, this threat to the department soon disappeared. Funding for energy projects was cut, especially in the area of conservation and renewable energy resources, but research funding for nuclear power was increased substantially. There appear to be no current plans for abolishing the Department of Energy.

4. The *Nuclear Regulatory Commission* (NRC) was created in 1975 to solve a dilemma. The old Atomic Energy Commission (AEC) had been charged not only with regulating nuclear energy but also with promoting it. This resulted in charges of conflict of interest and mismanagement. Opponents of nuclear power claimed that the promotional functions of the agency overshadowed the regulatory functions. These critics pointed out that the AEC had never denied a permit to any proposed nuclear power plant.

In 1975, the two functions of the AEC were divided. Promotional responsibilities were assigned to the Energy Research and Developmental Agency (ERDA) and later to the Department of Energy (DOE), while the regulatory functions were turned over to the Nuclear Regulatory Commission.

The NRC has the authority to issue and revoke licenses, levy fines or penalties for faulty operations and procedures, and even to order a shutdown of power plants in extreme circumstances. Prior to 1984, the NRC had never denied an operating license to a nuclear power plant, and critics argued that the NRC still saw itself as a promoter of nuclear power. But following the Three Mile Island accident in 1979, the NRC began taking a more critical attitude toward existing and proposed nuclear power plants. Some operating licenses have been suspended or denied, and at one power plant (the Zimmer plant in Cincinnati, Ohio), construction was halted because of serious safety and quality control problems. The NRC's performance is still controversial, but now the agency is attacked by representatives of the nuclear power industry as well as by critics of nuclear power.

5. The *Federal Power Commission* (FPC) is responsible for licensing privately owned hydroelectric power plants. The FPC plays a key role in decisions affecting total supply of electric power and the rate at which it is sold. Thus FPC regulates some activities of private electric utilities (IOUs), including interstate rates, accounting procedures, and so on.

6. The *Department of the Interior* (DOI) is the nation's principal conservation agency and is responsible for most of the nation's public lands and natural resources. The agency assesses energy and mineral resources and develops some of those resources while leasing other resources to the private sector. The U.S. Geological Survey, the Bureau of Reclamation, the Bureau of Mines, and the Office of Surface Mining are among the DOI's energy-related divisions.

7. The *Environmental Protection Agency* (EPA) is not strictly an energy-related agency, but many of its decision-making procedures have a significant effect on energy development. The agency requires the preparation of Environmental Impact Statements (EISs) before energy projects can be undertaken. In some cases, utilities have been forced to make design changes as a result of Environmental Impact Statements. For example, a major revision of the Alaskan Oil Pipeline was required as a result of the controversial 3500-page Environmental Impact Statement. The pipeline was not buried as planned, and underpasses were provided for animal migrations. Utility companies sometimes argue that environmental regulations cause time delays and increase power plant costs. But few would argue with the need to consider environmental concerns in the process of developing energy facilities.

8. The federal *Securities and Exchange Commission* regulates the corporate structure, financial dealings, mergers, and other business activities of major energy companies. Although the SEC does not directly establish energy policies, it has a significant influence on the structure of the energy industry. For example, in the early 1980s, several small oil companies became the targets of financial takeover attempts by larger oil companies. Some of these takeover attempts were blocked by the SEC, thus ensuring a somewhat more diverse range of oil companies in the United States. Without the SEC and its regulations, monopolistic structures could have arisen similar to the Standard Oil Trust of the late nineteenth and early twentieth centuries (see Section 3.3.1).

9. Many states and localities in the United States operate energy offices or agencies that establish and carry out energy policies at the state and local level. They also act as sources for dissemination of energy information, and in some cases they provide energy conservation services and fuel cost assistance to lower-income families.

Public Utilities Commissions (PUCs), *Public Service Commissions* (PSCs), and *Public Service Boards* (PSBs) are state agencies that control or regulate both public and private utilities. These agencies set the rate structures that utilities charge their customers. They also regulate other activities of IOUs, which affect investments, rates of return, the safety of systems, the development of all types of power plants, and so on. PUCs, PSCs, and PSBs have a strong and direct effect on energy production and development.

9.4 ENERGY POLICYMAKING

Energy supply and demand do not change instantaneously. The supply and consumption systems both have natural delays or "time lags." Power plants take years from conception to delivery; and even after many years, completion is not assured. During the early 1980s, more than 30 nuclear power plants were cancelled *after* construction was already under way. Oil fields, coal mine expansions, and hydroelectric projects all take time to put in service. Even small-scale wind and solar projects take time to come on line.

On the other side of the energy equation, the demand situation can be equally slow to respond. For example, the average fuel economy of American-made automobiles is expected to increase to 27.5 mpg by 1985 (from 18 mpg in 1978). However, this does not mean that the average automobile on the road will be capable of achieving this kind of gas mileage. Older cars die hard, and the less-affluent segment of the population cannot afford the newer, more efficient cars because of their inability to raise the necessary capital. As a result, the old gas-guzzlers may be on the road for a while yet.

Some theorists believe that changing demand patterns is a quicker process than changing supply options. Although it takes many thousands of individual purchasing decisions to cause a significant change in demand patterns, each decision requires only a short period of time to be carried out. For example, it takes only a few days for a family to make a decision about purchasing a new car, furnace, or air conditioner. If a large number of consumers make similar decisions, the overall effect can be very significant. Thus, if energy conservation becomes desirable at the consumer level, the cumulative impact of individual decisions can grow very rapidly. For example, in less than five years, the majority of homeowners in New England switched to partial or total use of wood heat. It would have taken over a decade to produce an equivalent increase in the nation's supply of nuclear-fueled electricity.

Most energy analysts agree that conservation is the most effective short-term measure for balancing energy supply and demand, both because of its low cost and because it is very fast compared to most supply options. But this does not mean that new energy resources are not worth developing. In the long run, it will be desirable to have a very broad range of energy resources to draw on, because future changes in energy supply and consumption are extremely difficult to predict.

In the past, energy planning has usually been done in a crisis atmosphere. The emphasis has always been on how to meet a current shortage or provide for projected consumption over the next decade. Against this backdrop, it has been difficult to develop longer-term, more fully integrated strategies. Long-range energy policy often consisted of science-fiction fantasies or projection of current trends into the indefinite future. But increasingly, it has been recognized that long-term energy policymaking depends on the immediate, short-term approaches we choose today. Thus energy policymakers are recognizing the need to plan the transition from short-term to long-term solutions.

9.4.1 Elements of Energy Policymaking

Basically, energy policymaking can be broken down into five stages:

1. *Energy forecasting.* This mathematically complex process requires foresight as well as a thorough knowledge of current trends. Computer modeling procedures can produce highly detailed predictions; however, those predictions are no better than the assumptions that underlie them. For example, early predictions of energy use were based on a simple projection of current trends. As a result, projections for the early 1980s proved to be much higher than the actual energy consumption. More sophisticated forecasting techniques are being developed today, taking account of changing energy resources, price fluctuations, and political conditions.

2. *Assessing energy supplies.* At present, the United States is experiencing a gap between domestic energy production and energy consumption. Since the 1950s, that gap has been bridged through the use of imported oil. But the Arab oil embargoes of the 1970s showed the unreliability of oil imports as a method of meeting energy needs. Other possibilities are now being investigated: synthetic fuels, renewable energy supplies, and advanced conservation technology.

3. *Limits on energy production.* In the past, it was assumed that technological developments would lead to ever-increasing availability of energy resources. New oil and gas discoveries, improved coal-mining techniques, and the development of nuclear fission all encouraged the assumption that energy would always be available in ever-increasing quantities. Today, energy policymakers recognize the limits of energy resources, as well as the time lag inherent in the construction of any new energy facilities.

4. *Ownership, control, and accountability.* Ownership of energy resources has traditionally been connected to land ownership. Landowners have been free to develop energy resources on their land, or to sell the energy development rights to other individuals or corporations. This has favored the rapid development of nonrenewable energy resources. However, of the new energy resources being looked at today, most are found on public lands, major rivers, or ocean beds. There is a strong national and international debate concerning who should receive the financial benefits of these public resources.

Recent trends have been toward wider ownership and development of energy resources. For example, the Public Utility Regulatory Policies Act of 1978 (PURPA) requires public utilities to purchase electric power from small producers at a price comparable to the cost of replacement power. This means, for example, that the owner of a small wind generator can be sure of a steady market for any excess power that the installation may produce. This has encouraged the development of thousands of small-scale wind, hydro, and photovoltaic systems owned by individuals or small companies.

5. *Research, information, and promotion.* The government's role in the development of new energy resources has gradually increased over the years. In the early twentieth century, the government's role was limited to regulating monopolies. Today, however, the government uses a variety of methods to promote and control energy development. Tax incentives, subsidies, rate structures, and penalties can be applied to achieve the desired energy production goals. For example, a nuclear power plant is eligible for tax investment credits, special insurance treatment, accelerated depreciation, and other favorable tax treatment. In addition, the operator benefits from large federal expenditures on research and development, fuel enrichment, and waste disposal. Some energy analysts claim that nearly half the cost of nuclear power development has been underwritten by the government. Most other energy resources also enjoy various direct and indirect subsidies.

Besides these broad areas of concern, energy policymakers must be concerned with the specifics of implementation: guidelines versus regulations; tariff policies versus import restrictions; distribution of subsidies among the various energy industries; establishment of agencies to direct energy policymaking. Energy policymaking is no longer a simple task.

9.4.2 Policy Alternatives

At present, U.S. energy policy is principally concerned with the problem of ensuring adequate supplies while reducing dependency on foreign sources of energy. Although the range of energy policies is theoretically unlimited, most policymakers favor one or a combination of the following alternatives:

1. *Deregulate energy production.* In many ways, a free marketplace is the ideal regulator. If demand exceeds supply, prices will rise, demand will be reduced, and production will increase. No special rules are required, no lobbyists can influence the decision makers, and no politicians can change the rules in order to gain popularity.

Unfortunately, the marketplace is not free. As was demonstrated by the case of the Standard Oil Trust, large corporations exert a disproportionate influence on the national and world economy. But even in a perfectly balanced marketplace, short-term profitability can override long-term considerations, leading to rapid depletion of energy resources and ultimately creating serious shortages.

Reliance on market forces will not necessarily reduce dependence on imported sources of energy. In fact, there is evidence to suggest that American dependence on imported oil was caused by the greater profitability of imported oil. Prices at the gas station for domestic and imported petroleum were the same, but the production costs of imported petroleum were much lower. As a result, virtually all the major oil companies shifted their emphasis from exploration in the United States to development of overseas production capability.

2. *Increase production of nonrenewable resources.* Given enough incentives and time, domestic energy production can be increased substantially. By opening up the continental shelves to oil production, encouraging production of shale oil and other unconventional energy resources, and speeding up the licensing of nuclear power plants, the government can greatly increase domestic energy production and reduce dependence on imports.

This plan has the capability of producing a dramatic increase in energy production, but at a large environmental and economic cost. Oil spills, atmospheric pollution, and radioactive leaks are likely to increase dramatically under such a plan, and the need to invest heavily in new facilities could create a shortage of available capital. Some economists believe that the extremely high interest rates of the early 1980s were caused in large part by heavy investment in energy facilities, especially thermal power plants.

3. *Increase production of renewable resources.* Solar and wind energy, geothermal heat, and hydro power have the potential to provide a large portion of our energy needs within the next 20 years. According to a number of recent studies, 12 to 28% of all U.S. energy needs in the year 2000 could be met by solar energy. Since the late 1970s, the federal government has encouraged renewable energy resources by providing a substantial tax credit for renewable energy installations, although the tax credits currently (1984) appear to be in jeopardy.

Most people would agree that renewable energy resources are desirable; however, in some cases the capital costs of producing renewable resources are very high. For example, the price of photovoltaic cells was $8.00 per peak watt in 1982—four times the DOE target price of $2.00. Although renewable resources tend to be environmentally benign, high capital costs present a serious obstacle to their development.

4. *Reduce consumption.* Energy consumption can be reduced through a system of incentives and penalties, or through the more drastic step of mandatory allocation of supplies (rationing). In recent years, the U.S. government has tried both approaches. A 15% tax credit is available for homeowners installing energy-conservation equipment, and penalties can be imposed on automobile makers who fail to meet fleet fuel economy standards. (In fact, this penalty has never been imposed, but the regulation is in place and could be used.) During the 1970s, the government imposed more direct restrictions, banning gasoline sales on Sunday and imposing temporary emergency energy standards for temperatures in public buildings.

BOX 9.4.1 CAN CONSERVATION HURT US?

Although conservation is inexpensive and environmentally benign, it can reduce the resilience of the energy-consumption system. When there is a large amount of waste in the system, sudden drops in energy supply can be handled with little inconvenience. But if energy use has already been reduced to very low levels, as in the Third World, disruptions in energy supply can cause serious problems. For example, if I have my house thermostat set at 75 °F, I can reduce my energy consumption without any problem if there is an oil shortage. But if I have already lowered my thermostat to 65 °F, I will have a hard time cutting down further.

This is not a plea for wasting energy. But we must recognize that strict energy-conservation standards must be accompanied by additional measures to ensure continuity in energy supplies.

5. *Do all of the above.* Although it is not possible to free the marketplace and impose gasoline rationing at the same time, it is possible to follow policies that attempt to reduce consumption and increase production of renewable and nonrenewable energy resources while relying as much as possible on market forces. This creates a variable mix of energy strategies that can be adjusted to fit any given situation. Over the years, U.S. energy policy has tended to follow this middle course.

Although this mixed approach to energy policy offers a measure of flexibility, it has its drawbacks as well. When energy policies are not clearly defined, there is more room for influence by special interests, general inefficiency, and duplication of effort. Nonetheless, any other approach will require a substantial national consensus on energy issues: something that may be a long time coming.

Government intervention in energy decision making should be and is generally resented on all sides. Oil company executives complain that government price restrictions were a major cause of the oil shortages of the 1970s. Advocates of nuclear power claim that government safety regulations have made it impossible for nuclear power plants to be built quickly and profitably. Opponents of nuclear power claim that the Nuclear Regulatory Commission has acted as a promoter rather than a regulator of nuclear power. Solar energy entrepreneurs object to what they consider disproportionate subsidies offered to competing energy resources. Conservationists claim that governmental controls have held energy prices far below their replacement values, thus undervaluing conservation technology.

But despite these general objections to government regulation and control, no one has offered an alternative system that would adequately consider social and environmental issues. Without government controls, it seems likely that those with the greatest wealth and power could impose their will on the rest of the population. Some may object that wealth and influence have a distorting influence on government policymaking, and it is true that democracy is inefficient and subject to a certain amount of abuse. But no system has yet been found that comes closer to respecting the will and general well-being of the public at large.

9.5 POLICY SUGGESTIONS

In a democratic system, it is entirely up to the political process to define what will constitute a reasonable or desirable energy policy. Table 9.5.1 presents a survey of our government's efforts in this area. But we believe that any sane energy policy, whether public or private, must recognize the need for sustainability: that is, we should provide for our energy needs in the long term as well as the short term. For once our present supply of energy is used, it is gone forever, never to be available again for our service. In keeping with this principle, we will suggest some specific policies that reflect the principle of sustainability. The exact means of implementing these policies is up to the policymakers, but we believe that voluntary approaches should be encouraged through the use of differential tax rates, rebates, and other economic incentives.

9.5.1 Energy Production

It seems unlikely that renewable energy resources can be successfully developed without some form of governmental subsidies. Fossil and nuclear energy resources have received enormous direct and indirect federal subsidies, and there is little doubt that continued federal support is essential to the success of the nuclear industry. For the sake of equity, similar incentives should be offered to developing other energy resources. These incentives can be in the form of direct research and development grants or indirect subsidies such as tax credits.

In addition to providing direct and indirect subsidies, government bodies need to concern themselves with the legal difficulties faced by developers of new energy resources. For example, developers of solar installations are concerned that their neighbors can plant vegetation or build structures which shade their collectors. This does not have to be a serious problem. Solar access can be protected under a system comparable to present-day transportation easements. As another measure to protect producers of new energy resources, small-scale power producers can be guaranteed purchase contracts at prevailing rates under programs similar to those provided by the Public Utility Regulatory Policies Act of 1978 (see Section 9.4.1).

TABLE 9.5.1 Important Energy Policy Decisions at National Level

Year	Description	Area Affected
1885	Interstate Commerce Act	Natural Gas
1910	Establishment of Bureau of Mines	Coal, fossil fuels
1911	Dissolution of the Standard Oil Company	Oil
1917	U.S. Army Corps of Engineers	Hydro, electric
1920	U.S. Federal Power Commission	Hydro, electric
1926	Depletion allowances for corporations	Fossil fuels
1930	Shale oil leasing restrictions	Synfuels
1933–1935	TVA, REA, HOT OIL Act	Electric power, oil
1938	Natural gas regulation	Natural gas
1942–1945	Wartime rationing, Synfuels Act	Oil, synfuels
1946	Atomic Energy Commission	Nuclear
1953	Outer Continental Shelf Lands Act	Oil
1954	Atoms for Peace	Nuclear
1957	Price-Anderson Act	Nuclear
1963, 1967	Clean Air Act	Environment
1970	EPA, CEQ, NEPA	Environment
1973	Mining Safety Administration	Coal
1973–1974	Petroleum allocation regulations	Oil
1974	Federal Energy Administration, ERDA	Energy research
1975	Nuclear Regulatory Commission	Nuclear safety
1975	U.S. Energy Policy Act	Conservation, energy
1977	U.S. Department of Energy	Energy
1977	Surface Mining Act, OSM	Coal, environment
1977	Federal Energy Regulatory Commission	Electric energy
1978	National Energy Conservation Policy Act, PURPA, Natural Gas Policy Act	Energy, renewables, fossil fuels, etc.
1980	Deregulation, windfall profits tax, Synfuel Corporation	Energy

9.5.2 Economic Analysis

Public utility commissions are charged with ensuring that energy projects are economically feasible before giving permission for them to be undertaken. A policy of marginal cost comparison will give a fairer estimate of the project's costs than other methods of cost comparison. Thus, before any power plants or other energy production projects are undertaken, it makes sense to determine whether a "least cost" energy strategy is being followed. Similarly, a comprehensive economic analysis must take into account the indirect costs of a project, such as medical costs from environmental pollution. Public costs, in the form of tax credits and other subsidies, should also be taken into account during the preparation of an economic analysis.

9.5.3 Energy Efficiency

Although conservation measures have been shown to be less expensive than energy supply alternatives, private investment in conservation technology has lagged behind investment in supply facilities. Governments can encourage energy efficiency by setting energy-efficiency standards in residential and industrial structures. Local ordinances can also require passive solar features (where appropriate) as a feature of all new residential and industrial construction. Energy-efficiency ratings and standards can also help to increase the energy efficiency of household appliances.

9.5.4 Energy Auditing

States and localities can initiate low cost/no cost programs of individual home visits by trained energy auditors to identify opportunities for energy conservation. Many states have already offered such programs with encouraging results. When combined with Community Action Programs, such as home weatherization and fuel assistance for elderly and low-income families, these energy audits can serve several important purposes at once. In some cases, these programs can be funded by public or private utilities, since investment in energy conservation technology is much cheaper than investment in additional generating capacity—in many cases less than one-fourth the cost.

9.5.5 Zoning and Taxing Policies

Local zoning and tax boards can adopt policies encouraging, or even mandating, energy efficiency and use of solar technology. In Davis, California, for example, a comprehensive local code mandates energy efficiency and passive solar features on all new construction, and encourages retrofit of existing dwellings. Through this plan, the city has been able to reduce energy consumption by 15%.

9.5.6 Transportation of Goods and People

Directly and indirectly, transportation accounts for 40% of gross U.S. energy consumption. Besides direct consumption of fuels to power vehicles, energy is used to build and maintain vehicles and roads.

Governmental policies, especially at the federal level, can promote and encourage energy-efficient means of transporting goods and people. For freight handling, the railroads are by far the most energy-efficient means of overland transport. Governmental programs similar to highway construction projects could be undertaken to repair and upgrade railway roadbeds in the United States. For transporting people over short distances, surface mass transit (high-speed rail) can be constructed or upgraded in metropolitan areas. Governmental programs are essential to the growth of rail travel in the United States.

The automobile will continue to play a major role in U.S. transportation. Here, energy efficiency can be encouraged through mandates, tax rebates on energy-efficient cars, and penalties on low-mileage cars. Increased fuel taxes can also help to lower fuel consumption.

State energy offices and motor vehicle departments can help in arranging car pools, van pools, and so on. Minibus service can be provided to rural areas and mass transit systems can be designed for urban areas. Special free parking areas can be provided for car-pool vehicles. Lower registration fees for energy-efficient, nonpolluting automobiles could also be initiated at the state level.

9.5.7 Industrial and Commercial Operations

Multipurpose, environmentally sound energy projects such as cogeneration (see Section 11.1) can be encouraged through regulatory practices and tax incentives. Accelerated investment tax credits are already in place to encourage more rapid replacement of aging industrial equipment. Together with energy-efficiency guidelines, these incentives can help pave the way for greater energy efficiency in industrial and commercial practices. Some states are experimenting with energy auditing services for the commercial and industrial sector. These services are similar to the residential programs but funded either by the state or by the beneficiaries. A volunteer program of this type conducted energy surveys of about 250 manufacturing plants in the state of Vermont, leading to savings of 20% on fuel costs and 10% on electric needs. State incentives and tax credits appear to be an effective tool for encouraging energy improvements.

These governmental actions could create some dislocations in the U.S. economy, especially as older, inefficient plants are closed to make room for newer, more efficient equipment. To minimize the effect of these dislocations, training can be provided for any workers displaced as a result of changes in national energy policy. Precedents for national worker retraining have already been established through the Comprehensive Employment Training Act (CETA) and other similar measures. The concept could form a part of any coherent national energy policy.

9.5.8 Public Sector

Public buildings such as schools, hospitals, and offices can be required to make energy-conserving improvements in order to upgrade their energy efficiency. Matching funds can be made available to assist public institutions in carrying out these improvements. Public buildings can be made to meet strict "energy standards," including the use of solar designs and natural lighting and ventilation.

State and local education departments, schools, and libraries can act as clearinghouses for energy-related information. They could provide "how-to" information, lectures, and audiovisual presentations on energy use and conservation. Schools and colleges can offer energy-related courses and programs. One of the main reasons for persistent waste of energy is a general unavailability of accurate energy information. Educational institutions can definitely help to overcome this problem.

Local and regional energy policies such as the ones outlined here are well within the capacity of most local, county, and state governments. In the short run, these programs cost money; but in the long run, they can lead to savings on road maintenance, environmental protection, and public assistance. Localities with programs such as these often have a significant advantage in attracting newer industries because of their reputation as "forward-looking" communities.

BOX 9.5.1 VERMONT: A CASE STUDY

Vermont is a small, rural state with about half a million people. The climate is on the cold side, with an average of 7500 heating degree-days per year. The state's energy conservation programs, directed by the State Energy Office, include HEAT (Home Energy Audit Team) service; RCC (Residential Conservation Corporation), a utility-funded project; CAP (Community Action Program) home weatherization services for elderly and handicapped citizens; HEAT II, a program to link energy audits to fuel assistance for low-income citizens; VIECAP (Vermont Industrial Energy Conservation Program), a volunteer program that draws on experts from industry and the business community to promote energy conservation in the industrial and commercial sectors. Other programs include the state energy office's Van Pool, Ride Sharing, and other information-sharing systems designed to increase public awareness (see the Governor's Report on Energy, September 1982). Together, these programs have helped to produce the following savings between 1975 and 1980:

1. Vermont's total energy consumption dropped by 3.8% despite a population increase of 8.8%.
2. Residential electric energy use per degree-day was reduced by 3.6%.
3. Gasoline consumption was reduced by 10%. (Clearly, the growing fuel economy of automobiles had a significant effect on gasoline consumption.)
4. Wood energy use was *increased* by 380%. By 1980, more than 60% of all dwellings in Vermont had wood stoves or furnaces.
5. Heating oil consumption was reduced by 33%, largely as a result of the increase in wood burning.

9.6 ENERGY AND FOREIGN POLICY

Although energy is often thought of as a domestic problem, it bears a close and important relationship to foreign policy. Even in the energy-rich United States, energy and foreign policy are closely connected because of our large energy imports. In countries such as Japan, which import nearly all of their energy resources, energy policy is clearly inseparable from foreign policy.

In general, nations establish their foreign policy in order to protect their "national interests." In today's highly mechanized industrial economy, energy supply is a highly important element of the national interest. In 1973 and 1978, oil shortages, gasoline lines, curtailment of Sunday sales, and rising prices were all attributed to the Arab oil embargoes, which affected less than 5% of the nation's energy supply. If *all* oil imports were lost, the economic effects could be even more disruptive. So maintaining a steady flow of energy imports is unquestionably in the "interest" of the United States.

Because the U.S. imports oil and gas from the Middle East, Venezuela, and Mexico, our relationships with those countries and regions need to be close and harmonious. But good diplomatic relations alone are not sufficient to ensure a steady flow of energy imports. More practical reciprocal arrangements are the real backbone of foreign energy policy. For example, in exchange for its substantial energy imports, the United States sends the following valuable items to oil-exporting countries:

1. Military assistance and weaponry
2. Technological assistance and equipment
3. Grains and other food products
4. Capital investments in factories and other production facilities

Thus the oil-exporting countries depend on us for economic and military assistance at the same time that we depend on them for our energy resources. These reciprocal needs and obligations have helped maintain a steady international flow of energy except in times of the most severe world tension.

Of the four general categories of exports mentioned above, military sales and other "national security" exports largely counterbalance the enormous outflow of dollars for oil imports. There has been no shortage of customers. Recently, the United States sold AWACS aircraft to Saudi Arabia, despite the objections of Israel, our closest ally in the Middle East. It is no coincidence that Saudi Arabia is the most oil-rich nation in the world. Other oil-exporting countries have also provided a major market for U.S. arms sales.

Energy technology has recently been used as an international bargaining chip by the Reagan administration, which tried to restrict the sale of equipment to the USSR for use in the trans-Siberian gas pipeline. The policy increased tensions between the United States and its European allies, many of whom hoped to benefit from the energy supplies to be provided by the pipeline.

Oil is not the only energy resource with international significance. Nuclear energy has important implications for foreign relations, both because of the need to import uranium and because of the threat of nuclear weapons proliferation. It would be quite possible for a nation to reprocess spent nuclear fuel and to use the recovered plutonium for nuclear weapons. For this reason, reprocessing technology is not exported by the United States, though this policy may change in the near future. Some other countries, including France, have already sold reprocessing technology to developing countries.

In 1974, India exploded a nuclear bomb made of plutonium from a CANDU reactor supplied by Canada. In response to this incident, several countries, including the United States, took steps to prevent future abuses of nuclear technology. However, the effectiveness of these measures is still very much in question.

Energy issues affect U.S. foreign policy indirectly as well. Imports of fuel-efficient Japanese cars have come under fierce criticism from U.S. labor leaders, who have urged restrictions on imports. There is a general consensus that Japanese cars succeeded in the American market largely because of their greater fuel efficiency. Again, we can see the relationship between energy and foreign policy.

Acid rain, which is largely the result of electrical power production, has recently caused a heated dialogue between the United States and Canada. The Canadian government claims that most acid rain in the eastern provinces of Canada is caused by air pollution originating in the United States. The U.S. government has consistently denied this claim. The dispute has caused an increase in tensions between the two countries and has even led to the suspension of some international fishing agreements. In North America and other parts of the world, offshore oil drilling has stimulated controversies over ownership of the continental shelf. This, too, has complicated the development of international fishing treaties. In all these cases, we can see that energy issues are instrumental in shaping U.S. foreign policy.

9.7 OPEC: AN INTERNATIONAL ENERGY CARTEL

In recent years, one of the most important energy organizations has been the Organization of Petroleum Exporting Countries (OPEC). Together, these nations account for approximately 50% of the world's total oil production and over 90% of the world's imported oil. Today, OPEC's membership consists Kuwait, Libya, Nigeria, Qatar, Saudi Arabia, United Arab Emirates, and Venezuela. Nine of these 13 nations are located in the Middle East region of the world. The Middle East contains the world's richest oil deposits. But the Middle East is also characterized by extremely volatile politics, with the constant potential for international conflict. This has had significant consequences for U.S. foreign policy and for world politics in general.

Nation	Annual Oil Production (MBD)
Algeria	1.0
Ecuador	0.2
Gabon	0.1
Indonesia	1.6
Iran	2.2
Iraq	2.5
Kuwait	1.65
Libya	1.75
Nigeria	2.0
Qatar	0.475
Saudi Arabia	10.0
United Arab Emirates	1.7
Venezuela	2.17

OPEC was originally formed in 1960 at a time when Persian Gulf oil producers were losing hundreds of millions of dollars in oil revenues each year due to falling oil prices. But the organization did not really emerge as a powerful economic force until the Arab oil embargo of 1973. Although not all OPEC members participated in this boycott, the resulting oil shortage and sharp rise in prices benefited all the members of the organization.

Following the Arab oil embargo, the OPEC members met regularly to set the prices they would charge their customers. At these meetings, the OPEC nations divided roughly into two groups: those who favored raising prices as high as the market would bear (principally Iran and Libya), and those who favored more moderate oil prices that would not disrupt the world economy (Saudi Arabia). Because of the large influence of Saudi Arabia, the world's largest oil producer, price increases were moderate, though steady, until about 1980.

It is important to note that the members of OPEC met openly for the purpose of setting their prices. If this were done by private corporations, it would constitute price-fixing, an illegal act. But no such restrictions apply to nations. The prices charged by OPEC members have been as much as 40 times the cost of production. And this, of course, has caused serious grievances in the less developed and Third World countries, where foreign exchange reserves have had to be seriously depleted in order to meet oil needs. The loss of foreign exchange forced some of these countries to turn to alternative sources of energy, including nuclear energy, at high prices. These technologies were supplied by developed countries, especially the United States, on a loan or grant basis rather than purchased for hard currency.

Third World countries were not the only ones hit hard by the rise in oil prices. Industrialized countries have also suffered considerable hardships. Many analysts attribute the worldwide economic declines of 1975 and 1980–1983 to the sharp rise in oil prices brought about by OPEC.

In response to this economic hardship, the industrialized nations of the world have soundly condemned the oil-exporting nations. Criticism from less developed nations has not been so public, largely because OPEC offers many poorer countries low-interest loans or outright subsidies amounting to billions of dollars per year.

Although the role of OPEC was crucial in bringing about the sharp price increases of the 1970s, other factors were also important. Multinational oil companies passed along and even inflated the price increases in order to maintain their own profit margins. At the same time, they cut back on their production of less profitable U.S. crude oil. In the political climate of the 1970s, criticism of the Arab countries was popular. But the role of the oil companies in maintaining a tight oil market was an essential element of OPEC's strategy.

During the 1970s, the U.S. government tried to influence the OPEC countries, largely unsuccessfully. A direct effect of the high oil prices was that the oil-exporting countries had large amounts of foreign exchange even after purchasing necessary supplies, equipment, food, and technical assistance. One politically destabilizing consequence of these tremendous surpluses was the exorbitant spending by Arab countries on military hardware. Under Shah Mohammed Reza Pahlavi, Iran built up one of the world's largest armed forces, using hardware and training purchased from the United States with oil revenues. Similarly, other oil-exporting countries developed prodigious military forces. To date, no member of OPEC has succeeded in purchasing nuclear weapons, although Libya's ruler, Moummar El Khadaffi has made it known that he will pay an unlimited price for nuclear armaments.

Besides buying military equipment and armaments, many OPEC member states are investing in the refining and other facilities necessary to transform crude oil into useful consumer products. Gasoline, jet fuel, heating oil, and other finished petroleum products offer a larger aggregate profit margin than unfinished crude oil, and the oil-exporting nations want a larger share of those profits.

Although OPEC was highly successful in raising oil prices during the 1970s, their strategy began to fail during the 1980s. This can be attributed to two primary factors, both of which can be seen as direct responses to the initial price increases.

1. *Conservation measures.* After the Arab oil embargo, automotive fuel economy began to improve steadily. In the United States, fuel-efficient imported cars began to replace the large domestic gas-guzzlers, and American manufacturers soon started introducing fuel-efficient cars of their own. Homeowners improved the insulation in their houses, causing a reduction in demand for fuel oil. Alternative sources of energy, such as solar space heating, grew in popularity, further reducing oil consumption. By

1983, U.S. oil imports had begun to decline significantly, causing a noticeable softening in the world oil market.

2. *Worldwide economic decline.* Worldwide economic conditions began to worsen in the mid-1970s, largely in response to the rise in oil prices and the disruptions in supply caused by the Arab oil embargo. As the world economy weakened, industrial consumption of petroleum products decreased. Tight national budgets led to reductions in foreign aid by the industrialized countries, with the result that underdeveloped countries were less able to buy expensive imported oil. In the early 1980s, the worst worldwide economic decline since the Great Depression led to sharply reduced worldwide demand for petroleum products. OPEC production fell from 30.928 MBD in 1979 to 26.89 in 1980, 22.61 in 1981, 17.89 in 1982, and 16.95 in mid-1983.

Once demand for petroleum products began to decline, OPEC lost much of its economic power. Prices began to fall on the "spot market" (international oil sales not covered by long-term contracts), and soon the official price began to fall as well. From a high of $40 per barrel in March 1981, the official OPEC price fell to $34 in 1982. At the same time, many OPEC countries were actually selling oil at prices as low as $30 per barrel. Then, at a special meeting at Geneva in January 1983, even this official price level was abandoned. OPEC's power to set world oil prices was broken, at least for the time being.

During the period of declining prices that began in 1981, the countries that had pushed most aggressively for price increases were now the first to cut their own prices. The "moderates," led by Saudi Arabia, called for production cutbacks to maintain the prices at a high level. Saudi Arabia even instituted unilateral production restrictions. But Iraq and Iran, who were at war with one another, ignored the call for cutbacks, and Libya signed a number of low-priced contracts with Asian countries.

Although many in the industrialized world initially concluded that the lower oil prices "served the Arabs right," it soon became apparent that a sudden drop in oil prices could have serious repercussions. Some oil-producing countries, such as Mexico, had established extremely high levels of foreign debt, basing their ability to pay on the high oil prices then in effect. When prices began to decline, these countries found that their ability to pay had decreased. For a time, there was a very real possibility that some countries would have to default on their foreign debts, causing enormous losses for the banks (primarily American) that had underwritten those debts. Once again, the established industrialized countries were hurt less than the developing countries, which had sought to finance their own industrialization with the revenues from their oil production.

In the light of OPEC's near collapse in 1982–1983, it is possible to make a few observations about the international petroleum market:

1. The dramatic price increases of the 1970s would not have been possible without the extremely high levels of demand which characterized the period leading up to the Arab oil embargoes.
2. Demand for energy, like demand for other useful commodities, can change in response to price increases.
3. Oil prices, like prices of other products, are more dependent on demand than on costs of production.
4. International economic cooperation is easier to achieve and more effective during times of prosperity than during times of hardship.
5. Abrupt price changes, whether upward or downward, are economically disruptive and cause problems worldwide.
6. Developing and Third World countries are the most likely to suffer in any petroleum crisis.

9.8 ENERGY AND WAR

According to Jacob Bronowski, war is possible only in societies with surpluses. When all the members of a society must work to provide for basic needs, no warrior class can develop, and potential raiding parties will not gain anything from their depredations. But after the invention of agriculture, food surpluses became commonplace, leading to the development of theft, and ultimately war.

Before the development of inanimate energy resources, humanity's energy came from food and from animal power (horses, cattle, and slaves). In most cases, the groups with the greatest surplus of these forms of energy prevailed. Thus the principle was established very early in human history that energy-rich societies could successfully wage war on those with fewer available energy resources.

With the advent of fossil fuels, the scope of war was broadened. Advances in energy technology were quickly incorporated into military strategy, so war became steadily more destructive. Energy-rich societies continued to have the upper hand in warfare. In the U.S. Civil War, the North, with its coal mines, steel industry, and railroads, overcame the South, which lacked these important energy resources.

Throughout history, energy-affluent societies have tended to wage war on low-energy societies with a high degree of success. The period of European colonialism was characterized by European conquest of less industrialized regions of the world. The colonial powers had an insurmountable advantage because of their enormous energy resources and the formidable war machines which harnessed those resources.

During the early twentieth century, the military was the driving force behind the development of technology. Aircraft, submarines, communications equipment, and rocketry were all improved greatly during the two world wars.

By the end of World War II it had become clear to the world that victory in war depended to an enormous extent on the development of advanced technology for releasing energy rapidly and explosively. The bombings of Tokyo and Dresden were crippling blows to Japanese and German civilizations. But the final and most devastating conversion of energy to warlike purposes came at the very end of the war.

BOX 9.8.1 NUCLEAR WEAPONS CHART

How much is enough? The dot in the center square represents all of the firepower exploded in World War II, including Hiroshima and Nagasaki—the equivalent of 3 megatons of TNT. The other dots represent the number of World War II equivalents that now exist in the form of nuclear weapons—18,000 megatons, or the firepower of 6000 World War II's. These dots equal the firepower of nearly 1 million Hiroshima bombs—4 tons of TNT for every man, woman, and child on earth. The circle in the upper left-hand square represents one Poseidon submarine, which has the explosive capability of almost 9 megatons, or the equivalent firepower of 3 World War II's. This is enough to destroy over 200 cities in the USSR. Soviet firepower could do the same to us. The circle in the lower left-hand square represents one Trident submarine, which has the explosive capability of 24 megatons, or the equivalent firepower of 8 World War II's. (Reproduced with permission from Jim Geier and Sharyl Green, Burlington, VT.)

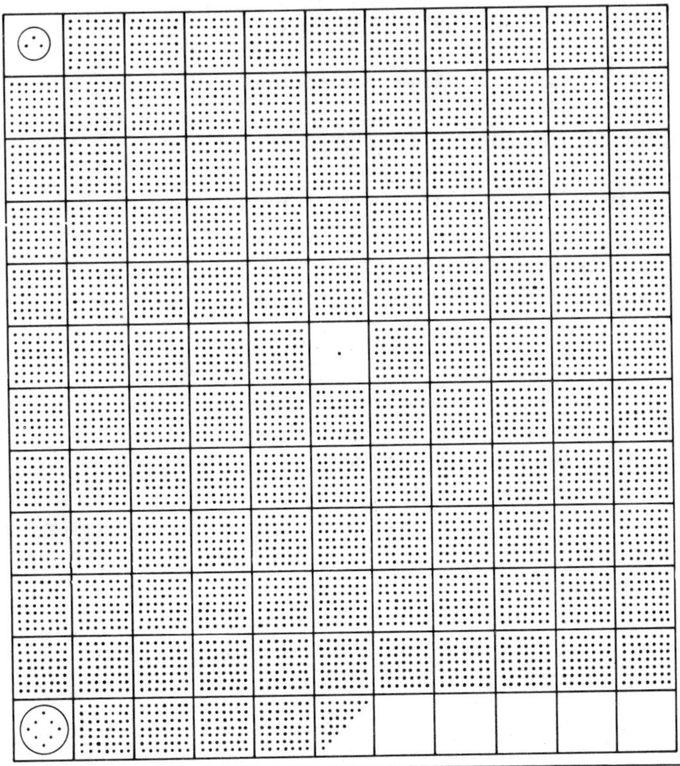

Working in secrecy for a period of five years, a team of engineers developed a method for applying advanced principles of subatomic physics in order to release some of the energy stored in the form of mass. The destructive power of the resulting explosion was unprecedented in human history. The cities of Hiroshima and Nagasaki were leveled, and hundreds of thousands of people perished.

Following the end of World War II, military engineers continued to apply energy technology to the development of weapons. More powerful and more efficient nuclear fission weapons were developed. Then, in the early 1950s, the principle of nuclear fusion was applied to weapons, resulting in even more powerful weapons. (As we noted in Chapter 3, *peaceful* use of nuclear fusion is still decades away, although we have had fusion weapons for 30 years.) The increase in explosive power has continued to the present. Today, the combined explosive power of the United States and the USSR amounts to the equivalent of 18 billion tons of TNT: 6000 times as much energy as was exploded in all of World War II, including Hiroshima and Nagasaki.

Nobody is pleased by the terrible increases in nuclear weapons. Attempts are being made to limit nuclear arms, with limited success. Less developed nations seek nuclear weapons capacity, and the great powers continue to build up their arsenals. Everyone wants to stop the buildup, but no one knows how to do it. In many ways, it seems that the very availability of an energy resource dictates that it will be applied to warfare.

9.8.1 Going to War for Energy Resources

Besides applying energy technology to enhance their effectiveness in waging war, societies and nations have long used war as a means of acquiring energy and other natural resources. In the first primitive wars, raiding parties seized stocks of food from their peaceful neighbors. Later, the principle was extended to include seizing territory. By acquiring territory, a society increases its potential for agricultural production. In both cases, the principle was the same: by going to war, it was possible to increase the society's supplies of valuable resources.

As warfare and foreign policy developed and became more sophisticated, the European colonists sought *access* to natural resources rather than outright conquest. Rather than concerning themselves with seizure of vast territories, industrialized powers sought control of vital shipping points, ports, and other methods of access to natural resources. Treaties and contracts replaced seizure as the method of acquiring the resources for industrial development. But the potential for war was always present. During the late 1970s, this fact came home very strongly in the United States. It was revealed that the Carter administration was prepared to go to war if necessary to ensure the security of the Persian Gulf oil-producing region. This is still part of U.S. foreign policy.

Figure 9.8.1 Mushroom cloud rising from a 1954 thermonuclear weapon test in the Pacific testing area. Many such detonations over the past years have released alarming quantities of radiation in the atmosphere. Environmental, social, and economic consequences of an all-out nuclear exchange would be catastrophic. (*Source:* Lookout Mountain Air Force Station.)

9.8.2 Energy Facilities as Strategic Targets

One consequence of centralized, technically sophisticated energy systems is that the energy facilities themselves may become strategic targets. For example, during a recent war between Iran and Iraq (still in progress as of mid 1984), Iranian oil fields were the target of Iraqi attacks. In a related incident, Israeli planes bombed an Iraqi nuclear power plant, claiming that the plant was being used to produce materials for nuclear weapons. In both cases, an attack on energy facilities was seen as a way of harming the adversary.

In large-scale war, central power stations, especially nuclear power plants, would almost certainly be the targets of military attacks. Some military analysts have even suggested that the U.S.'s entire electrical power grid could be shut down by a single thermonuclear weapon detonated in space above the United States. In light of this increased vulnerability, it has been suggested that a switch to decentralized energy facilities would be in the strategic interest of the United States. It would certainly be difficult for an adversary to shut down a power system based on millions of small-scale hydro, wind, and photovoltaic installations—much more difficult than attacking a few dozen nuclear power plants.

Energy and war do not necessarily have to go hand in hand. Following World War II the United States exported vast amounts of energy technology to assist in the rebuilding of Europe and Japan. More recently, under the Atoms for Peace program, the United States has supplied less technologically developed nations with the potential of producing their own electric power from uranium. (Critics of this program point out, however, that some of the recipient nations have diverted this technical assistance to warlike purposes.) Other energy technologies, including hydro and other electrical power technologies, are commonly offered by industrialized countries to less developed countries. In principle, there is no reason why nations cannot cooperate to develop their energy resources in a way that will contribute to the greater benefit of humankind.

Energy surpluses can be used for economic, cultural, or leisure purposes—or they can be used for purposes of destruction. Energy can be a powerful force for international cooperation or a force for uncontrolled destruction. The same reactions that power our electrical plants can be used to blow them up in a matter of seconds. Nuclear power plants can produce power to light our dark nights, and nuclear weapons can bring darkness in broad daylight. In either case, energy is the driving force, and one that must be reckoned with.

The "superpowers," the United States and the USSR, are truly "super" in their energy-conversion capacities. Their ability to command enormous energy resources brings with it a responsibility to use those resources wisely. The greater the power of an engine, the greater the need for caution. While walking, a person may sing, look around, enjoy the weather, and even close his or her eyes. A bicyclist, an automobile driver, and a truck driver need to be progressively more careful and attentive. The same can be said about nations with vast energy resources at their command. Our greatest strength lies in our ability to employ energy conversions, and our greatest weakness lies in our fallibility, our limited (although very impressive) reasoning power, and our limited perceptions. Through a greater awareness of our own limitations, we may be successful in harnessing the enormous energies that threaten to destroy us. Surely this, the ultimate energy issue, is the ultimate political issue as well.

STUDY QUESTIONS

1. What are the common forms of ownership in the energy industry? List specific companies that fall into each category.
2. Is the electric utility company that serves your area publicly or privately owned? What are their rates? Compare their rates with those of neighboring utilities.
3. Why is it necessary to have regulated monopolies in the electric and gas energy industries and not in the oil or coal energy industries?
4. Which of the following governmental policy measures are the best for dealing with energy problems? Justify your choices.

Direct subsidies
Tax incentives
Regulations
Penalties for failing to go along with policies
Public ownership of energy companies

5. If you were responsible for establishing U.S. energy policy for the next 30 years, what objectives would you try to accomplish? Which energy resources would you wish to develop to achieve those objectives? Explain why those resources are most appropriate for achieving your objectives.
6. What specific policy steps would you institute to facilitate the development of the energy resources you chose in Question 5? What political and social obstacles would you anticipate in carrying out your plan?
7. Is democracy the most effective governmental system for dealing with the complexity of energy issues? Do you believe any changes should be made in our system of government to deal more effectively with the energy crisis? If so, what changes should be made?
8. What methods exist for eliminating the "energy gap" between U.S. production and imports? In your judgment, which of these are sound alternatives? Justify your choices.
9. Does your state or local government offer any free energy services? Make a list of those services and consider how you might be able to take advantage of them.
10. The city of Davis, California, has passed a number of laws regulating energy consumption. For example, buildings are required to meet strict energy-efficiency standards. Is this an undue restriction of people's right to live as they please? If so, what steps, if any, should governments take to encourage energy conservation?
11. Are any highly industrialized nations members of OPEC? What does this suggest about the nature of the organization?
12. If the United States became an oil-exporting country, would it be a good idea to join OPEC? What would be the benefits and drawbacks of OPEC membership?
13. How many countries have nuclear weapons? How many of those countries have nuclear power plants? Do you believe that export of nuclear power can lead to the spread of nuclear weapons?
14. What controls, if any, should be imposed on exports of nuclear technology to reduce the threat of proliferation?

FURTHER READINGS

CLARK, WILLIAM, and JAKE PAGE. *Energy Vulnerability and War: Alternatives for America.* New York: W. W. Norton, 1981.

COLLINGRIDGE, DAVID. *The Social Control of Technology.* New York: St. Martins Press, 1982.

COMMONER, BARRY. *The Politics of Energy.* New York: Alfred A. Knopf, 1979.

Cook, Earl. *Man, Energy, Society.* San Francisco: W. H. Freeman, 1976.

Energy Research and Development Administration, Bicentennial Project. *Energy History of the United States.* Washington, DC: ERDA, 1975.

Engler, Robert, ed. *America's Energy.* New York: Pantheon, 1980.

Lovins, Amory. *Soft Energy Paths: Toward a Durable Peace.* New York: Harper & Row, 1977.

Lovins, Amory B., and L. Hunter Lovins. *Brittle Power.* Andover, MA: Brick House, 1982.

______. *Energy/War: Breaking the Nuclear Link.* New York: Harper & Row, 1980.

Nash, Hugh, ed. *The Energy Controversy: Amory Lovins and His Critics.* San Francisco: Friends of the Earth, 1979.

Scheffer, Walter F., ed. *Energy Impacts on Public Policy and Administration.* Norman, OK: University of Oklahoma Press, 1976.

Steinhart, John S., et al. *Pathways to Energy Sufficiency: The 2050 Study.* San Francisco: Friends of the Earth, 1979.

10

Personal Awareness

Whenever we make decisions such as what sort of house to build, what kind of heating system to install in the house, which appliances to buy, whether to drive a small car or a large car, even when to watch TV, we are making decisions that have a significant impact on energy needs, both at the local and the national level. By adding to (or reducing) the demand for a particular type of product or service, we influence manufacturers and distributors, who adjust their production to respond to demand.

For example, in the 1970s, when the demand for small, fuel-efficient automobiles jumped and the demand for large gas-guzzlers dropped, automobile manufacturers got the message. Today, the automobile buyer can choose from a wide variety of small, fuel-efficient automobiles. Fifteen years ago, the choice was limited to a few imported cars and one or two domestic models. Although governmental regulations had some influence on the trend toward smaller automobiles, customer demand was the main motivating force. Thus decisions by individual consumers indirectly led to a fundamental restructuring of the largest industry in the United States. Other individual choices can have equally significant effects on other areas of the energy economy.

10.1 FACTORS IN PURCHASING DECISIONS

What factors do consumers consider important when they make their choices? First, of course, they take into consideration the factors that will directly affect them at the time of the purchase. These include the following:

Need for the product
Initial purchase cost
Appearance and surface features
Luxury and status considerations
Convenience and ease of operation

In many cases, the purchaser is concerned only with these factors. For a home builder who is planning to sell the house, these initial considerations will be the main guide to the decision-making process. But for a consumer who will have to pay the costs of operating and maintaining the heating system, additional factors must be considered, such as the following:

Operating costs
Reliability and life expectancy of the product
Repair and maintenance costs

In addition to these considerations, a few customers are also concerned with the broader implications of their purchasing decisions. They see their purchases not only as a business transaction but also as a personal statement. These consumers are likely to be concerned with such matters as the following:

Energy consumption
Energy efficiency
Environmental impacts of the product
Social and political effects of the product

Consider, for example, the purchase of a home-heating system.

Purchase cost. A wood stove is probably the least expensive heating system that can be installed in a house. But wood heat usually requires a backup system. Besides wood, electric baseboard heating is the least expensive heating system to install in a house. No ductwork or chimney is required, and the baseboard units can be installed using a minimum of labor. Other inexpensive systems include space heaters and wall furnaces. Oil- and gas-fired central furnaces are more expensive, and active solar heating systems are the most expensive of all. Solar heating systems also require some kind of backup.

Appearance and surface features. Electric baseboard systems and central furnaces are usually the most visually attractive alternatives. Wood stoves and space heaters often take up living space. Solar heating systems can be very attractive if they are well designed, but some people object to the appearance of solar heating systems, especially roof-mounted active systems.

Luxury and status considerations. Solar heating carries an air of status unmatched by other heating systems. Fireplaces and electric heating systems (especially heat pumps) are often considered luxurious. Space heaters and wall furnaces are usually considered low-status heating systems.

Convenience and ease of operation. Thermostatically controlled heating systems are clearly more convenient than systems such as wood stoves or simple space heaters. Electric baseboard systems are especially convenient because the user can set each room at a desired temperature rather than using one central thermostat for the entire house. Automatically controlled active solar heating systems are as convenient as any thermostatically controlled heating systems, but passive solar systems require a substantial amount of owner attention. Wood furnaces require even more owner attention, and wood stoves are probably the least convenient of all heating systems.

Based on these considerations, a builder who is planning to sell the house would be likely to install electric baseboard heating, since it is the cheapest system initially and will appeal to consumers because of its convenience and ease of operation. In fact, electric baseboard heating is the most common system in such applications, followed closely by gas furnaces in areas where natural gas is readily available. But from the user's point of view, electric baseboard heating is not so advantageous because of long-term costs.

Operating costs. Clearly, a passive solar heating system is the cheapest to operate. After that come active solar heating systems, with wood, gas, and oil heat following in order. The operating cost of an electric heat pump will vary depending on geographical location. In the South, a heat pump will be quite economical to operate, but in the North, it may be more expensive than oil heat. Baseboard electric heating is the most expensive to operate in any climate.

Reliability and life expectancy. In general, simple systems have a smaller probability of breaking down. Thus systems with complex controls, such as active solar heating systems, are the most prone to breakdowns, and the simplest systems, such as passive solar heating systems or wood stoves, have the fewest breakdowns. In terms of life expectancy, strong materials usually lead to a longer life; thus a heavy, expensive cast-iron wood stove will last longer than a lighter, cheaper sheet-steel stove. Usually, a customer must pay more in order to buy a product with a longer life expectancy.

Repair and maintenance costs. Repair and maintenance costs are usually related to initial cost and to reliability. Thus a highly reliable product will require few repairs and little maintenance. But when maintenance is needed, an expensive system will usually cost more to repair than will an inexpensive system. Thus an active solar heating system is much more expensive to repair than an electric baseboard system or a wood stove.

Based only on the considerations listed above, a consumer would be likely to rule out active solar heating systems entirely. Passive solar heating systems might be considered in the case of new construction, where the heating system could be integrated into the construction at minimal cost. Electric baseboard heat might not be the system of choice because of its high operating costs, although in warm climates or areas with low electric rates, the operating costs might not be prohibitively high. Of the remaining choices, wood is the least expensive (at least in rural areas where wood is plentiful); thus many consumers choose wood heat as the primary heating system and use electric baseboard heating as a backup system because of its low installation costs. In fact, this is the rule in large areas of rural America.

Most consumers are concerned primarily with short-term savings. Raising the necessary purchase capital is usually a problem, and long-term projections of energy costs and environmental effects are not always reliable. Nevertheless, a consumer purchase does have far-reaching effects, especially when the same purchasing decisions are being made by large numbers of people. If we consider long-term, global factors in reaching a purchasing decision, we come up with yet another approach.

Energy consumption. Clearly, solar heating systems have the edge here. Although solar heating does require energy, that energy is entirely renewable and would be wasted if it were not used immediately. Wood heat comes in a close second because wood is also a renewable resource. But wood is a stored form of solar energy; thus in terms of global energy consumption, it is preferable to use solar energy for heating and save the wood for higher-level tasks such as carpentry or biomass conversion. Other energy sources, such as fossil fuels and electricity, are much less desirable in terms of global energy consumption.

Energy efficiency. In terms of energy efficiency, electric resistance heating would erroneously appear to be the system of choice because of its high efficiency at the point of end use. But if we consider the whole *system* efficiency, including generation and distribution of electricity, we find that electric heat is the least efficient of all the alternatives (see Chapter 5). Electric heat pumps have a higher efficiency rating, especially in warm areas. But electricity is also a very high quality form of energy (see Chapter 5), which means that it is not a good choice for low-grade energy tasks such as home heating, even if it is used with a high degree of efficiency. The clear winners in terms of matched end use and overall energy efficiency are solar heating systems. They may use only 30 to 50% of the incoming solar energy to heat the house, but this must be compared with the much lower fraction of the incoming solar energy—

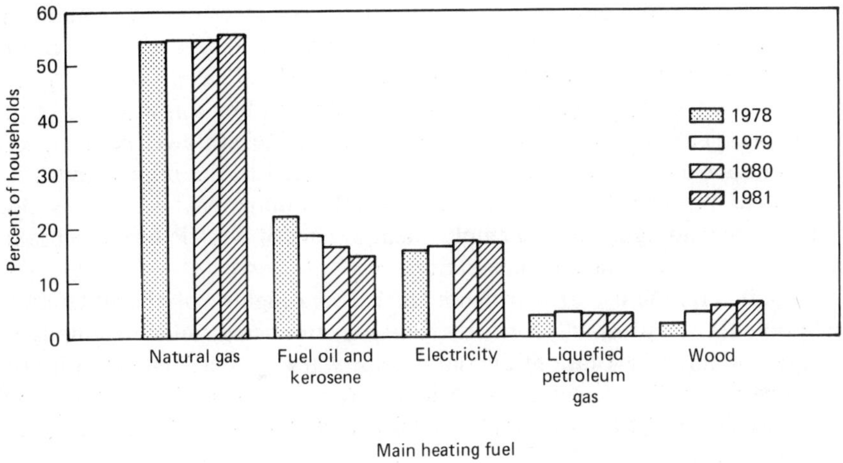

Figure 10.1.1 Distribution of households by main heating fuel, 1978–1981. (*Source: Monthly Energy Review,* Energy Information Administration, Department of Energy, Sept. 1, 1983.)

anywhere from a fraction of 1% to 10%—which is used by conventionally heated houses. Taking all energy resources into account, solar-heated houses are the most energy efficient.

Environmental impacts. As was shown in Chapter 7, different energy resources have widely differing impacts on the environment. If we consider only local environmental effects, solar heating is the most environmentally benign method of heating, with electric heating coming in a close second. Wood heat and direct combustion of coal would come in last. But if we consider large-scale environmental impacts, the story is somewhat different. Solar heat is still the best and coal burning is still the worst, but electricity becomes significantly less attractive. Although power companies tend to claim that electricity is "clean," the methods of generation are not. Most thermal power plants use coal, oil, or uranium fuel, all of which have adverse environmental consequences. Coal-burning power plants are believed to be largely responsible for acid precipitation, and the environmental effects of nuclear power, in the event of a mishap, could persist for thousands of years. Hydro power presents land-use problems. So in environmental terms, solar heat is the clear choice, followed by clean fossil fuels, electricity, and wood and coal. (Some have theorized that a wind-powered heat pump would be very clean and use only renewable energy resources, but the technology for such a heating system is still in its infancy.)

Social and political effects. In assessing the social and political implications of choosing a certain heating system, we need to consider a range of issues ranging from employment to the effects on the world balance of trade. Of the heating systems under consideration, solar heating provides the largest number of jobs because the manufacture and installation of solar heating equipment is a labor-intensive sector of the economy. Similarly, wood heat provides a fairly large number of jobs. Of the other energy types, natural gas production probably provides the fewest jobs, with oil and coal providing a few more. In the case of electricity, the situation is not clear. Fossil-fueled power plants create a fair number of jobs if we include the jobs created in the mining industry. But nuclear power plants create very few jobs compared to other energy-producing technologies. So the employment impact of electric heat is variable depending on the method of generation.

Politically, the use of petroleum has led to a highly volatile situation in the Middle East, with small, formerly rural societies being suddenly inundated with great wealth. Revolutions and large-scale wars have broken out, with the superpowers showing an increasing inclination to involve themselves in area conflicts. Thus the use of oil heat can be said to jeopardize both national and international security.

Most natural gas and coal is produced domestically, so the political disruptions caused by use of those fuels is considerably less than that caused by oil burning. However, coal mining historically has imposed terrible social costs on coal-mining regions such as the southwestern Appalachian mountains. Uranium mining has caused serious health consequences for Navajo and other native American communities located near the mines. Wood burning is less socially disruptive, although it does cause increased health problems associated with air pollution. Of all the choices for domestic heating, solar energy probably produces the fewest adverse social and political consequences.

For the homeowner who looks beyond the immediate local consequences of his or her selection, solar heating is the clear choice. Opinion polls have repeatedly shown that people regard solar energy as the most desirable form of energy for home heating. However, few choose to install costly solar heating systems, despite the fact that the initial costs will be recovered eventually through reduced fuel bills. Generous governmental subsidies have begun to increase the popularity of solar heating systems, although many consumers do not benefit from the tax credits because they have difficulty raising sufficient capital in the first place. For most people, personal choices are weighted heavily toward immediate economic problems rather than the long term. We can only hope that public awareness will grow, and that individual energy decisions will be based on a more complete understanding of their consequences.

10.2 MONITORING ENERGY CONSUMPTION

Personal awareness of energy begins with taking a look at where we are consuming energy, how much, and in what form. Just as we calculate our weekly or monthly income and expenditures, we need to calculate our expenditures of energy. That will allow us to discover possibilities for savings (of money as well as energy) and to monitor our progress in conserving energy.

As a first step in monitoring personal energy use, it is useful to draw up an estimate of all the household's various energy requirements. Using these estimates, it is then possible to estimate total household energy use for a particular period of time. These estimates can be compared with actual energy use. If the two figures differ significantly, the estimates of energy use can be reexamined and brought in line with the actual consumption figures. Once household energy use has been estimated with a reasonable degree of accuracy, opportunities for conservation should be easier to pinpoint.

10.3 ESTIMATING PERSONAL ENERGY USE

Energy consumed by individuals or households can be grouped into the following categories:

Space heating and air conditioning (HVAC)
Hot water
Lighting
Food preparation and storage
Transportation
Entertainment (radio, TV, video games, etc.)
Cleaning, lawn care, maintenance, and repairs

Which of these categories accounts for the largest energy use in your household? Is it transportation? Or space heating, cooling, and hot water? There is no answer that applies to every household. If a household is located in a temperate climate and has two wage earners, both commuting to work, transportation will probably be the main cause of energy consumption. In a cold climate, however, or where the wage earners do not drive to work, transportation will be less significant. To assess a household's energy consumption, it is necessary to examine not only the energy efficiency of the house and the energy-using appliances, but also the life-styles of the residents. Here are some techniques that may be helpful in estimating energy use.

10.3.1 Home Heating

In many households, home heating is the largest single factor in family energy consumption. Calculating home-heating requirements for design purposes can be a very complex procedure requiring detailed knowledge of home construction and heat-loss principles. But heating requirements can be estimated fairly well based on the quality of house construction. Although this is not a perfectly precise process, it yields reasonable results.

The first step in this process is to estimate the house's *heating coefficient*. This is a measure of the amount of energy required (in Btu) to heat 1 ft^2 (floor area) of the house by 1°F for one day. This heating coefficient can be estimated

with a reasonable degree of accuracy by consulting Table 10.3.1. These estimates can be refined somewhat through the use of more precise quantitative estimating techniques, but they are adequate for making a preliminary estimate of energy use for home heating.

Once the approximate heating coefficient has been determined, it is possible to estimate the annual heating load through the use of *heating degree-days* (see Table 10.3.2). Each degree-day represents a temperature difference of 1°F between a predetermined interior temperature (usually 65°F) and the average outdoor temperature for one day. Thus, if the average temperature on a given day is 45°F, the number of degree days for that day will be 65 − 45, or 20 degree-days. Weather bureaus maintain records of degree-days, including monthly and annual totals. To determine the building's heating load for a certain period of time, simply multiply the heating coefficient by the house's floor area and then multiply the result by the number of degree-days.

TABLE 10.3.1 Heating Coefficients

Description of House Construction	Heating Coefficient (Btu/(°F/day/ft²)
No insulation in side walls; attic poorly insulated; single glass in windows; no insulation in basement or crawl space	20–25
Moderately insulated (2 in. of fiberglass, old, settled cellulose insulation, board insulation, or poured vermiculite); attic moderately insulated (3–6 in. of fiberglass or cellulose); storm panes on most windows; no insulation in basement or crawl space	16–20
Adequately insulated walls (3½ in. of fiberglass or poured cellulose); attic insulated (6+ in. of fiberglass or cellulose); all windows double-glazed or thermopane; floor insulation above crawl spaces	12–16
Highly insulated walls (6 in. of insulation or 3½ in. with insulating sheathing such as styrene board or foil-faced urethane); attic extremely well insulated (12+ in.); some windows triple-glazed; few windows on north side; well-insulated around foundation or above crawl spaces	8–12

TABLE 10.3.2 Degree-Days for Selected U.S. and Canadian Cities

State or Province	City United States	Average Winter Temperature[a] (°F)	Yearly Degree-Days (rounded off)
Alabama	Mobile	59.9	1,560
Alaska	Fairbanks	6.7	14,280
	Juneau	32.1	9,075
	Anchorage	23.0	10,860
Arizona	Phoenix	58.5	1,765
Arkansas	Little Rock	50.5	3,220

TABLE 10.3.2 Degree-Days for Selected U.S. and Canadian Cities (Cont.)

State or Province	City United States	Average Winter Temperature[a] (°F)	Yearly Degree-Days (rounded off)
California	Los Angeles	60.0	1,350
	San Francisco	55.0	3,000
Colorado	Colorado Springs	37.3	6,425
	Denver	40.0	6,000
Connecticut	Hartford	37.3	6,235
District of Columbia	Washington	45.7	4,225
Florida	Jacksonville	61.9	1,240
	Miami	72.0	200
Georgia	Atlanta	51.7	2,960
Hawaii	Honolulu	74.2	0
Idaho	Boise	39.7	5,810
Illinois	Chicago	38.0	6,000
Indiana	Indianapolis	39.6	5,700
Iowa	Des Moines	35.5	6,590
Kansas	Wichita	44.2	4,620
Kentucky	Lexington	43.8	4,680
Louisiana	New Orleans	61.5	1,300
Maine	Portland	33.0	7,500
Maryland	Baltimore	46.0	4,300
Massachusetts	Boston	40.0	5,640
Michigan	Detroit	37.1	6,300
	Lansing	34.8	6,910
Minnesota	Minneapolis	28.3	8,400
Mississippi	Jackson	55.7	2,250
Missouri	St. Louis	44.0	4,600
Montana	Missoula	31.5	8,130
Nebraska	Lincoln	38.8	5,875
Nevada	Las Vegas	53.5	2,710
New Hampshire	Concord	33.0	7,380
	Mt. Washington	15.2	13,810
New Jersey	Newark	42.8	4,600
New Mexico	Albuquerque	45.0	4,350
New York	Buffalo	34.5	7,060
	New York City	42.8	4,900
North Carolina	Raleigh	49.5	3,400
North Dakota	Fargo	24.8	9,230
Ohio	Cleveland	37.2	6,350
	Cincinnati	45.1	4,410
Oklahoma	Tulsa	47.7	3,860
Oregon	Portland	47.0	4,400

TABLE 10.3.2 Degree-Days for Selected U.S. and Canadian Cities (Cont.)

State or Province	City United States	Average Winter Temperature[a](°F)	Yearly Degree-Days (rounded off)
Pennsylvania	Philadelphia	43.0	5,000
	Pittsburgh	40.0	5,500
Rhode Island	Providence	38.8	5,950
South Carolina	Columbia	54.0	2,500
South Dakota	Rapid City	33.5	7,350
Tennessee	Memphis	51.0	3,200
Texas	Dallas	55.3	2,370
	Houston	61.5	1,300
Utah	Salt Lake City	38.4	6,050
Vermont	Burlington	29.4	8,270
Virginia	Norfolk	49.2	3,420
Washington	Seattle	46.0	4,800
	Spokane	36.5	6,650
West Virginia	Charleston	44.8	4,475
Wisconsin	Milwaukee	32.6	7,640
	Madison	30.9	7,850
Wyoming	Cheyenne	34.2	7,380
	Canada		
Alberta	Calgary	−10.9°C	9,700
British Columbia	Vancouver	2.4°C	5,510
Manitoba	Winnipeg	−18.3°C	10,680
New Brunswick	Moncton	−7.9°C	8,720
Newfoundland	St. John	−3.8°C	8,990
Northwest Territory	Aklavik	−30.0°C	18,000
Nova Scotia	Halifax	−3.2°C	7,360
Ontario	Ottawa	−10.9°C	8,750
	Toronto	−4.4°C	6,830
Prince Edward Island	Charlottetown	−6.7°C	8,165
Quebec	Montreal	−8.9°C	8,000
	Quebec City	−11.6°C	9,000
Saskatchewan	Regina	−17.3°C	10,800
Yukon	Dawson	−28.6°C	15,070

[a]Average values shown for winter temperatures are for population centers rather than airports or weather stations. U.S. temperatures are measured in degrees Fahrenheit; Canadian temperatures are measured in degrees Celsius. The degree-days are calculated by conversion factor.

Sources: Based on data from U.S. Weather Bureau and Canadian Government Department of Transportation, Division of Climatology. For more complete information, the reader is referred to the charts in *Environmental Control Principles* (New York: American Society of Heating, Refrigerating and Air-Conditioning Engineers, Inc., 1977), App. A, pp. 461–467.

BOX 10.3.1 ILLUSTRATIVE EXAMPLE

Suppose that your house is reasonably well insulated: 3½ in. in the walls, storm windows, some insulation around the perimeter, good attic insulation. You calculate the probable heating coefficient to be 12 Btu per degree-day per square foot of floor space. The floor area of the house is 1000 ft^2 and the number of degree-days for one year is 7000. (To estimate the degree-days for your locality, consult Table 10.3.2.) This will give the following energy-use figure:

$$12 \times 1000 \times 7000 = 84{,}000{,}000 \text{ Btu}$$

This is the rough equivalent of 800 gallons of fuel oil burned with an efficiency of 75%, or 102,500 ft^3 of natural gas burned with an efficiency of 80%, or 6 cords of hardwood burned with an efficiency of 50%, or 25,000 kWh of electricity used for resistance heating. For heat-pump calculations, multiply the resistance heating figure by the heat pump's average *coefficient of performance* (see Box 10.5.1).

BOX 10.3.2 COMPARATIVE HEATING COSTS

Here is a comparison of the costs of typical heating systems used in a cold climate. These estimates were developed by the Vermont State Home Energy Audit Team (HEAT). According to HEAT, an average house in a cold climate uses approximately 80 million Btu every year for space heating. At current fuel prices (1982), to keep this house at a uniform 65°F throughout the house, the following costs will be incurred.

System	Operating Costs	Comments
Electricity	$2120	Winter rates. Substantial reductions are possible with zone or individual room heat. Also, reduced rates are available during off-peak hours.
Kerosene furnace	1480	
LP furnace	1280	
Oil furnace	1080	
Gas furnace	944	Natural gas is not widely available in New England.
Kerosene space heat	936	
Coal stove	680–792	Depending on efficiency.
Wood stove or furnace	544–592	Depending on efficiency.

Source: Vermont State Energy Office

10.3.2 Water Heating

The amount of energy used for water heating depends on the number of people in the household, the frequency and duration of showers, baths, and dishwashing, the amount of laundry done per week, and the temperature of the hot water. Top-loading washing machines use twice as much water as the front-loading type. Each time a top-loading machine is filled, depending on the size, between 12 and 25 gallons of water is used. The average U.S. resident uses about 600 gallons of hot water per month. If the water heater is set at 140°F (a typical figure), this will require approximately 500,000 Btu of energy per month. Depending on the method of heating water, this figure can be converted into kilowatthours, gallons of fuel oil, or cubic feet of natural gas, using the conversion factors listed in Appendix A.

10.3.3 Lighting

Calculating energy use for lighting requires careful observation. Over a period of a week, keep records of which lights are turned on, and how many hours they stay on. Then note the *wattage* of each light. Multiply the wattage of each light by the length of time it stays on per week. This will give the number of *watthours* of electricity used by that light. Adding up all the individual figures will give the total watthour consumption of your household for lighting.

BOX 10.3.3 ILLUSTRATIVE EXAMPLE

A kitchen light is turned on for a total of 36 hours in one week. The lamp is a three-bulb incandescent lamp containing three 60-W bulbs. Thus the power rating of the lamp is 180 W. Multiplying this by 36, we find that the lamp uses approximately 6500 watthours per week for this light fixture. This translates into 6.5 kWh per week.

$$180 \text{ W} \times 36 \text{ hr} = 6480 \text{ watthours} = 6.48 \text{ kWh}$$

Adding this to the energy consumption of the other light fixtures in the household, we can compute the house's total energy use for lighting.

10.3.4 Food Preparation and Storage

Energy used for cooking and food storage depends largely on personal habits, the size of the family, the type of appliances used, and other factors.

The energy efficiency of a refrigerator/freezer is determined by size, insulation, method of defrosting, location, temperature setting, and the amount of empty space in the compartments. Manufacturers of cooking appliances are required to calculate the average annual energy consumption of their products. This information should be available from an appliance dealer.

Microwave ovens, toaster ovens, and regular stoves have widely different energy requirements. The same food preparation, done using different appliances, will require widely varying amounts of energy. In general, use of the smaller appliances will save energy. But for purposes of a preliminary estimate, the manufacturer's estimate of annual energy use should be a good starting point.

10.3.5 Transportation

Calculate your car's fuel economy. This is fairly easy to do; just keep track of how much gasoline or diesel fuel you buy in a month's time and also calculate how many miles you drove in that month. Then divide the number of miles by the number of gallons. Using this gas mileage figure, you should be able to calculate your transportation energy use by simply reading the car's odometer. (Note: The EPA mileage ratings for cars are not reliable figures for calculating actual energy consumption; they should be used only to compare the relative economy of different automobiles.) Automobile energy use should be calculated for an entire year's time, because automobile use fluctuates greatly from season to season. A yearly total will compensate for these seasonal fluctuations.

Trips on airplanes or other forms of public transportation are not part of household energy use. But as we will show in Section 10.5.6, some forms of public transportation are much more efficient than automobile travel; thus overall energy consumption can be reduced in some cases through the use of public transportation rather than private automobiles.

10.3.6 Entertainment

Since the invention of the radio, electronic entertainment devices have played an ever-increasing role in people's lives. Television, video games, and recently home computers, have made their way into a very large number of households. The energy consumption of these entertainment devices varies dramatically, from low-powered, highly efficient audio equipment to "instant-on" color television sets, which use a considerable amount of electricity even when they are not being watched, because the cathode heater is on all the time. To determine the amount of energy consumed by electrical appliances, determine the wattage (often printed on the nameplate, found on the side or back panel) and multiply by the average number of hours the appliance is used. Table 10.3.3 lists the energy requirements of typical household appliances. By using these figures, you will be able to calculate the electrical consumption of most of your home appliances.

TABLE 10.3.3 **Typical Household Appliance Loads**[a]

Appliance	Average Power (W)	Annual Energy Use (kWh/yr)
HVAC system (climate control)		
Air conditioner, window type	1,400	600
Dehumidifier	260	375
Fan, attic	370	300
Fan, exhaust or circulating	90	50
Fan, furnace	280	400
Fan, movable	175	140
Heat lamp, infrared	250	15
Heat pump system (heating and air conditioning)	11,800	16,000
Heater, portable	1,400	200
Heater, baseboard-type system	15,000	17,000
Humidifier, room	60	60
Humidifier, large	175	160
Oil burner	265	410
Air conditioning, central	5,000	2,500
Kitchen		
Broiler	1,450	100
Can opener	50	5
Carving knife	90	10
Coffee maker	900	110
Deep fryer	1,450	80
Dishwasher	800	250
Blender, food	500	25
Mixer, food	125	15
Drain food waste disposer	450	30
Frying pan	1,200	200
Grill	1,150	40
Hot plate	1,250	90
Oven-range	5,000	1,200
Roaster	1,350	200
Toaster	1,150	40
Waffle iron	1,125	25
Living room		
Radio	30	30
Hi-fi system, tube type	115	160
Hi-fi system, solid state	30	60
Television, black and white	250	375
Television, color	350	500
Electric clock	2	18
Food storage		
Freezer, 15-ft^3 chest type	350	1,200
Refrigerator, standard, 14 ft^3	300	750
Refrigerator-freezer	325	1,150
Refrigerator-freezer, frostless	360	1,600
Lights		
60-W bulb	60	75
100-W bulb	100	110

TABLE 10.3.3 Typical Household Appliance Loads (Cont.)

Appliance	Average Power (W)	Annual Energy Use (kWh/yr)
Fluorescent fixtures, 4-ft size	50	150
Three-way light fixture	200	280
Laundry room		
Clothes dryer, gas	400	150
Clothes dryer, electric	4,850	1,000
Iron	1,000	100
Washing machine, automatic	400	100
Washing machine, old-fashioned	200	60
Water heater, 40 gal	4,500	4,500
Bathroom		
Toothbrush	7	5
Hair dryer	390	20
Shaver	15	20
Light fixtures	100	100
Exhaust Fan	200	60
Cleaning department		
Vacuum cleaner	750	100
Floor polisher	300	15
Bedroom/family room		
Electric blanket	200	150
Sewing machine	100	15
Shop/basement		
Water pump, ½ hp	460	230
Shop drill, ¼ in., ⅙ hp	250	10
Skill saw, 1 hp	1,000	70
Table saw, 1 hp	1,000	60
Lathe, ½ hp	460	10

[a]The numbers shown (rounded off for clarity) are average values and should be taken only as a rough guide to actual energy consumption. The actual annual use will depend on factors such as your life-style, habits, and the actual size and models of the appliances. Many new appliances are required to have clear labels indicating their power and energy consumption. These are called "energy guides." These guides also give estimated energy costs per year at common electric rates.

Sources: Data based on the following sources: Cornell University Extension Service; National Center for Appropriate Technology; and the State Energy Offices of Vermont and New York.

10.3.7 Cleaning, Lawn Care, Maintenance, etc.

A century ago, most cleaning chores were done by hand. But since the invention of the automatic washer and dryer, vacuum cleaner, power lawn mower, and electric hedge trimmer, most cleaning and maintenance chores involve some kind of power equipment. Appliances that use electric resistance heat, such as electric clothes dryers, use substantial amounts of energy. Those that operate with motors use less electricity but can contribute to peak demand. Electrically powered appliances should be treated like home entertainment appliances for purposes of energy-use calculations. For gasoline-powered machines, gasoline consumption can be measured over a period of time in much the same way as calculating automobile fuel economy.

10.3.8 Total Energy Use

After calculating approximate energy use in each of these categories, it is necessary to see how these figures relate to actual energy consumption. Grouping energy consumption by type of energy used (i.e., all electrical energy grouped together, all uses of fuel oil combined, etc.), see what the total figures will be for each type of energy. Once that is done, it will be a simple matter to test those figures against actual energy consumption.

10.4 MEASURING ENERGY CONSUMPTION

Since all energy sources for a house are purchased (with the exception of solar energy), energy suppliers will provide a bill indicating the amount of energy sold. To keep track of overall energy use, keep all records of fuel purchases. This will tell you how many gallons of oil, cubic feet of gas, kilowatt hours of electricity, or cords of wood you used during the year. Comparing this with the estimates you made earlier, you should be able to get a good picture of your overall energy consumption.

In some cases, the estimated and actual use figures will not agree. If the differences are 10% or less, you may assume that your estimates are essentially correct. (*Note:* This does *not* necessarily mean that you are using energy as efficiently as possible. Suggestions on reducing your energy use will be offered in Section 10.5.) In the case of larger discrepancies—especially 20% or more—it will be necessary to search for the source of the discrepancy. If you use 20% *less* fuel than your estimate, perhaps you are already practicing some conservation measures, and you should not worry too much. But if your energy use *exceeds* your estimates, you may want to take a careful look at your patterns of energy consumption.

10.4.1 Possible Sources of Discrepancies

If your energy use for heating, cooling, and air conditioning are significantly higher than your estimates, two areas are likely to be at fault: inadequate insulation or excessive use of energy. Both these problem areas can be difficult to locate, but a few standard methods may be helpful.

1. *Detecting inadequate insulation.* Walk through your house on a cold day touching the outside walls. Are there areas of the walls which are much colder than the average? These may be areas where insulation has settled or was never installed. Are there spots of water condensation on the walls? If so,

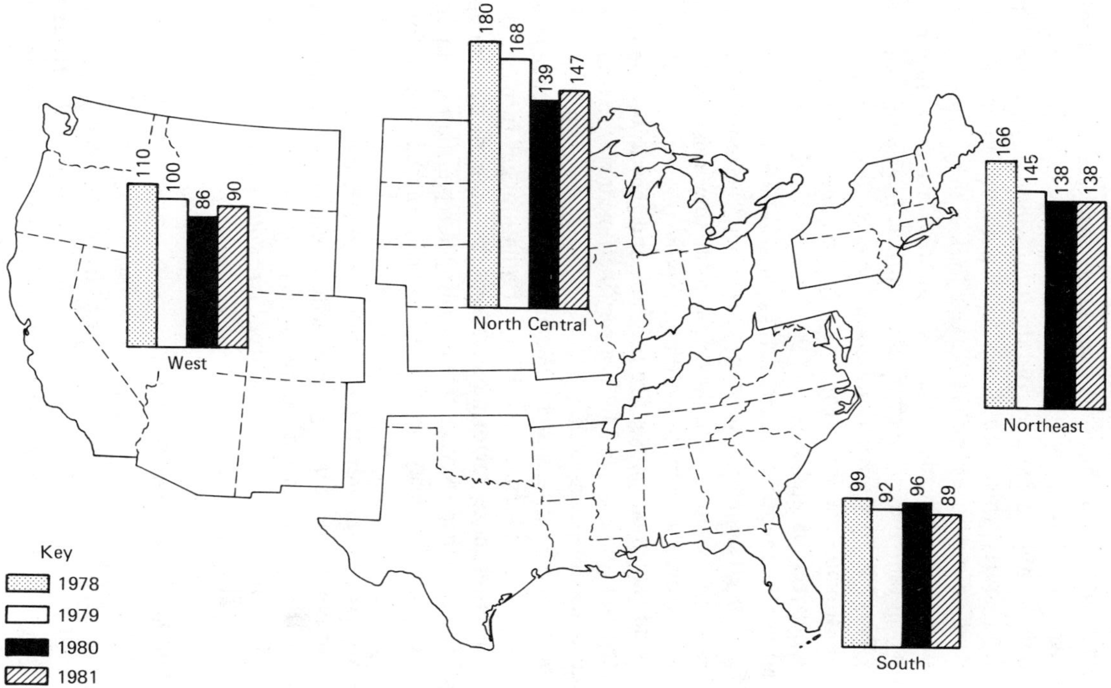

Figure 10.4.1 Average household energy consumption by region, 1978–1981 (Million Btu). (*Source: Monthly Energy Review,* Energy Information Administration, Department of Energy, Sept. 1, 1983.)

this is another likely spot of excessive heat loss. How about the attic insulation? Is there as much as you thought? Check for settling and for areas where the insulation was improperly installed or removed during repairs. Any of these problems can lead to greater heat loss than expected.

2. *Finding leaks.* If the house is adequately insulated, the heat may be escaping through cracks in the walls or around windows. If there are major leaks, you should be able to feel the cold draft on windy days. If your house is not particularly drafty, this is not likely to be a major source of heat loss.

3. *Checking your personal habits.* Take a look at the thermostat. Where is it set? If your thermostat reads 75 °F, this is almost certainly a major reason for using more heat than the estimate prepared earlier. In general, you can add approximately 2 to 3% to your heating costs for each degree you raise your thermostat. The estimates developed earlier apply to a house with the thermostat set somewhere between 65 and 70°F.

If you adjust the thermostat often, this can also be a source of trouble. Some people, in an attempt to save energy, set their thermostats at uncomfortably low settings. Then, when they feel cold, they raise their thermostats—often just before the furnace would have gone on anyway. After a few minutes, they find that their houses are too hot, and they lower the thermostats, starting the cycle once again. This is a very inefficient way to heat a house, and it can cause discomfort as well.

4. *Checking your heating system.* We have suggested efficiency levels of around 75% for a fuel oil furnace and 80% for a natural gas furnace. If your furnace is not meeting those performance standards, your fuel consumption may be significantly higher. A visit from a furnace repairman or an energy auditor can save you both energy and money if your furnace is not performing efficiently.

5. *Checking your estimates.* If you are unable to locate the source of your excess energy consumption, you may have made an error in your calculations. Have you chosen the wrong heating coefficient for your house? Have you made a mistake in estimating the house's floor area? Are your degree-day figures accurate? Are your calculations correct?

10.4.2 Revising Your Estimates

If you have still been unable to locate the source of discrepancy, even after going through all these steps, your energy use is probably being affected by a factor that you have not considered. Perhaps your personal habits have a greater effect than you imagine, or there may be a defect in the construction of your house. These sorts of discrepancies can be very difficult to locate. If you are attentive, you may be able to discover an opportunity for significant energy conservation. In any case, you should revise your energy estimates so that they are more in line with your actual measured energy consumption.

10.5 OPPORTUNITIES FOR ENERGY CONSERVATION

Once you have estimated your energy use, you may want to evaluate the possibility of lowering your personal energy consumpton. Conserving energy can have significant benefits, including the following:

Lower energy costs
Greater personal comfort
Reduced labor (especially if your house is heated with wood)
Possible reductions in future energy price increases
Conservation of natural resources

As we have explained in Chapter 4, energy conservation does not mean "being too cold in winter and too warm in summer," as many people seem to believe. Intelligently chosen and applied, energy-conservation measures can actually *increase* comfort levels in the home.

Some conservation measures, such as improved insulation, require no changes in life-style, whereas others can require fairly substantial changes. We are not suggesting that all the measures listed should be adopted. But all of them deserve serious consideration. Studies have shown that energy consumption in the United States can be reduced by up to 50% with no reduction in our standard of living. Indeed, some European countries have a higher standard of living than the United States with substantially lower energy consumption per capita (see Section 8.2). Personal energy conservation can make a significant contribution toward the goal of national energy self-sufficiency.

A word about costs. Some conservation measures, such as lowering thermostats or turning off lights and appliances, cost nothing at all. In fact, they begin saving money at once. Other conservation measures, such as replacing old, inefficient appliances, can have a very high initial cost and a longer payback period. These replacements are not usually carried out until the appliance is old and worn out anyway. Cost effectiveness is an important consideration in energy conservation, as it should be.

In the rest of this chapter, we will offer a number of suggestions for reducing personal energy use. Most of the energy-conservation measures described will have very short payback periods, often less than six months. When possible, we will give a rough estimate of the payback period for each energy-conservation measure discussed.

When assessing a conservation option, it is important to consider the question of *net* energy conservation. By replacing an older automobile with a newer model, you may be able to save some household energy. But a new car requires a large amount of energy for its manufacture and delivery. Unless the energy savings from the new automobile are fairly substantial, it hardly makes sense to replace the old one if it still works adequately. In other cases, replacing an old, inefficient product makes very good sense in terms of net energy savings. We will identify any conservation measures where net energy savings can be a problem.

10.5.1 Space Heating and Cooling

In northern regions of the United States, space heating is a major area of household energy use. A typical house consumes approximately 80 million Btu for heating alone. In the southern regions of the United States, heating requirements are considerably lower, but most houses have some kind of cooling system to provide comfort during the hot summer months. In the mid-Atlantic states, the Midwest, and large areas of the West and Southwest, temperatures are high in the summer and cold in the winter. In these areas of the country, both heating and cooling are major energy users in large numbers of households. The principles of conserving energy for heating and for cooling are quite similar, so the following discussion is applicable to all geographical areas.

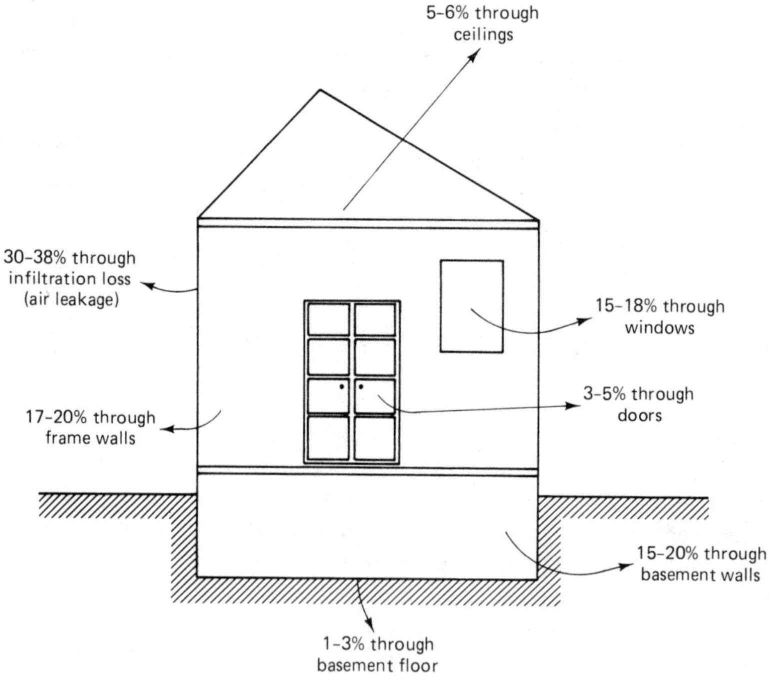

Figure 10.5.1 Heat-loss pathways in a typical insulated house. Infiltration occurs through cracks in walls, windows, foundations, doors, joints, chimneys, and air vents.

Getting the most out of your HVAC system

Sizing. When installing a HVAC (heating, ventilating, and air conditioning) system, it is desirable to select a system that is the right size for the house. Large or oversized systems will be less efficient because of frequent on and off cycling. The efficiency of any HVAC system is lower during the start-up and shutdown phases of operation than when the system is running steadily. Thus oil and gas furnaces, wood or coal stoves, and air-conditioning systems should be selected carefully. Their ratings will be stated in terms of output (in Btu per hour) or heating capacity (in cubic feet or cubic meters).

Needless to say, the system's output must be sufficient to keep the building warm or cool enough in times of extreme weather. But in general, it is best to select the smallest system that will do an adequate job of heating or cooling the house. This not only saves energy; it also saves money. Smaller systems are invariably cheaper than larger systems, all other factors being equal. Also, a larger system tends to be noisier than a small system and can cause overheating in parts of the house. If you find that your heating system goes on and off too often, the system may be oversized. Often a smaller nozzle for the furnace will solve the problem.

Efficiency. Not all HVAC systems are created equal. Whether the choice is a furnace, a wood stove, a heat pump, or an air conditioner, efficiency and economy of operation will vary among types, manufacturers, and models. Once the appropriate type of system has been selected (see Section 10.1), the purchaser should try to determine whether the product has an efficiency rating or, in the case of a heat pump, a coefficient of performance (COP) (see Box 10.5.1). Dealers often do not have such information or are reluctant to provide it; in fact, many dealers will insist that efficiency or coefficient of performance is not important. Certainly, efficiency is not *everything*. Convenience, reliability, and price are all important considerations. But energy efficiency is an important factor and should be considered in any purchase of a HVAC system. Prudent consumers will do a bit of research and obtain comparative figures before settling on a system.

Maintenance. For optimum performance, the system should be in excellent working condition. Moving parts should be lubricated, heat exchanger surfaces should be clean and free of deposits, and combustion chambers, motors, and belts should be tuned for maximum efficiency. An annual check-up and continued maintenance by trained personnel is a good idea. Utility companies and many state governments offer free energy audits to the public. These energy audits usually include efficiency tests of HVAC equipment.

Location. The location of HVAC systems also plays an important part in the efficient use of energy. Heating systems should be centrally located so that heat distribution is even and efficient. Chimneys and flue pipes should also

BOX 10.5.1 HEAT PUMPS: COEFFICIENT OF PERFORMANCE

Heat pumps do not convert electric energy directly into heat; rather, they use the kinetic energy of electricity to move heat from one place to another: in winter, from the outdoors to the indoors, and in summer from the indoors to the outdoors. A refrigerator is a familiar example of a heat pump.

When the temperature difference is small, using electrical energy to move heat is much more efficient than converting electrical energy directly into heat. The *coefficient of performance* (COP) describes the ratio between the amount of heat obtained from the heat pump and the amount of heat that would be obtained through direct conversion (resistance heating).

$$\text{COP} = \frac{\text{heat transferred by heat pump}}{\text{energy input}}$$

Thus, if the coefficient of performance is 2.5, the heat pump will produce 2½ times as much useful heat (heat moved to the desired space) as an electric resistance heating system, for the same amount of electrical energy.

The COP of a heat pump will vary from one climate to another. In extremely cold weather, a heat pump may actually be *less* efficient than electric resistance heating (the COP will be less than 1.0). For this reason, it is important to determine the COP for a heat pump *in your particular climate* before deciding to make a purchase.

be located in the center of a house as a built-in thermal mass. If they are located on outside walls, efficiency can be reduced because of heat losses from the stack. In the case of wood heat, corrosion and creosote buildup are more of a problem when the chimney or flue pipe is located on an outside wall. This is because the low temperature of the flue increases condensation on the inside, thus increasing creosote buildup and corrosion of metal pipes.

Home weatherization. In older houses, infiltration can be a major source of heat loss. Heated air can escape through cracks around doors, windows, electrical outlets, and the joints of walls. In some cases, the walls themselves may have cracks that permit the escape of heated air. Reducing this unwanted infiltration can be a highly cost-effective way of reducing energy consumption and costs.

Methods of Weatherizing. Caulking around window and door frames will reduce infiltration significantly at negligible cost. Weather-stripping loose windows and doors will also prevent the escape of heated air. Adding storm windows will cut down on infiltration as well as reduce transmitted heat loss through the windows. Some people tape polyethylene sheeting over windows during the winter to reduce both infiltration and transmitted heat losses.

Materials. Most hardware stores carry a large selection of caulking compounds, weather stripping, clear plastic sheeting, and tape. Materials should be chosen carefully, not only on the basis of low cost, but also to ensure good performance. For example, if plastic sheeting is simply taped onto a window frame with masking tape or duct tape, the paint on the window frame can be damaged. However, special weather-stripping tape reduces this problem at a relatively low cost. Hardware suppliers or energy auditors should be able to describe the advantages and drawbacks of the various weatherizing products. How-to books and pamphlets are available from the U.S. Department of Energy, state energy offices, and other sources. These publications describe products and methods for weatherizing. (See the list of sources at the end of the chapter.)

Costs. Weatherizing is generally one of the most cost-effective ways of reducing energy consumption. Payback periods are usually less than one year except in the case of storm windows, where the high cost may result in a payback of up to five years. A second advantage of weatherizing is that it greatly reduces uncomfortable drafts, thereby encouraging lower thermostat settings in the house. Thus, weatherizing makes especially good sense as a personal energy-conservation measure.

Humidity and HVAC. During cold weather, indoor air can become uncomfortably dry. Adding humidity to the air often makes a significant difference in comfort and health levels. 65 °F (18 °C) will feel much more comfortable with adequate humidity in the house. More moisture in the air will not only reduce your heating bill, but also prevent chapped skin, dry, sore throats, and bleeding noses.

To add humidity to a house, several choices are available. Commercial humidifiers are both effective and expensive. A large, flat pan of water placed on a radiator or wood stove can do nearly as good a job. Drying laundry in the house can add humidity to the air, and venting an electric clothes dryer inside the house (using appropriate filtering materials) can provide both heat and humidity.

During the summer, high humidity contributes to personal discomfort. Weather bureaus calculate a *temperature-humidity index* to account for the effect of humidity. Lowering the humidity can make as much difference as lowering the air temperature. Often a simple dehumidifier can eliminate the need for air conditioning while preventing mold and mildew inside the house.

Indoor air pollution. Although it is not as widely publicized as outdoor air pollution, indoor air pollution is quite a serious problem. Most indoor pollution comes from sources within the structure itself (see Table 10.5.1). Naturally, the concentration of pollutants is greatest near the source. But the limited air movement inside a house increases the concentration of pollutants. In pursuit of energy efficiency, many householders are making their houses "tighter," thus cutting down on ventilation. This reduces the rate at which pollutants can be exhausted from the dwelling place, leading to an even greater concentration of indoor air pollutants. The result can be a general feeling of unease, or even perceptible health effects.

TABLE 10.5.1 Sources of Indoor Air Pollution

Pollutant	Source
Radon and radioactive materials	Masonry, natural soil
Carbon monoxide	Stoves, space heaters, attached garages
Formaldehyde vapor	Foam insulation, paneling, carpets
Asbestos and other fibers	Insulation, filters, construction materials
Nitrogen oxides	Space heaters, stoves
Particulates	Wood stoves, cleaning, general repairs
Aromatic compounds	Cleaning, electronics
Microorganisms (bacteria, viruses, and fungi)	Air-cleaning filters, heat exchangers, humidifiers, air conditioners

Formaldehyde, a serious irritant and a potentially cancer-causing substance, is considered one of the worst indoor pollutants. Formaldehyde vapors are given off by many commonly found products—foam insulation, construction boards (plywood, particle board, and paneling), furniture, textiles, and other products. During the 1970s, thousands of houses were insulated with urea formaldehyde, a highly insulating foam. In many cases, the foam was installed incorrectly, and formaldehyde fumes escaped into the houses, causing nausea and general discomfort. Removing the foam has proved to be an expensive and complex process costing an average of more than $10,000 per dwelling.

Asbestos fibers float in indoor air due to the presence of asbestos-based roofing, flooring, some acoustic tiles, insulation, filters, and other common household items. These suspended particles can lodge in lungs and on the skin, causing irritation and possibly cancer. Many schools and other public buildings have asbestos-based tiles on their ceilings. These ceilings must be replaced or coated with sealant in order to reduce the concentrations of asbestos fibers in the air.

Chemical fumes are released by hundreds of consumer products, especially cleaning and deodorizing fluids. Spray insecticides and pest strips release highly toxic vapors into the air to be inhaled by inhabitants of the house. Hair sprays, perfumes, and many other cosmetic products also release fumes which can become concentrated in well-sealed houses.

Combustion products such as carbon dioxide, carbon monoxide, and oxides of nitrogen are released during most types of combustion. Stoves, heaters, and other appliances are the major producers, but automobiles in attached garages can also be a source of these combustion products. Gas stoves are also high on the list of indoor air pollution sources.

Suspended particles can reach very high densities inside houses. Dusting and vigorous activity can stir up particles. Vacuum cleaners are especially serious offenders because they blow air into the room at high rates of speed. This moving air can easily stir up dust from horizontal surfaces and blow it into the air. House dust can carry spores, bacteria, and other irritating or harmful substances.

Environmental laws do not apply to the atmosphere inside a household, but the indoor environment can be a source of danger to residents of a tightly sealed house. Most building codes include ventilation requirements for commercial buildings, but few such standards have been developed for residential housing. Products such as air-to-air heat exchangers allow for sufficient ventilation at the same time that they reduce heat loss due to the escaping air. But heat exchangers are currently available in only a tiny percentage of new housing. Awareness of the problem is the householder's best defense.

Insulation. Regardless of the heating or cooling system used, every household can benefit from the proper installation of good-quality insulation in proper amounts. A well-insulated house will not only keep heat in during the winter; it will also keep heat out during the summer. For this reason, insulation is important not only in cold climates but also in areas with hot, humid summers.

All exterior walls, ceilings, and floor spaces should be insulated. How much? It depends on the climate. The latest recommendations for new construction are listed in Table 10.5.2. Higher values are for colder climates.

TABLE 10.5.2 Recommended *R* Values in Homes

Exterior walls	*R*-14 to *R*-22 (3–6 in. of insulation; *R*-22 requires 6-in. wall studs)
Attic/ceilings	*R*-33 to *R*-38 (9–12 in. of insulation)
Floors	*R*-11 to *R*-22 (2–6 in., depending on the method of installation

Types of Insulation. The marketplace offers a wide choice of insulating materials. Materials currently available include rock wool, glass fiber, cellulose fiber, vermiculite, perlite, polystyrene boards and beads, polyurethane board, and isocyanurate boards. Urea-formaldehyde foam was used for a number of years because of its high insulating value and its ease of installation, but it is no longer recommended because of health hazards associated with its use.

Insulation comes in a variety of forms. Blankets and batts of insulating material are especially suited for frame walls, floors, and ceilings. Loose fill is good for attic floors and hard-to-reach or irregular-shaped areas. Blown-in insulation is suitable for retrofitting older frame walls with air spaces between the inner finish surface and the outside sheathing. It is also suitable for attics and hard-to-reach spaces. Rigid insulation is suitable for basements, masonry walls, underground insulation, and retrofitting of lath and plaster interior walls. Specially shaped insulating products are also available for water pipes, heating

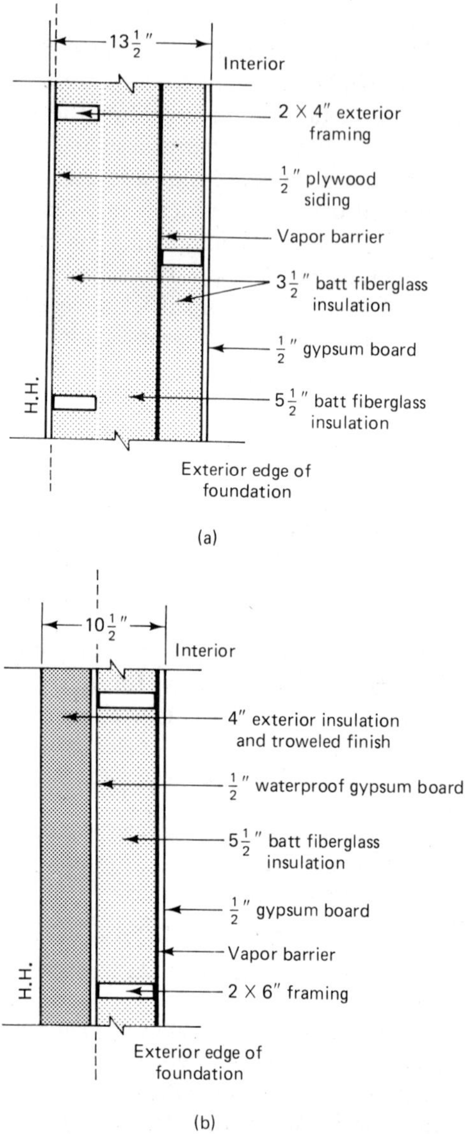

Figure 10.5.2 Superinsulated wall options: (a) double-stud wall, (b) single-stud wall with exterior insulation system. [Reproduced with permission from *Superinsulation: A Housing Trend for the Eighties* (Butte, MT: National Center for Appropriate Technology, 1980).]

BOX 10.5.2 *R* VALUES AND HEAT LOSS

The amount of heat transferred from a higher-temperature region to a lower-temperature region depends on the conductivity of the material separating them. The *U value* of a building component (walls, ceilings, floors, windows, doors, etc.) is a measure of how much heat that component will conduct from the cooler to the warmer side. The units in the British system are Btu per hour per square foot per degree Fahrenheit. This means that if a surface has a U value of 2 and there is a temperature difference of 10 °F between the two sides, the surface will conduct 20 Btu per hour per square foot from the warmer side to the cooler side. In the Système International, the units are watts per square meter per degree Celsius.

Obviously, the smaller the U value, the better the material functions as an insulator. It does not matter whether we are trying to save on heating or on cooling. In either case, a lower U value means energy savings.

For purposes of calculation, U values can be somewhat clumsy to work with, especially for calculating the conductivity of a wall made up of several different materials. For this reason, a second measure, called the *R value,* is commonly used. The R value is a measure of how well a surface resists the flow of heat. The R value is the reciprocal of the U value, and vice versa:

$$R = \frac{1}{U} \qquad U = \frac{1}{R}$$

Most common building materials, especially insulating materials, have R values listed on the package. The table below lists some R values for typical building components. Building supply dealers should be able to provide more detailed information on materials that are not listed in this table.

R Values of Typical Building Components

Material	R Value[a]
Wood (per inch)	1.2
1/2-in. plywood	0.6
Building paper	0.1
Lapped bevel siding (wood)	0.8
Shingles (wood)	0.8
Stucco (per inch)	0.2
Fiberboard sheathing (1/2 in.)	1.2
Concrete (per inch)	0.0
Brick (per inch)	0.2
Gypsum board (1/2 in.)	0.4
Carpet with fiber pad backing	2.0
Carpet with foam rubber padding	1.2
Extruded polystyrene sheathing (1/2 in.)	2.5
Vermiculite, loose fill (per inch)	2.0
Perlite, loose fill (per inch)	2.7
Fiberglass insulation (per inch)	3.3
Rock wool (per inch)	3.3
Polystyrene rigid board (per inch)	3.4
Cellulose fill (per inch)	3.7
Urethane foam (per inch)	5.3
Urea-formaldehyde foam (per inch)[b]	4.5
Isocyanurate rigid insulation	7.0

[a]To convert these R values to the R values used in the Système International, multiply by 0.176. Thus the R value of wood is $1.25 \times 0.176 = 0.22\ \text{m}^2 \times$ degree Celsius per watt.

[b]Not recommended because of toxic vapors.

To find the R value of a surface, add the R value of all the components. Figure 10.5.3 shows how this is done for a typical wall. This number can then be converted to a U value by calculating the reciprocal. For example, if a wall has a total R value of 22, the U value is 1/22, or 0.045. When multiplied by the area of the wall, this will yield the heat loss through the wall per hour per degree Fahrenheit (see Figure 10.5.4). When these figures are added up for all the surfaces of the house and added to the heat losses due to infiltration, the resulting number is the house's total heat loss per hour per degree Fahrenheit.

In an earlier section of this chapter we showed how to estimate a house's seasonal heat loss based on *degree-day*. But the R-value method yields a heat-loss figure per *degree-hour*. To convert from degree-hours to degree-days, simply multiply by 24. Example: The wall in Figure 10.5.4 has a total heat loss of 6.345 Btu per hour per degree Fahrenheit. This means that the heat loss per degree-day equals the following:

$$\begin{aligned}\text{heat loss per degree-day} &= 24 \times 6.345 \\ &= 152.28 \text{ Btu}\end{aligned}$$

In a climate with 7000 degree-days (a relatively cold climate), this wall's total heat loss for one year will be roughly equal to the following:

$$\begin{aligned}\text{heat loss per heating season} &= 7000 \times 152.28 \\ &= 1{,}065{,}960 \text{ Btu}\end{aligned}$$

A word of caution. These quantitative estimates are deceptively precise. Other factors also contribute to a building's heat loss. For example, all buildings lose heat by *infiltration* (heated air escaping through cracks), which is very difficult to calculate accurately. Wind speed, thermostat settings, type of foundation, and quality of construction can also affect the amount of heat loss from a building. These calculations provide only a rough estimate of a house's heat losses.

ducts running through unheated areas, and water heaters. In general, it is easy to find the proper insulating materials for any job.

Installation. In general, insulating materials should not be exposed to the living space. The insulating properties are created by many isolated air pockets between tiny particles or fibers of the insulating material. If the insulating material is exposed to the air, the insulating value will be reduced. Fiberglass strands can also break off and float in the air in the living space, causing irritation and even respiratory problems. Many insulating materials, especially polystyrene, are flammable and give off toxic fumes during combustion. Obviously such products should not be exposed to the living space. All these problems can be avoided if the insulation is sealed behind a barrier surface such as plywood, gypsum board, masonry, plaster, etc. Attic insulation can be exposed to the enclosed space as long as the attic is not used as living quarters.

When walls are insulated, special attention must be paid to the problem of moisture buildup. Condensation inside the walls can reduce the R value and ultimately cause deterioration of the insulating materials. Installation of a vapor

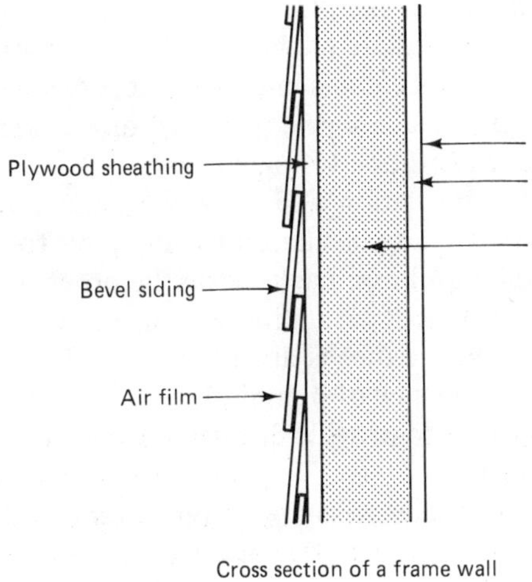

Component		R value
Air film, interior		0.68
Gypsum board or (lath and plaster)		0.32
$3\frac{1}{2}''$ glass fiber		11.00
Sheathing, plywood		0.50
Siding, bevel		0.80
Air film, exterior*		0.17
Total R value	≅	13.5

*Depending on the wind speed

Figure 10.5.3 Calculating the *R* value of a typical wall (cross section of a frame all).

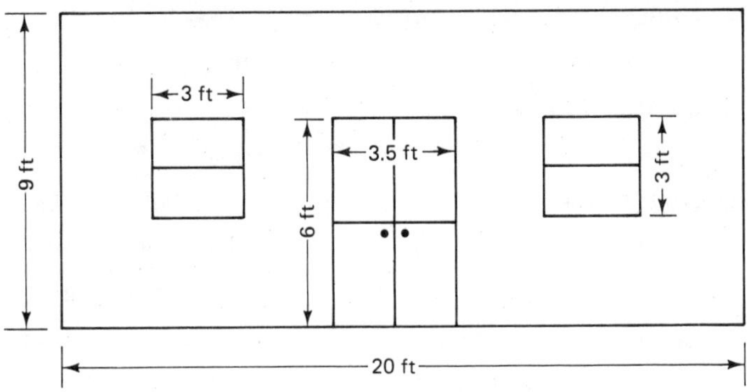

Net wall area = wall area (20′ × 9′)
− door area (6′ × 3.5′)
− window area [(3′ × 3′) + (3′ × 3′)]
= (180 − 21 − 18) ft^2
= 141 ft^2

Heat loss through the wall (per °F per hour)
= area of the wall × U
= 141 ft^2 × 0.045
= 6.345 Btu/°F/h

Figure 10.5.4 Computing heat loss through a wall.

barrier, facing the living space, will help prevent such problems. (Energy auditors tell us that too often, the vapor barrier on fiberglass roll insulation is installed facing the wrong way. Make sure that you pay attention!) No vapor barrier should be installed between attic insulation and the heated living space.

Costs and Payback. Depending on the particular application, insulation can have either a short or an intermediate payback period. Insulation of water heaters, hot water pipes, and furnace ducts in crawl spaces can provide a payback of less than one year. This means a rate of return on your investment of more than 100%. No bank can hope to match this rate!

Floor, basement, or foundation insulation also has a very short payback period, usually less than three years. Ceiling insulation usually pays for itself in two or three years, and wall insulation has a payback on the order of three to five years. If a house is poorly insulated, insulation will have a very short payback. If insulation is added to a house that is already insulated, the payback is somewhat longer. (Remember the law of diminishing returns? See Section 5.8.) Energy auditors should be able to provide detailed payback information for any proposed insulating projects.

It is important to realize that the energy savings from several different measures are not additive. For example, if you carried out five different insulating and weatherizing methods, each of which reduced your energy use by 10%, the total energy savings would not be 50% but less—actually around 40%.

original energy consumption	=	100%
savings from first insulating project	=	0.1 × 100%
	=	10%
energy use after first project	=	100% − 10%
	=	90%
savings from second project	=	0.1 × 90% = 9%
energy use after second project	=	90% − 9% = 81%
savings from third project	=	0.1 × 81% = 8.1%
energy use after third project	=	81% − 8.1% = 73.1%
savings from fourth project	=	0.1 × 73% = 7.3%
energy use after fourth project	=	73.1% − 7.3% = 65.8%
savings from fifth project	=	0.1 × 65.8% = 6.5%
energy use after fifth project	=	65.8% − 6.5% = 59.3%
total energy savings, all projects	=	100% − 59.3% = 40.7%

10.5.2 Water Heating

Water heating consumes a significant portion of the average household's energy use, and opportunities for saving energy in this area are substantial. In the average household, water heating is the second largest domestic user of energy, excluding the automobile. Energy use for water heating can easily be cut in half with no reduction in comfort or convenience. Those who are willing to modify their personal habits slightly can save even more.

Choosing a water-heating system. If you are planning to install a new water-heating system, it makes sense to consider a range of possibilities. Selection of a water-heating system can make a significant difference in energy consumption.

1. *Tank water heaters* are the most common water-heating systems. They maintain a reservoir of water, anywhere from 20 to 80 gallons, at a desired temperature. Tank heaters are widely available using a range of fuels, including electricity, gas, fuel oil, and propane. They can easily be integrated with solar water-heating systems.

2. *Tankless water heaters* heat water on demand but do not maintain a reservoir of heated water. They use less energy than tank systems because they do not use energy to heat water that is simply being stored. If they are properly sized, tankless water heaters can provide an unlimited supply of hot water, unlike tank heaters. They are generally considered to be the ideal backup system for solar water-heating systems.

   Tankless water heaters are not as widely available as tank heaters, although they can be ordered in all parts of the country. As a rule, they are fueled by electricity or natural gas. In areas with very hard water, the water lines in tankless heaters can become obstructed by minerals. In these areas, it is necessary to flush the heater periodically with special solvents to dissolve mineral deposits. But if these precautions are followed, tankless heaters can be used anywhere.

3. *Solar water-heating systems* vary from manufacturer to manufacturer and from installation to installation. The payback periods can vary from five years to indefinite, depending on the location and orientation of the house, the alternative fuels, and patterns of family water use. Clearly, solar water-heating systems can save large amounts of energy, but careful cost analysis is essential to produce satisfying results.

4. *Wood-fired water-heating systems* are available in many areas of the country. The most common type of system is a water jacket which can be installed in the back or side of a wood stove. Water can be circulated from the water-heating tank to the stove jacket, or the stove can be used to preheat the water before it enters the tank. Wood-fired tank water heaters are also available. In these systems, the owner builds a fire about half an hour before the hot water is needed.

All water-heating systems must have pressure-relief valves to prevent water-line breaks or explosions. Any commercially available water-heating system will have this safety feature built in. But do-it-yourself installers sometimes neglect this essential safety feature, with unfortunate results.

Patterns of water use. An easy place to start conserving is with the thermostat setting. Many water heaters are routinely set at 160 °F (70 °C). This is a very high setting. 120 °F (50 °C) is certainly adequate for all purposes except operating certain types of automatic dishwashers. Also, if the domestic hot water system is tied in with the space-heating system, a high temperature setting will have to be maintained in both systems during the winter to maintain a high enough temperature in the space-heating system. But in all other situations, there is a substantial potential for savings from simply lowering the thermostat. By experimenting with different thermostat settings, you should be able to find a setting that is high enough for your needs, yet lower than the so-called "normal" setting. Lowering the water temperature from 160 to 120 °F (from 70 to 50 °C) can reduce energy consumption for heating water by as much as 40%. 120 °F (50 °C) is plenty hot. At that temperature, water can scald human skin.

Further energy savings can be realized by installing flow restrictors on shower heads and faucets. No decrease in performance will be caused by these measures; in fact, the reduction in demand caused by flow restrictions will actually increase the length of time a person can spend in the shower before the hot water runs out (see Box 10.5.3). Hot water leaks should be stopped as soon as possible. Dripping faucets can waste 60 gallons of hot water per week.

Insulation. Tank water heaters are already insulated, but usually not adequately. Their efficiency can be improved greatly by installing an insulating blanket around the tank. Many electric utilities sell these blankets at a greatly reduced cost or give them away free to customers. In other areas, it may be necessary to purchase an insulating blanket or fabricate one from leftover wall or attic insulation. In each case, the paycheck period is quite short, usually less than six months.

Instructions for making and installing water heater blankets are widely available from utilities as well as state and local energy agencies. Most books on home energy conversation include instructions on installing water heater jackets. It is also a good idea to insulate hot water pipes, especially if they run for a long distance through an unheated space. Insulating water pipes will help prevent freezing during cold weather, and it can increase the temperature of the water at the tap, thus allowing the water temperature control to be turned down by a few degrees.

10.5.3 Cooking and Kitchen Energy Use

Food preparation is a major energy-consuming activity. Some foods are heated several times: once during processing, again during packaging, and a third time before they are served. All these steps require energy use. Refrigerators and freezers are also major consumers of energy. The kitchen is an important place to practice energy conservation.

BOX 10.5.3 ENERGY AND LIFE-STYLE IN THE BATH

Personal hygiene is one of the most important uses of hot water. Disease and infection can be reduced significantly through regular bathing. A good hot bath can also be very relaxing after a hard day's work. But it is not necessary to use enormous amounts of water simply to get clean. Bathing styles can be suited to the occasion, with possible savings in hot water use. Here is a table of the amount of hot water, and the amount of heat energy, required for a variety of bathing techniques.

Shower Styles

Personal Preference	Hot Water Required (gal)	Energy per Person [thousand Btu (kcal)]	Comments
Big tub bath	25–30	30–35 (7500–8500)	Can provide space heating if the water remains in the tub long enough to transfer heat to the room
10-minute shower	15–20	18–25 (4500–6000)	
5-minute shower	5–6	6–7 (1500–2000)	
Sauna with cold shower	—	8–10 (2000–2500)	Therapeutic
Sauna with hot shower	5	15–20 (4000–5000)	Can provide space heating
Shower for two	20–25	10–12 (2500–3000)	May not save energy!
Sponge bath	1	1–1.5 (250–300)	Cleans adequately; needs warm room
Swim (unheated water)	0	0	Summer only; good exercise
Solar shower	0	0	On sunny days
Hot tub (five persons)	45–50	11–16 (2700–4000)	Sociable
Indian shower	2–2.5	2.5–3 (400–500)	Uses a bucket of water and a cup; no plumbing required

Adapted with permission of Macmillan Publishing Company from Michael Corr and Dan MacLeod, "Home Consumption as a Function of Life Style," in *Energy and Human Welfare: A Critical Analysis,* edited by Barry Commoner et al., Vol. III: *Human Welfare: The End Use for Power*.

Equipment. In the kitchen, the type of equipment used largely determines the amount of energy consumed. Small ovens, microwave ovens, pressure cookers, and stovetop cooking can each save from 9 to 50% of the energy used, compared with less efficient methods. Microwaves will not save energy if a whole meal is to be cooked, but they are highly efficient for warming up a portion or two.

Pilot lights can consume up to 40% of the total energy used by a gas stove. This is not only uneconomical, but it can also be a nuisance in the summer or in inadequately ventilated houses. Pilotless gas stoves are widely available today, and existing gas stoves can have their pilot lights turned down or converted to electric ignition. However, it is important to remember that pilot lights were originally introduced as a safety feature. Before turning off any pilot lights, you should make sure that children will not be able to turn on the burners accidentally. Be aware that gas is a highly combustible substance and act accordingly.

"Frost-free" refrigerators can save the time and nuisance of manual defrosting, but with a noticeable penalty in energy use: typically 50% more than manual-defrost models of similar size. Much of this excessive energy use is caused by the *mullion* heaters, which heat the areas around the doors to prevent condensation. These heaters are unnecessary and should be disconnected. Newer frost-free refrigerators have a manual switch (typically called an "energy-saver switch") which allows the mullion heater to be turned off.

Chest-type freezers are more efficient than upright freezers. Because cold air tends to sink, upright freezers allow cold air to fall out into the room whenever they are opened. Any type of refrigerator or freezer will use less energy if it is kept full and opened as little as possible. If you are in the habit of standing with the refrigerator door open, staring inside and wondering what to eat, you are probably not really hungry.

Cooking techniques. Methods of food preparation can reduce energy use significantly. Most vegetables taste better and are more nutritious if they are slightly undercooked rather than overcooked. In summer, salads are delicious and much more energy efficient than roasts and heavy cooked meals—keeps the house cooler, too. Preplanning meals and thawing frozen food beforehand (in the refrigerator) will save energy.

Using appliances efficiently can also save energy. Preheating an empty oven is not always a good idea. Many foods can get a head start during the preheating period. It is also a good idea to cook an ovenful of food rather than just one or two items. The amount of energy used is the same for one item or for five. Opening the oven door too often will let out a large amount of heat. It is also true that a watched pot takes longer to boil, especially if the chef keeps lifting the lid.

10.5.4 Cleaning

Clothes washers take on the average 30 to 35 gallons (110 to 130 liters) of water for each wash load. A warm wash with a cold rinse will do the job fine in most cases, and a cold wash is preferable for many delicate fabrics. A full load of clothes uses the same amount of energy as half a load, unless the washer has a special low-water setting. Presoak cycles are not really necessary except in the case of extremely heavily soiled clothes.

Electric or gas clothes dryers are mostly a waste of energy. Clothes can be dried equally well with solar clothes dryers (clotheslines), and in times of inclement weather, clothes can be hung in a basement. During the winter months, this will add moisture to the air in the house, reducing the need for a humidifier.

Many automatic dishwashers have an automatic drying cycle. This cycle is not at all necessary. Dishes will dry perfectly well without it. Most newer dishwashers offer users the option of turning off the automatic drying cycle; homeowners should take advantage of this feature. Dishwashers should not be operated unless they are full. A half-empty dishwasher uses as much energy as a full one, but it washes only half as many dishes. For small jobs such as washing a few breakfast dishes, washing dishes by hand can save a considerable amount of energy compared to running a dishwasher—as long as the person washing the dishes does not let the hot water run continuously.

10.5.5 Miscellaneous Household Energy Use

"Instant-on" color televisions use more energy than conventional TV sets because the cathode heater is on all the time, using valuable electricity. This "instant-on" feature can be disconnected easily. Incandescent lights use more electricity than fluorescent lights for the same light output. Conversion kits are available to convert standard incandescent lamps to fluorescent. Solid-state electronic equipment uses much less energy than older, vacuum-tube equipment, and requires less frequent servicing. These are only a few examples of opportunities for energy savings which are often overlooked. With a little attention and imagination, you should be able to find quite a number of other ways to save energy in the home.

10.5.6 Transportation

In many households, transportation is the largest user of energy. Nationwide, transportation offers more potential for energy conservation than any other area except space heating and cooling. On the household level, possibilities for energy conservation can be equally significant.

Automobiles. By now, most people in the United States are inclined to buy small, fuel-efficient automobiles, even if this means purchasing imported

European or Japanese models. American automobile makers, aware of this trend, have worked hard to reduce the size and weight of their passenger cars. However, the new, smaller cars cost more to purchase than used gas-guzzlers. Even used economy cars tend to cost more than used large cars because they are considered more desirable. This creates a serious problem for lower-income Americans, who would like to purchase smaller cars but cannot afford the high prices. High interest rates add to the problem, because larger interest payments offset the fuel savings obtained by operating an energy-efficient car. But in the long run, fuel-efficient cars make good economic sense as well as good energy sense.

Gas mileage can be improved in any car if it is kept in top running condition. Tires should be inflated to the recommended pressure, engines should be tuned, and wheels and axles should be balanced and well lubricated. This will ensure that the car gets maximum performance and fuel economy.

Driving Habits. Planning trips in advance can also save a significant amount of fuel. It makes sense to combine errands rather than making one trip to the grocery store, another to the hardware store, a third trip to pick up the kids from nursery school, and so on. This will save time as well as energy. Anticipating stops, accelerating gradually, lowering speeds, and driving smoothly will also help save gasoline.

For travel to and from work, car pooling and van pooling is becoming more popular. If two or three people can travel in the same car, not only will energy be saved, but the trip will be more sociable, highways will be less crowded, parking will be easier, and wear and tear on automobiles will be reduced. Cars will be available at home for necessary errands rather than tied up in the company parking lot. Insurance companies frequently offer reduced premiums for people who car-pool to work rather than driving their own automobiles every day.

Some companies take car pooling one step further and provide van pools for their employees. In a typical van pool system, several employees (typically six to eight) share rides in a van which is provided by the company. The riders are not responsible for payments, insurance, or repairs. The company benefits because van pool riders show a pattern of lower absenteeism and less fatigue from long-distance commuting. Energy savings is far from the only benefit of van pooling.

Alternatives to traveling by automobile. Roughly 85% of all personal travel in the United States is done by automobile. But while most people travel everywhere by driving their cars, alternatives are available. For travel between cities, buses and trains are safer and more energy efficient than individual automobiles. A 300-mile bus or train trip is tiring, but it does not require as much concentration as a 300-mile automobile trip. Many people can relax and doze off on long-distance bus or train trips. But after a long trip by automobile, most drivers are completely exhausted and ready for bed. For shorter distances, rapid-transit systems, buses, or subways are available in most cities and suburban areas. When carrying a full load of passengers, all these means of transportation are considerably more energy efficient per person than are one or two people traveling in a car.

For those in good physical shape, *bicycling* offers a good alternative to automobile travel. The bicycle is considered to be the most energy-efficient mode of transportation ever devised by human beings. A bicycle trip of a few miles can be good exercise and will save plenty of gasoline. Except in large urban centers, travel by bicycle is more enjoyable than driving (and in those areas, mass transit is clearly the preferred method of travel). Bicycle riding does require an extra measure of caution, but it pays off in better health as well as reduced gasoline bills. Motorized bicycles offer a good alternative to bicycles for those who lack the stamina for bicycling the necessary distances, and they offer excellent fuel economy. Again, caution is necessary to prevent safety problems.

Walking represents the oldest and the best alternative to the automobile. For short trips to work or the corner grocery store, a walk provides fresh air and relaxing, invigorating exercise, as well as fuel savings.

As it is plain to see, there are many alternatives to driving. Walking, bicycling, buses, subways, trains, car pools, van pools, and motor bicycles all offer significant energy savings. In many instances, a simple telephone call can be the fastest, easiest, and most energy efficient of all. Business, banking, and short conversations with friends and relatives can all be accomplished over the telephone, usually with a significant savings in time and energy.

10.6 RESOURCE CONSERVATION

Recycling and reuse. We can all be more careful in our personal use of materials. Paper, glass, aluminum, and other valuable resources can be recycled or reused many times. For example, using a knapsack or a string bag for shopping trips, as is done in most other parts of the world, saves the energy costs of unnecessary paper and plastic bags. Most Americans have huge stacks of paper bags behind the refrigerator or in a closet, and no more are needed. Some food stores are trying to encourage conservation by charging their customers extra for paper bags and marking down their produce to reflect the cost savings.

Many consumer products, including food, are heavily packaged. In many cases, the extra packaging is unnecessary and wasteful. In some areas, particularly New England, *bottle bills* have been passed to reduce the use of throwaway containers. The consumer pays a deposit, typically 5 or 10 cents, for each bottle. The deposit is refunded when the bottle is returned. In addition to increasing recycling and the use of returnable bottles, these bills have significantly reduced roadside litter in states where they are in effect.

Returnable bottles not only reduce roadside litter but also save consumers money. To confirm this, simply go to a supermarket and compare the prices for beverages in returnable bottles with the prices for throwaway containers. Remember to subtract the cost of the deposit from the price of the beverage in returnable bottles. In the great majority of cases, the beverage will be less expensive in returnable bottles.

In addition to recycling paper, glass, and metal, many people recycle other materials as well. Vegetable garbage can be composted rather than thrown down the garbage disposal or hauled out to the curb. Grass clippings and tree leaves can also be composted. This will produce valuable garden material as well as energy savings. Old crankcase oil can be recycled. Many garages accept oil for recycling. Scrap paper can be used to start fires in wood or coal stoves. Some waste paper can be reused, too. For example, most of the first draft of this book was written on discarded computer printouts or office paper. The unused side was perfectly clear and more than adequate for early manuscript drafts.

10.7 LIFE-STYLE CHANGES

Most of the suggestions in this chapter require very slight, if any, changes in life-style. Conserving energy is mostly a matter of awareness rather than a matter of extra work. We do not mean to suggest that people should abandon their energy-intensive laborsaving devices and return to a life of hard labor. But consider the following example: a suburbanite mows his half-acre lawn with a small tractor-type mower. Then, when he is finished, he drives 5 miles to a gym, where he lifts weights and does exercises. If he had used a regular hand mower in the first place, he could have had his exercise and done his mowing at the same time. Not only would he have saved energy, but he would have saved time as well. Finally, there is a special satisfaction in manual labor which is not accomplished by pushing buttons.

In the final analysis, personal energy use is based on ethical as well as rational grounds. In making our personal decisions about energy use, we should ask not only how much energy we want to use, but in a more basic way, ask ourselves what kinds of lives we want to live. There is a special kind of satisfaction that results from adjusting one's life to the natural ebb and flow of energy. For many, this, rather than saving money, is the fundamental motivation for conserving energy.

SELECTED DOE PUBLICATIONS THAT HELP SAVE ENERGY AND MONEY

DOE OPA-0022, *Energy Savings Through Automatic Thermostat Controls*
Controls that save energy by automatically setting back thermostats.

EDM-1050, *Heat Pumps*
Heat pumps—operation, selection, and new developments.

DOE/CS 0158, *Heating With Wood*
Everything you need to know about wood heating.

DOE/OPA-0018, *How to Improve the Efficiency of Your Oil-Fired Furnance*
Beneficial effects of regular servicing of oil-fired furnaces on energy usage and the environment.

DOE/OPA-0021, *Insulate Your Water Heater and Save Fuel*
Value of insulating water heaters with a glass fiber insulation kit sold in local stores.

DOE/CS-0017, *Insulation*
Reducing energy waste in heating and cooling homes by installing adequate insulation.

DOE/CS-0006, *Lighting*
Energy-efficient uses of incandescent and fluorescent lamps.

DOE/OPA-0033, *Put the Sun to Work Today*
Solar energy—its applications, heating and cooling technologies, and factors to consider when buying home solar equipment.

DOE/OPA-0037, *Tips for Energy Savers*
Practical ways to save energy in your home, garden, and car.

For single copies of these or other free publications, or for information on a specific energy topic, write to DOE Technical Information Center, P.O. Box 62, Oak Ridge, Tennessee 37830.

STUDY QUESTIONS

1. Watch five television commercials chosen at random. If possible, watch during prime time (evenings). What do those advertisements say about energy use in the United States?
2. Write a letter to the chairman of the board of a corporation suggesting energy improvements in a particular consumer product. How does the company respond?
3. Visit an appliance store and determine the energy consumption of five different refrigerators. How easily did you find the information? How helpful was the clerk? Do you think most people ask about energy efficiency when they shop for appliances?
4. Energy consumption of large appliances such as refrigerators can vary widely. The difference in operating costs between the most and least efficient models can amount to $100 or more per year. Why is there so much difference between different models? Why are the less-energy-efficient models still on the market?
5. Can you think of a situation where you made a conscious effort to save energy? How much energy do you think you saved? What would be the overall effect if everyone saved a comparable amount of energy?
6. Can you think of a situation where you wasted energy—that is, where you used energy unnecessarily? How much did you waste? What would be the overall effect if everyone wasted that much energy?
7. What is the largest consumer of energy in your house? What is its efficiency? How much energy does it use in a year? How can it be made more energy efficient?
8. Look around your house, apartment, or dormitory room and find the places where heat escapes. What could be done to reduce that heat loss? Are there any actions you could take to help reduce that heat loss?

9. Why is 65 °F used as the reference point for calculating heating degree-days? Is this a good choice? What temperature should be used for calculating *cooling* degree-days for air conditioning?
10. If you live in a state that does not have a bottle bill, visit your local grocery store and compare prices for beverages sold in returnable and non-returnable bottles.

PROBLEMS

1. Calculate the energy consumption and energy cost of a refrigerator, an electric heater, and a 100-W light bulb over a period of one year. The following information should help you with the calculation.

   Power rating of the refrigerator: 300 W

   Average on-time for the refrigerator: 10 h/day

   Power rating of the electric heater: 1500 W

   Average on-time for the heater: 7 h/day during heating season

   Length of heating season: 150 days

   Average on-time for light bulb: 5 h/day

   Cost of electricity: 7.5 cents/kWh
2. The refrigerator in Problem 1 is a manual-defrost model. A frost-free model would use 1800 kWh per year. What are the annual energy and money savings from using the manual defrost refrigerator?
3. Calculate the average kWh consumption of a hot water heater. The average temperature of incoming water is 45 °F. The hot water temperature is 160 °F. Assume that the average consumption is 60 gallons of hot water per day. The water heater tank loses 40% of its heat due to insufficient insulation. At 7.5 cents/kWh, what is the annual cost of operating the water heater?
4. How much energy and money would be saved if the water temperature in Problem 3 were reduced to 120 °F? Assume that the tank still loses 40% of its energy.
5. How much energy and money would be saved if the water heater in Problem 3 were insulated so that it lost only 15% of its energy?
6. An electric clothes dryer has a power rating of 5000 W. It takes 1 hour to dry a load of clothes.
   - **(a)** How much would be saved if the clothes were dried using a solar clothes dryer (the old clothesline in the backyard)?
   - **(b)** How much would the savings amount to if the electric dryer were a high-speed model with a wattage rating of 8400 and a drying time of 45 minutes?
7. Assume that your dormitory room contains four electronic appliances: an electronic calculator, an electric clock, an electric toothbrush, and an electric hair dryer. How much energy would you save if you completely stopped using each of these appliances? Are the savings significant?
8. Can you read your electric meter? See the explanation given in Figure P.8 if you are not sure how to do it.

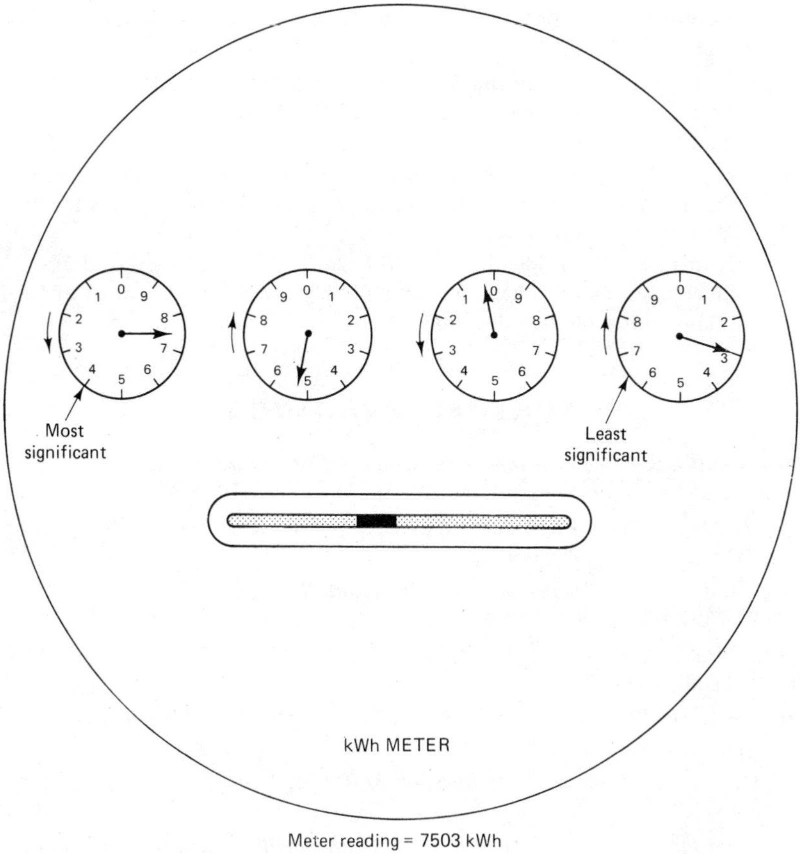

Figure P.8 How to read an electric meter. To read the electrical consumption, note the pointer position on each dial. If the pointer is in between two numbers, use the smaller number. For example, the meter above reads 7503; this means 7503 kWh. Be sure to note the direction of rotation on each dial.

- **(a)** Read your meter accurately for a period of 1 week. What is your energy consumption for that week?
- **(b)** Estimate your electrical consumption for a month, and also for a year.

9. Calculate the heat loss through the floors of a house 30 ft by 24 ft. The R value of the floors, including insulation, is 14.5. Assume a climate of 5000 degree-days.

10. (a) Calculate the energy lost from the ceiling of the house in Problem 9. The ceiling area is also 30 ft by 24 ft. The R value of the ceiling is 14.

(b) How much energy would be saved by adding:

(1) A 3½-in. layer of fiberglass insulation (R-11)?

(2) A 6-in. layer of fiberglass insulation (R-19)?

11. Calculate the energy loss through the windows of the house in Problem 9. There are six windows with glass area 24 in. by 36 in. and four windows 30 in. by 36 in.

There is also a sliding glass door 72 in. by 72 in. Assume that all the windows are double-glazed, with an *R* value of 1.7.

12. The walls of the house in Problem 9 are 8 ft high. Except where there are windows or glass doors, the walls have an *R* value of 14. What is the energy loss through the walls?

13. **(a)** Using the information in Problems 9 through 12, calculate the total heat loss for the house in Problem 9, assuming that infiltration accounts for 25% of the total heat loss.

    (b) If this heat is provided by fuel oil, burned with an efficiency of 70%, how much does it cost to heat the house for 1 year? The heat content of 1 gallon of fuel oil is 140,000 Btu, and fuel oil costs $1.25 per gallon.

FURTHER READINGS

BROWN, LESTER R., ET AL. *Running on Empty: The Future of the Automobile in a Petroleum-Short World.* New York: W. W. Norton, 1979.

CENTER FOR SCIENCE IN THE PUBLIC INTEREST. *99 Ways to a Simple Lifestyle.* Garden City, NY: Doubleday/Anchor, 1977.

HAYES, DENIS. *Energy: The Case for Conservation,* Worldwatch Paper 4. Washington, DC: Worldwatch Institute, 1976.

MAHON, HAROLD, ET AL. *Efficient Energy Management.* Englewood Cliffs, NJ: Prentice-Hall, 1983.

MORRISON, JAMES W. *The Complete Energy-Saving Handbook for Homeowners.* New York: Harper & Row, 1979.

NASH, HUGH, ed. *Progress As If Survival Mattered.* San Francisco: Friends of the Earth, 1977.

Save Energy, Save Dollars: Efficient Energy Management. Ithaca, NY: New York State Co-operative Extension, 1977.

THURMANN, ALBERT. *Handbook of Energy Auditing.* Atlanta, GA: Fairmont Press, 1979.

11

Integrations

In the preceding chapters we have tried to identify some of the major energy issues. For purposes of clarity, we have looked at those issues using six different methods of analysis. But in doing so, we may have overlooked the most important perspective of all: the ways in which the various analytical methods interact.

When analytical principles are applied across disciplinary boundaries, it is often possible to perceive connections that were hidden before. For example, ecologists have found that the greater the diversity of organisms found in an ecosystem, the greater the resilience of the ecosystem in the face of adverse conditions. During the 1970s, for instance, a lack of genetic diversity is believed to have contributed to the devastation caused by the corn blight that destroyed 15% of the U.S corn crop. Many researchers believe that the Irish potato famine of the mid-nineteenth century resulted from a similar lack of diversity.

A related principle has been observed in economics. In the southern Appalachian region of the United States, a large percentage of the work force has traditionally been employed in the coal-mining industry. When demand for coal declines, as it did in the 1960s, the economy of the coal region can nearly collapse. Much of the same thing happened during the early 1980s in Detroit and other areas heavily dependent on the U.S. auto industry. But regions with a more diverse economic base, like New England, have been more successful in withstanding economic hardship.

The same principle applies, in a somewhat different form, in the area of energy supply. Where a range of energy resources is available, disruptions in the supply of one resource can cause only minor hardship. But where one resource has dominated the energy picture, a break in the supply of that resource can create serious problems. The oil shortages of the 1970s are a case in point.

In this chapter we show how the analytical principles developed in the previous chapters can be brought to bear on a handful of energy topics. Our purpose is not to provide a definitive synthesis of the various perspectives; rather, we hope to convey a sense of how the various perspectives work together, laying the groundwork for future integrations.

11.1 COGENERATION AND TOTAL ENERGY SYSTEMS

Cogeneration is a general term for electrical generating technologies that perform other energy-related functions at the same time. Whenever electricity is generated, there is some leftover energy that cannot be converted to electricity. In most cases this means that some energy is wasted, usually in the form of low-grade heat. Cogeneration reduces this waste energy by using it productively for other purposes, such as space heating and industrial processes. Waste heat from power plants can also be used to heat greenhouses and seed beds.

Magnetohydrodynamics is one proposed high-technology application of cogeneration principles (See Figure 11.1.1). In an MHD generator, a very high-temperature stream of gases is used to produce electric power, and the exhausted gases are then used to drive a mechanical turbine. This combined generating cycle is expected to generate electricity with an efficiency of 50% or even more.

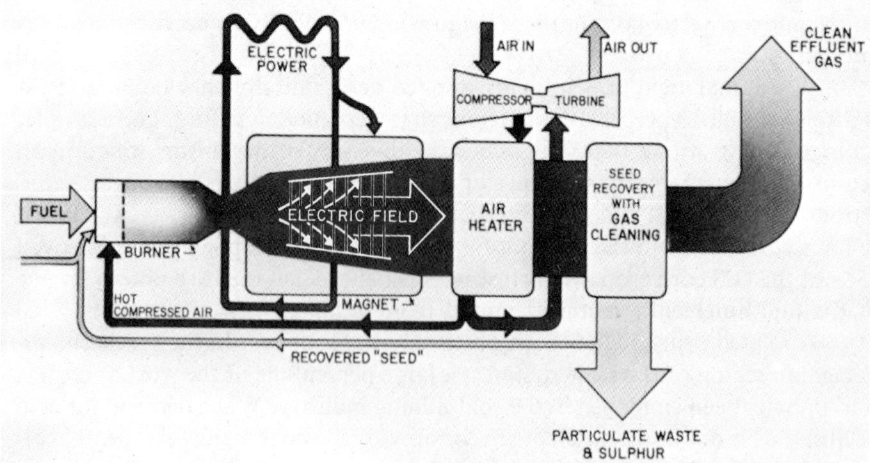

Figure 11.1.1 Diagram of a magnetohydrodynamic generator. (*Source:* Department of Energy, Washington, D.C. Courtesy of Avco Everett Research Laboratory.)

In principle, the waste heat from the second stage of the cycle can be used in lower-temperature applications as well.

Cogeneration is not a new technology. In 1922, 20% of the U.S. electric supply was produced in cogenerating stations. But by 1975, cogenerators produced only 5% of U.S. electricity. As cheap thermal power plants were developed, cogeneration lost favor with utilities. Pricing policies actually discouraged industries from installing cogeneration capacity. In Europe, however, cogeneration has been more widespread. Cogeneration produces 27% of all electrical generation in West Germany, and 20% in Great Britain. One factor that may lie behind this unequal development is the relatively low price of energy in the United States.

In one sense, cogeneration is simply an application of the "matched end-use" principle discussed in Chapter 5. The high-temperature steam is used for the high-grade task of producing electricity, and the low-grade steam is used for lower-grade tasks. It can also be seen in terms of the nutrient recycling described in Chapter 7, where the wastes from one organism are not simply discarded, but can be used by other organisms in an ecosystem. But perhaps the most important feature of cogenerating systems is that they are *integrative*: they combine two energy-conversion techniques into a larger, more effective system.

11.1.1 Conditions for Cogeneration

Opportunities for cogeneration arise wherever steam is being generated for industrial or heating purposes, or wherever electric power plants are located near facilities that can make constructive use of low-temperature steam. But cogeneration is not quite as simple a matter as adding on a device to draw off the waste heat from an existing thermal power plant. Cogeneration changes the operating characterisitics of a thermal power plant by reducing the temperature difference across the steam generator. This means that the efficiency of the electrical generating component of a system with cogenerating capacity is somewhat lower than in a thermal power plant without cogeneration capacity. But the overall efficiency of energy use is much higher—at least 50%, compared to 30 to 40% for a conventional thermal power plant.

In general, large (500- to 1000-MW) power plants are not good choices for cogenerating facilities even though they are technically suited. Most of the needs for space heating and steam occurs in cities and other heavily industrialized areas. But large thermal power plants cannot easily be located in cities. They require large areas of land which is hard to come by in cities. Also, the emissions from coal plants, or the threat of radiation releases from nuclear plants, make them environmentally unsuited to urban environments. But long-distance transport of steam is very inefficient because the temperature drops rapidly. Thus the efficiency advantages of cogeneration would be lost because of the distance between a large thermal power plant and the region where the steam can be used.

On the other hand, if several smaller cogenerating facilities are distributed throughout the area of electrical use, rather than being concentrated in a central generating station, the efficiency advantage of cogeneration increases even more. Not only is the additional heat and process steam available where it is needed, but electrical losses due to transmission and distribution are also reduced substantially. Thus, by matching the scale of the electrical production with the need for waste-heat products, it is possible both to achieve somewhat higher overall conversion efficiency and also reduce losses due to electrical transmission.

Cogeneration is ideally suited for smaller power plants serving local, widely distributed facilities: factories, hospitals, schools, and apartment buildings. The cogenerating facilities can be owned individually by the consumers or leased from utility companies. In both cases, the result is the same: a large network of many small, local generating facilities, each providing residual steam which is used for other purposes by the institutions operating the generating facilities.

11.1.2 System Characteristics of Cogeneration Networks

Like other energy systems, cogenerating networks have their own special operating characteristics.

1. Land-use impacts of electrical power generation are reduced. Since the new generating facilities are installed in factories, schools, hospitals, and other institutions, very little new land needs to be set aside for the new facilities. The transmission and distribution lines also tend to be in place, so not much additional land is required for transmission facilities.
2. If a cogenerating station goes out of service, the loss of electrical power is not as great as the loss that would be experienced if a large power plant went out of service. If a plant goes out of service permanently, another plant can be expected to take its place if the plant was economically feasible in the first place.
3. The cogenerating facility pollutes the region where the power is used rather than a remote rural region.
4. The number of cogenerating plants can change fairly rapidly in response to the need for power. As additional cogenerating sites become economically feasible, they can be developed rather quickly.

These operating characteristics can be compared to those of a biological organism that occupies a formerly uninhabited ecological niche (See pp. 293-294). The total volume of space occupied has not changed; the loss of that organism will not cause significant disruptions in the ecosystem; waste products of the organism become part of the local environment; and an increase in the availability of niches will produce an almost immediate increase in the number of organisms occupying those niches.

11.1.3 Cogeneration and Political Accountability

Because of their small size and diversity, cogenerating plants present fairly serious problems for public agencies charged with assuring continuity of power production. Regulated utilities are primarily in the business of producing electrical power, and they must respond to the requests of the political regulatory bodies. But in many cases, cogenerating facilities are operated by institutions whose primary concerns lie in other areas. If the requirements of utility regulatory boards become too stringent, cogenerators can choose to suspend their cogenerating activities. This reduces the possibility for political accountability and control.

One environmental consequence of this lack of control is that cogenerating facilities may actually create more pollution than central power plants for an equal amount of power generation. Accurate monitoring of emissions from a central power plant is well within the capacity of governmental agencies. But if the pollution is being created by hundreds or thousands of small cogenerating facilities, monitoring is more difficult. And because electrical generation is only one of the functions being performed by the institution, it is very difficult to determine how much of the pollution has been produced by electric power generation and how much is being produced by other functions. Given the current structure of government agencies, it seems likely that widespread cogeneration would present significant regulatory problems. Voluntary restraints by plant operators could help to avoid these difficulties.

11.1.4 Choice of Fuels

Theoretically, cogeneration can be accomplished using any fuel that is capable of producing high-temperature steam. Fossil fuels, solar energy, biomass fuels, or nuclear energy would all seem to be possible fuels. But in fact, not all these fuels are good choices. The choice of an appropriate fuel depends on the nature of the tasks being performed, the availability of fuels, and environmental, social and political constraints. Let us see how these factors affect the choice of a fuel for cogeneration.

For efficient electrical generation, steam must be produced at very high temperatures. This requires a fuel that is capable of providing those temperatures economically and predictably. But cogeneration usually takes place in urban areas; thus the chosen fuel must also be environmentaly benign. Finally, any new energy-conversion technology should take account of future energy supplies. If fuel will not be available over the lifetime of the plant, or if prices will make the fuel economically uncompetitive, alternative fuels should be chosen.

Based on these criteria, solar energy does not appear to be an appropriate fuel for cogeneration. First, the supply of solar energy is irregular and unpredictable. Second, to produce high temperatures, it would be necessary to provide a large absorbing area: too large for urban areas, where land is precious. Finally, solar energy can be used more economically for other tasks, such as providing low-grade heat for buildings and domestic hot water.

Coal appears to be a good fuel for cogeneration because it is well suited for the production of high temperatures. Supply lines are well established, coal will probably continue to be an abundant fuel for many years into the future, and electrical generation is an especially appropriate use for coal. However, environmental considerations require that we should be very careful how we use coal. First, most cogeneration facilities will be located in urban areas where air quality is already a significant problem. Unregulated coal burning could add an intolerable burden to the already heavily laden atmosphere. Second, even if emissions can be reduced to an acceptable level, there are still problems associated with the production of the resource. These include mine safety, land use, and reclamation of surface mine sites. If all these problems can be adequately addressed, coal may very well be an appropriate fuel for cogeneration systems.

Petroleum is a good fuel choice for cogeneration if we consider only its environmental effects and burning qualities. However, the supply of petroleum is subject to disruptions caused by international tensions, and the price is far from stable. Perhaps more serious, petroleum is clearly in very limited supply. Because of these facts, petroleum is perhaps the least desirable fuel for future electric generating capacity. Its use should be reserved for higher-level tasks such as transportation and petrochemicals.

Uranium is a very high-energy fuel, capable of producing extremely high temperatures. Its effectiveness for electrical production has been clearly demonstrated in the nuclear power plants which are found all over the world. However, uranium is not at all a suitable fuel for cogeneration. We have already mentioned the problem of locating nuclear plants in urban areas. Other problems include the danger of steam contamination. Even in plants with two separate cooling loops (see Section 3.6.6), contamination of the secondary coolant is not uncommon. Clearly, contaminated steam must not be used for other industrial processes or space heating. Finally, nuclear energy appears to be falling behind most other fuels in economic efficiency. For the foreseeable future, nuclear energy will not be used as a fuel for cogeneration.

If we overlook the problem of supply, natural gas is the ideal fuel for cogeneration. Its burning qualities are excellent, it is a very clean fuel, and it is lower in price than most competing fuels. The most serious drawback of natural gas is the limited supply. World supplies of natural gas are being depleted at an alarming rate. However, methane produced from crop residues is nearly identical to natural gas and could easily be substituted for natural gas using existing transportation and distribution pipelines. If such a substitution could occur without serious disruptions in price or supply, a mixture of natural gas and methane could produce an ideal, and ultimately renewable, energy resource. On the other hand, if biofuels are not developed quickly, a drop in supplies of natural gas could cause serious disruptions in electrical supply.

The burning characteristics of wood are certainly adequate for the production of cogenerated electricity, but wood will probably not be widely used in the near future. Wood is a very bulky fuel, and its use requires considerable attention. Wood chips and fuel pelletization schemes have simplified wood handling to a degree, but storage of fuel and disposal of wastes are still cumbersome and time consuming. In urban areas where cogenerating facilities are most likely to be located, wood smoke presents additional environmental problems. For many of the institutions that are likely to undertake cogeneration projects, these obstacles will prevent wood from being considered seriously as a fuel.

11.1.5 The Overall Picture

As we look at cogeneration from a variety of perspectives, we can see that its use implies much more than a simple increase in the efficiency of electrical production. If we turn toward cogeneration, we are also making an implied choice in favor of the following:

1. Decentralized power production
2. Reduction in government control of the energy industry
3. Greater private control over energy production facilities
4. Less separation between energy production and consumption
   a. More environmental effects at the point of end use
   b. Possibly greater concentration of jobs in urban centers

Cogeneration is not the only energy system with such far-reaching social and environmental consequences. Central power plants have equally far-reaching, if somewhat different effects. The point is that every energy choice, whether for central power generation, cogeneration, or conservation, represents a whole range of social, political, environmental, and personal choices.

11.2 ENERGY AND ARCHITECTURE

Architecture plays a vital role in shaping the living and working patterns of a society. Architecture can do more than put a roof over our heads; it can make us more productive, creative, vigorous, and self-assured. Studies have shown that the design of a workspace affects the productivity of workers. Control of lighting, placement of walls, level of noise, and use of color all affect our relationship to our workplace. But architecture also determines how we relate to our environment. The nature of this relationship has a strong influence on our use of energy.

During the mid-twentieth century, improvements in heating, ventilating, and air conditioning (HVAC) technology permitted the construction of buildings that were almost completely isolated from their natural surroundings. Thus the same design could be used in every type of climate, from the deserts of the Southwest to the frigid northern plains. We are all familiar with the glass-and-steel skyscrapers that dominate the landscapes of most urban centers, the identical tract houses we see in suburbs from Minneapolis to San Diego, and the squat, windowless metal warehouses on the outskirts of cities and towns everywhere.

Although these "universal" designs succeeded in providing space for families and businesses, they did so at a significant cost in energy. Many of the large buildings constructed during the 1960s and 1970s feature inoperable windows—necessitated by the complex heating and ventilation systems. The huge windows and the lack of thermal mass combine to produce large temperature differentials on sunny days. On occasion, one portion of the building may require air conditioning while another section is being heated. In the tract houses built during the same periods, inadequate insulation and hasty construction resulted in large energy expenditures for heating and air conditioning. In the construction of factories and warehouses, energy efficiency was usually subordinated to considerations such as purchase cost and speed of construction.

11.2.1 The Growing Interest in Energy-Efficient Architecture

With the increase in energy prices of the 1970s, architects began to emphasize climate-responsive, environmentally sensitive buildings, rather than using one design for all climates. In doing so, they began to consider questions such as the following:

Is the design energy efficient?
Is it economical and affordable?
Is it convenient, simple, and accessible?
Is it compatible with the climate?
Does it blend with the surroundings?

These principles are not mutually exclusive. A climate-responsive structure tends to blend with its surroundings and offers greater energy efficiency than a climate-independent structure. These principles are not new: over 2000 years ago, Socrates noted the basic principle of solar gain in buildings. Dwellings designed to fit the regional climate have traditionally been built in all areas of the world, from the cliff dwellings of the American Southwest to the igloos of the North, from the well-ventilated reed houses of tropical Africa to the wind-resistant saltbox houses of colonial New England. In China, Greece, South America, India, and the Middle East, climate-responsive dwellings have been built for centuries without any formal "theoretical" knowledge. Today, increasing numbers of architects are turning toward simple principles of energy-efficient building design:

Low-energy design. Extensive insulation, tight construction to minimize infiltration, reduction of perimeter wall areas, and shielded entrances all help to minimize energy use. Where reduced infiltration threatens to cut down on ventilation, air-to-air heat exchangers permit increased ventilation without excessive heat gain or loss.

Passive solar design. Solar gain from the south side can easily be incorporated into a building design. In many climates, passive solar heating can serve as the main source of heating. See Table 11.2.1 for an overview of solar's advantages over other building types in energy terms. For a fuller discussion of passive solar heating principles, see Chapter 4.

Earth-sheltered design. Some architects are using the insulating and moderating properties of the earth to reduce heat loss. Buildings can be built into hillsides, set into the ground, or sheltered with artificial "berms" piled up around their outside walls.

Natural design. Natural lighting can easily be provided by properly located windows, lending a more cheerful feeling to the interior while cutting down on lighting and heating requirements. Proper placement of operable windows can also contribute to natural ventilation.

Landscaping. Proper placement of trees and shrubs can help to reduce heat loss by breaking the force of winter winds. Shade trees can significantly reduce the need for summer air conditioning.

TABLE 11.2.1 Building Types and Use of Solar Energy

Home Type	Direct Solar Gain (%)	Energy Reduction (%)	Solar Energy Storage (%)	Conventional Fuel (%)
Conventional	5–10	0	0	95
Well-insulated	10	26	0	90
Sun-tempered	25	50	0	75
Passive solar	75	84	50	25

Source: The First Passive Solar Homes Awards. By Franklin Research Center, Philadelphia, for the U.S. Department of Housing and Urban Development, Jan. 1979.

There is tremendous room for improvement in the energy performance of buildings. About one-third of all U.S. residential structures have *no* insulation and another 50% are underinsulated. In Europe, approximately 50% of all structures have no insulation. By simply bringing existing buildings up to standards, it should be possible to save large amounts of energy. For a sampling of current standards, see Table 11.2.2.

11.2.2 Government Actions

There are many steps that governments can take to improve the situation. Some of these are already being considered widely.

TABLE 11.2.2 Annual Home Heating Costs[a] (1979 dollars)

Structure or Standard	Annual Cost
U.S. average house, 1978	$680
French building code, 1974	500
U.S. building standards, 1978	360
Swedish building code, 1977	230
California building code, 1979	220
Town house with retrofit, Twin Rivers, New Jersey	95
Saskatchewan Conservation House	20
Village House I, passive solar	15

[a]Assumes similarly sized houses using oil heat in a similar climate.

Source: Reproduced with permission from Christopher Flavin, *Energy and Architecture: The Solar and Conservation Potential,* Worldwatch Paper 40 (Washington, DC: Worldwatch Institute, 1980).

1. *Retrofitting existing structures.* Energy audits and Community Action programs for poor and economically deprived households are available in most states of the United States. Other beneficial programs include low-interest loans for energy-efficient improvements and tax incentives for proven energy-saving technologies. These programs can be highly beneficial if continued and expanded.
2. *Establishing governmental building standards for energy efficiency.* The U.S. government is in the process of developing a set of Building Energy Performance Standards (BEPS) for all public, commercial, and industrial construction. These standards will require much more stringent attention to energy efficiency in large construction.
3. *Providing tax incentives, preferential mortgage treatment, and other "help" for solar designs.* The U.S. government and many states offer such tax incentives. Some banks take the reduced heating costs into account when preparing mortgage loans for solar houses. If extended, these practices could encourage greater energy efficiency.
4. *Considering energy efficiency in the design of public housing and other integrated community facilities.* Pedestrian needs, accessibility of facilities, proximity to services, and suitability to the community are other important considerations in the design of public buildings.

Architecture is both an art and a science. As such, it is essentially an integrative pursuit. Traditionally, architects have been important in shaping the social consciousness of their communities. Today, through greater awareness of energy efficiency and the possibilities of designing to suit local climates, architects can have an equally profound role in bringing about a future of energy consciousness.

11.3 ENERGY AND TRANSPORTATION

As we noted in Chapter 2, transportation is a major user of energy. In 1982, 19 quads of energy were devoted to the transportation sector in the United States—about one-fourth of total U.S. energy consumption.

The transportation needs of a society are dependent on factors such as city planning, economic conditions, job availability, population distribution, and personal preferences. In turn, the system of transportation strongly affects economic, environmental, and social conditions.

In the United States, the automobile is the primary means of passenger travel, both locally and over long distances. In metropolitan areas, mass-transit systems and buses are available. Railroads are used for short-distance commuting between suburbs and urban centers, but their role in intercity travel is not very significant. Airlines are used for long-distance intercity travel, and buses transport passengers over intermediate distances. Goods are transported mainly by truck and rail, with some use of air freight. Other modes of transport in the United States include corporate aircraft and, to a lesser degree, bicycles and other low-energy systems.

The automobile is the heart of the U.S. transportation system. Over 120 million automobiles are registered in the United States: more than one for every two people. Automotive travel accounts for roughly 90% of all personal transport in the United States. Americans use their cars to shop, go to work, and visit their friends. Dependence on the automobile is so great in the United States that when the government developed plans for evacuating cities in the event of nuclear war, private automobiles were to be the main means of transportation.

11.3.1 Reasons for U.S. Dependence on Automotive Transport

There are clear reasons for the tremendous popularity of the automobile in the United States. Cars are extremely convenient compared to any other method of travel. Petroleum has been abundant and relatively inexpensive, and personal income has been high enough for most Americans to afford automobiles. At the same time, patterns of urban development have favored the automobile, heavy government spending on highways has increased the convenience of automotive travel, and social pressures have made automobile ownership an important item of personal status. Finally, once people have made the necessary expenditure to own a car, they are likely to use the car for all their transportation needs.

North America is relatively unique in its dependence on the automobile for public transportation. In Japan and most European nations, there is greater reliance on public transport systems and small, fuel-efficient automobiles. In Asia, Africa, and other developing parts of the world, the automobile has not taken over human transportation as it has in the United States, for simple reasons of economics. The majority simply cannot afford the high prices.

11.3.2 The Automobile and Energy Use

America's dependence on the automobile has had a strong effect on the use of energy. In 1981, an average U.S. automobile used 581 gallons to travel 9056 miles, with an average fuel economy of 15.5 miles per gallon. This represents a reduction in consumption from 1973, when the average was 763 gallons, 9992 miles, and 13.1 miles per gallon. But it is still much higher than the consumption that could be achieved with new, fuel-efficient automobiles.

When we compare the automobile with other means of transport, its inefficiency becomes even more apparent. Fully loaded city buses or trains can get 150 passenger-miles per gallon of fuel. A loaded intercity bus can carry a passenger 200 miles on a gallon of fuel, and an express train can do twice as well as an intercity bus (400 passenger-miles per gallon). Ridership in the United States does not reach capacity levels, but even so, trains and buses are more efficient than automobiles for transporting passengers.

From the perspective of end-use analysis, much automotive travel is not well matched to the need being served. Social critics often comment on the inappropriateness of using a 1500-kg machine to transport one person, a bottle of milk, and a dozen eggs for a distance of a few blocks. In fact, however, this is not the result of intentional mismatching; rather, it results from the tendency to use the automobile for all transportation needs.

11.3.3 Consequences of Automobile Use

Environmentally, our heavy reliance on the automobile has had significant consequences. Automobiles and trucks use more land than any other means of transport for carrying comparable quantities of passengers and freight. In cities, the high level of land use is especially apparent. Most modern U.S. cities devote 50 to 55% of their land area to automobile use. In Los Angeles, the figure is approximately 65%. Air pollution from automotive exhaust is the main cause of air quality problems in urban centers. Lead emissions from automotive exhaust has contaminated the soil in some cities so badly that it is unsafe to raise vegetable crops for human consumption.

The social costs of reliance on the automobile are extremely high—over 50,000 deaths and 4 million injuries each year from automobile accidents. Superhighways cut through many cities of the United States, producing urban deterioration and reducing property values. Automobile-related crimes constitute a significant percentage of all crime in the United States. Some social theorists even believe that the privacy of automobile travel has led to a sense of fragmentation in American communities, a lack of neighborliness, and important changes in the values of American youth.

Political costs of automotive travel are also substantial. The task of building and maintaining roads places significant pressures on governmental institutions already hard-pressed for tax revenues. Increased reliance on imported oil has had a significant effect on U.S. foreign policy. Recently, the U.S. government recognized this problem by establishing a "strategic petroleum reserve" to reduce the possibility of other countries using petroleum for political purposes.

11.3.4 Evolution of Transportation Systems

If we simply look at a transportation system at one moment in time, we may be baffled by some of its more obvious inefficiencies. From our perspective, the enormous gas-guzzlers of the 1960s seem to be a historical aberration. But if we look at the development of the transportation system in the United States, we can see how they fit into the overall evolution of that system. Once we understand the patterns of that evolution, we may very well find cause for optimism.

Throughout nearly all of its history, the United States has been largely a rural society. Land has been abundant, and vast undeveloped regions have been found along the western frontiers. The emphasis has been on development of resources, and the greatest financial rewards have been available to those who have been able to exploit previously undeveloped resources and land areas. With the population distributed over vast areas, mass transportation has been impractical.

At the same time, technological advances have been enthusiastically embraced in the United States. The tinkerer, the backyard inventer, has been one of the country's folk heroes. Thomas Edison, the Wright brothers, and Henry Ford are held up as models to be emulated by those with the intelligence and the opportunity. Careful planning, complex analysis, and concern for safety have rarely been celebrated. In this climate, the automobile represented a perfect integration of the need of the American people for highly individualized transportation and their strong interest in technological innovation.

As automotive technology continued to develop, American transportation developed along with the automobile. The greatest periods of American economic expansion occurred after the automobile was firmly in place as the preferred means of transportation. This meant that when new enterprises were formed, or new areas of the country were urbanized, the automobile influenced the shape of development. Systems of national highways linked major cities, with expressways reaching from the urban centers into the suburbs. Outlying districts became "bedroom communities" with no industrial or commercial facilities. Central shopping districts were replaced by shopping centers with vast parking lots. Soon, demographic patterns had adapted to the automobile so thoroughly that even in heavily populated suburbs it was impossible to manage without an automobile.

By the middle of the century, many people spent significant fractions of their lives in automobiles, commuting to and from work, shopping, and driving into the country for recreation. The car had ceased to be a simple means of transportation and had become a second home. As this happened, more of the conveniences of home were installed in automobiles: reliable heating systems, air conditioning, comfortable seats, room to move around. By 1956, cars had become so large that one of the authors of this text overheard a Britisher referring to an American station wagon as a "blooming furnished room."

This expansion of the automobile's functions, and of the automobile itself, was accomplished at a noticeable penalty in fuel economy. But this, too, was in keeping with the spirit of the times. New energy sources were being discovered every few years. Prices of electricity were dropping due to efficiency improvements and mass production of electrical generating machinery. Kitchens were being filled with energy-consuming, laborsaving devices. By the 1950s, nuclear electricity was on the horizon and "all-electric" houses were being promoted. In a sense, the energy appetite of the large automobiles was almost seen as a virtue rather than a drawback.

11.3.5 The Petroleum Shortage and Beyond

Earlier in this chapter, we mentioned that diversity contributes to the resilience of ecological systems. The same applies to transportation systems. As long as petroleum was abundant and inexpensive, the lack of diversity in American transportation was an advantage. But with the supply interruptions of the 1970s, it became a major problem. Long gasoline lines formed, disrupting traffic flow patterns in many urban areas. Public transportation, which had been neglected for decades, was unable to offer commuters a reasonable alternative to their oversized, inefficient cars.

Fortunately, there was still diversity in other parts of the world. Japan, in particular, was able to introduce a number of automobiles that were suited to the particular needs of the American people during the 1970s. Soon, Japanese cars occupied a large share of the American car market. Some analysts believe that the key to the Japanese success lay in their ability to offer inexpensive, high-quality, fuel-efficient cars during a time of fuel crisis. There can be no doubt that this contributed significantly to their success. But perhaps equally important was the diversity offered by the Japanese car makers. Some of the cars were tiny, fuel-efficient commuting vehicles. Others were small trucks, suitable for carrying small loads. Small station wagons were available for family use. Some companies even offered four-wheel-drive fuel-efficient cars for people who lived in remote rural areas.

The success of the Japanese automobiles changed the entire structure of the American transportation system. Interestingly enough, car ownership did not decrease; in fact, the number of two-car families increased significantly. But the two cars were usually quite different in size and function. A typical suburban family might own a large American station wagon for shopping trips and around-town use, and a small imported car for commuting. In many families, third and fourth vehicles began to appear: a pickup truck for hauling, and perhaps a motorcycle for pleasant weather. Bicycles also began to increase in popularity; in fact, during the late 1970s, bicycle sales in the United States exceeded automobile sales. Experimental vehicles such as electric cars have also appeared, although the technology for many of these vehicles is still in the developmental stage. In the future, however, alternatives to the petroleum-fueled automobile may become more attractive.

In response to the increasing diversity of vehicles, highways are beginning to be used in more diverse ways. Many highways now have bicycle lanes, especially in California. Bus "express" lanes have been added to many urban freeways. Differential tolls have been introduced: cars with few passengers pay more than fully loaded cars. One increasingly common feature of urban expressways is the presence of large parking areas at the outskirts, where commuters can park their cars, often free of charge, and ride buses or trains to the city centers.

11.3.6 Effects of Diversification in Transportation

As the American transportation system has continued to evolve since the fuel crises of the 1970s, other changes have appeared in American society at large. Urban planners are beginning to experiment with new ways of structuring downtowns. City streets have been blocked off and turned into "street malls." A few cities have experimented with traffic-free city centers. Minneapolis has a system of covered walkways running above the city streets. For nearly the first time in half a century, major cities have begun to build new mass-transit systems.

As transportation systems continue to evolve, we can expect additional changes in American life. The current emphasis on physical fitness may lead to a wider use of bicycles, especially as safer, more practical designs are developed. Housing patterns may change from the familiar pattern of one-family houses on suburban lots. Youth activities may become less dependent on the automobile. Indeed, there is reason to believe that among many American youth, computers may have already replaced the automobile as the main form of recreation.

As we look at how the American transportation system has evolved since the mid-1970s, we can see a pattern of increasing diversity and technological development. These changes have already helped to reverse the pattern of steady increases in petroleum consumption. There is reason to hope that other beneficial changes may occur as the system continues to evolve.

11.4 ENERGY AND AGRICULTURE

Agriculture is not usually regarded as an energy-conversion enterprise. In fact, however, farmers are engaged in the business of converting energy from one form to another. The principal agricultural conversion is from solar energy to useful forms of chemical energy. The two principal products of agriculture are chemical energy for human consumption (food) and chemical energy for animal consumption (feed). To an increasing extent, though, farmers are producing a third type of chemical energy: fuel for heating and driving machinery. Firewood, wood chips, pellets, and feedstocks for biomass conversion are the principal fuel products of today's agriculture.

Viewed as a system of energy conversion, agriculture should conform to the standards of analysis established earlier for energy systems. Sustainability, diversity, environmental impact, economic feasibility, social effects, and the role of government must all be considered as we look at systems of agriculture.

11.4.1 Energy Resources for Agriculture

In most discussions of energy and agriculture, the important contribution of renewable energy resources is either overlooked or dismissed lightly. Analysts are concerned primarily with the amount of human, mechanical, and chemical energy used in crop production. However, the principal—and irreplaceable—energy input is solar energy. It is impossible to practice agriculture on any significant scale without sunlight. Other renewable energy inputs include:

Winds (assist in the transpiration of moisture)
Rainfall
Carbon dioxide (CO_2)
Soil (includes fertilizer derived from plant and animal wastes)
Human and animal labor
Action of soil microorganisms and macroorganisms (earthworms, nematodes, etc.)
Pest control by natural predators
Pollination by insects

For centuries, these renewable resources were the basis of all agricultural production. But during the past few decades, farmers have begun using nonrenewable resources in increasing quantities. These resources fall into four basic categories:

1. Fuel to operate mechanical equipment
2. Fuel to pump water for irrigation
3. Synthetic fertilizers
4. Synthetic pesticides and herbicides

Each of these nonrenewable resources is designed to substitute for one of the natural, renewable resources listed above. Tractors and other mechanical equipment replace human and animal labor. Irrigation replaces rainfall. Synthetic fertilizers replace soil microorganisms and organic wastes. Synthetic pesticides replace natural predators, and herbicides replace the labor of weeding and cultivation.

11.4.2 Changing Agricultural Methods

The introduction of nonrenewable resources changed the character of farming. The use of tractors and other mechanical equipment increased the area that could be cultivated by a single person. This led to a steady increase in the size of farms, together with a significant reduction in the number of agricultural workers. It also contributed significantly to the amount of capital required for an average-size farm. As capital requirements increased, farm size had to increase even more to pay for the machinery (see Figure 11.4.1).

The introduction of mechanical power for irrigation permitted the introduction of agriculture into arid areas such as the desert Southwest. These areas are usually very sunny, so they rapidly became some of the most important agricultural regions in the country. In many cases, the newly cultivated regions produced food more cheaply than it could be produced in other regions. Thus the focus of agriculture shifted from the eastern states, with their abundant rainfall, to the arid western states. Many eastern states began to import large quantities of food, even though they had previously been agriculturally self-sufficient.

The use of synthetic fertilizers allowed farmers to plant new hybrid strains

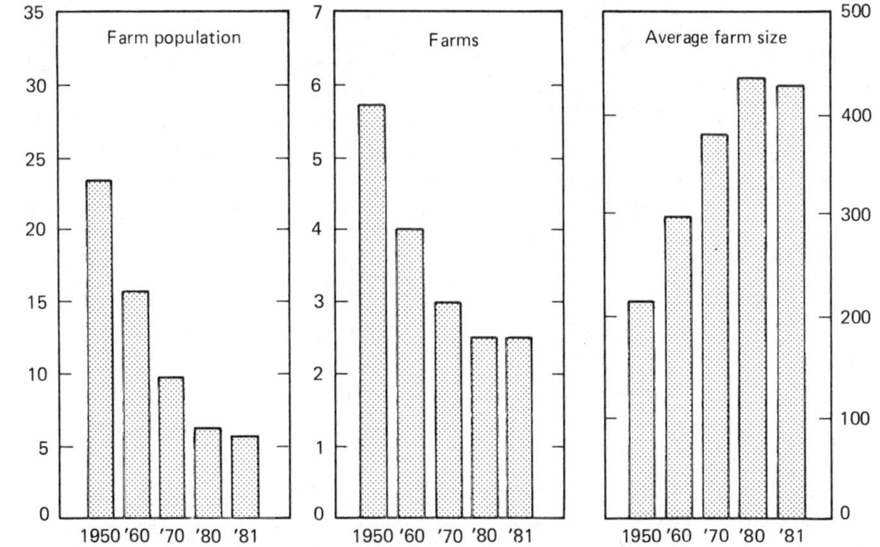

Figure 11.4.1 Farm population, number of farms, and average size 1950–1981. (*Source: Statistical Abstracts of the United States,* U.S. Department of Commerce, Bureau of the Census, 1983.)

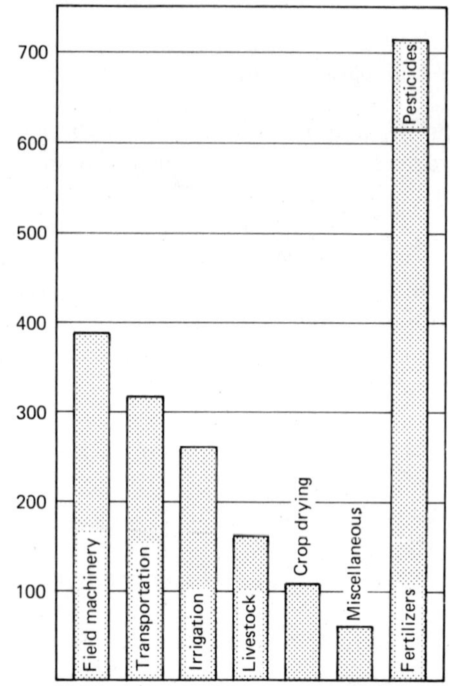

	Trillion Btu	Percent
Field machinery (till, plant, cultivate, applications, harvest)	391	30
Transportation (pickups, on-farm hauling, business auto trips)	317	25
Irrigation	261	20
Livestock, dairy, poultry	161	12
Crop drying	107	8
Miscellaneous	61	5
Total fuel and electricity	1298	100
Indirect energy: manufacture of fertilizers and pesticides	716	
	2014	

Figure 11.4.2 Energy used in agricultural production. (*Source:* Reproduced with permission from *Energy Use and the Food System* Special Circular 246, The Pennsylvania State University, College of Agriculture Extension Service.)

of crops, producing higher yields per acre of land. It also enabled many farmers to discard the traditional practices of crop rotation, planting of cover crops, and composting. As soil conditions changed, the mix of synthetic fertilizers could be changed to suit the new conditions. Even worn-out soil could now be used to produce crops. In some instances, soil was eliminated entirely: crops were grown in sterile sand, fed by a nutrient solution of synthetic fertilizer.

Chemical pesticides allowed farmers to kill insects and undesirable weeds quickly and effectively, with a minimum of labor. In some cases, such as "no-till" corn farming (where heavy application of herbicides replaces normal working of the soil), weeding and cultivation were eliminated altogether. This permitted farmers to farm larger areas of land, or farm the same area of land in less time.

As farmers practiced energy-intensive agriculture, they discovered that it was necessary to keep increasing the amounts of energy inputs. For example, the increase in farm size meant that farmers needed larger machinery with higher fuel consumption. The westward shift in agriculture increased the demand for irrigation. This led to lower water tables and higher energy requirements for irrigation. Use of energy for transportation also increased because food had to be shipped longer distances to reach eastern markets. The use of pesticides led to the development of resistant strains of insects, requiring the application of ever-heavier doses of pesticides.

In many cases, these effects reinforced one another. For example, the use of chemical pesticides killed many of the microorganisms that had previously contributed to soil fertility. To make up for the loss of these nutrients, it was necessary to use even higher levels of synthetic fertilizer. Applying the fertilizer required more use of farm machinery and larger expenditures for equipment and supplies. So the net effect of these changes in agriculture was a steadily increasing demand for energy.

11.4.3 The Role of Government

Federal and state governments have been instrumental in promoting the development of energy-intensive agriculture. Through the system of land-grant colleges and extension services, state governments provide farmers with a number of services, including research publications, soil analysis, and consulting services. In most cases, those services are oriented toward energy-intensive, mechanized agriculture. Extension agents tend to dismiss low-energy agricultural methods as primitive and old-fashioned.

In the area of agricultural research, state government agencies have also worked to promote the spread of mechanized agriculture. Research at land-grant colleges is often sponsored by manufacturers of farm machinery. Researchers have worked on projects such as developing tomatoes that can be picked more easily by mechanical harvesters.

The federal government, too, encourages large-scale, energy-intensive farms. For example, the government sells water for irrigation at very low cost to assist farmers in the arid Southwest. The water is made available to large users as well as small, although the original program specified a limit on the size of farms that were to be eligible for the program. Critics of the program charge that the low cost of the water discourages farmers from installing water-conserving irrigation systems. Other federal programs have also been oriented toward large farms and heavy use of nonrenewable energy resources.

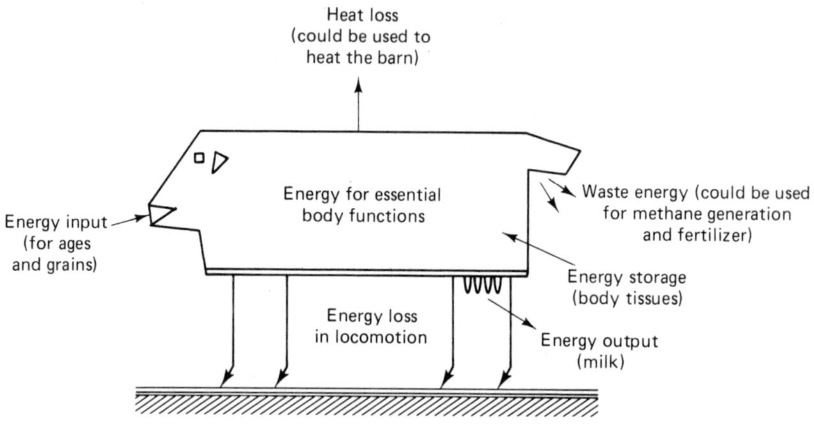

Figure 11.4.3 Dairy cow as an energy system.

11.4.4 The Energetics of Agricultural Systems

Before we can fully understand the importance of these changes in agricultural practices and energy consumption, it is necessary to look at the energy flow in agricultural systems. This will help identify areas of energy loss and also suggest possibilities for greater energy efficiency.

Solar energy provides the primary energy input to every agricultural system, even those where large amounts of fertilizer are being used. Of this solar energy, only a tiny fraction—on the order of 1%—is converted to plant matter. Thus 99% of the system's energy is lost immediately.

Not all the plant matter of a food crop is directly edible by human beings. In many cases, the food material is only a fraction of the plant. This introduces further losses into the system. Rather than 1%, we are left with only a fraction of a percent of the original solar energy input.

If the crops are fed to animals rather than directly consumed by people, additional losses are introduced into the system. Only about 10% of the food energy consumed by animals is actually converted to animal tissues. The rest is wasted. This means that for each additional step in the food chain, 90% of the energy value is lost.

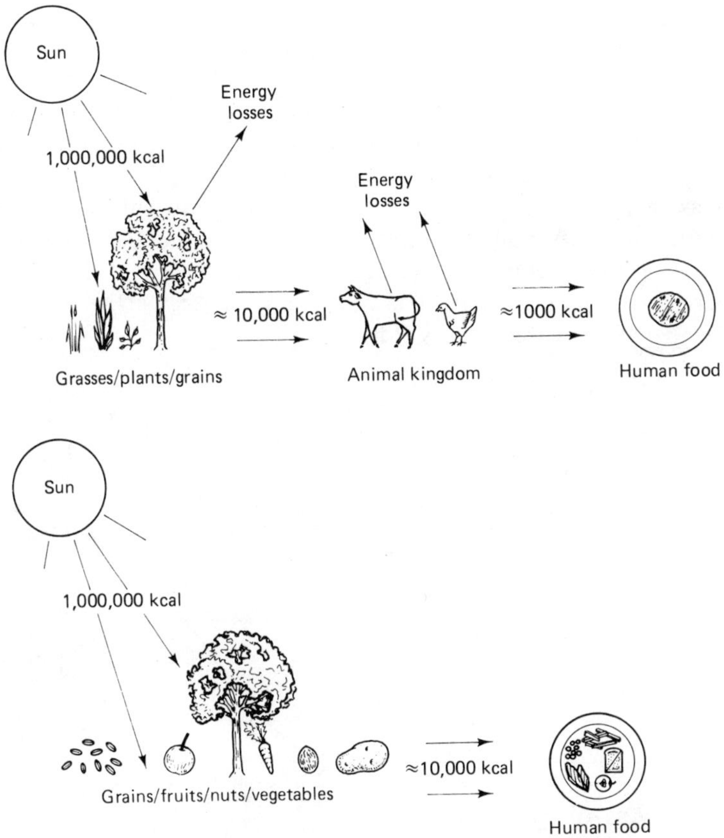

Figure 11.4.4 Energy losses in the food chain.

11.4.5 Energy Efficiency in Agriculture

Because the natural energy inputs in agriculture are difficult to quantify, meaningful energy efficiency figures are not readily available. In a sense, crop yields per acre represent such a measure of efficiency because the rate of solar energy per acre remains relatively constant from year to year. However, since the supply of sunlight and rain is not directly controllable by the farmer, it seems more reasonable to look at the efficiency for those inputs which are controllable, primarily petroleum, fertilizer, and pesticides. A useful tool in performing this analysis is called the "energy ratio" or "energy efficiency ratio." In this type of analysis, the energy value of the foods produced (in calories) is compared with the human-controlled energy inputs. This ratio can be expressed mathematically as follows:

$$\text{energy ratio} = \frac{\text{food output}}{\text{energy input (excluding natural inputs)}}$$

It is important to recognize that this ratio can be greater than 1 because it does not take into account the amount of solar energy that goes into the production of food.

Figure 11.4.5 shows some typical energy ratios. In general, the higher the energy ratio, the better the system energetics. As the figure makes clear, modern agriculture is very inefficient in its use of artificially supplied energy resources. As long as energy prices were low, this presented no difficulties. The low cost of the additional energy input was overshadowed by the higher income from increased crop yields. But as petroleum prices have increased, fertilizer prices have increased proportionately. Today, increased energy efficiency can be as cost-effective as increased crop yields.

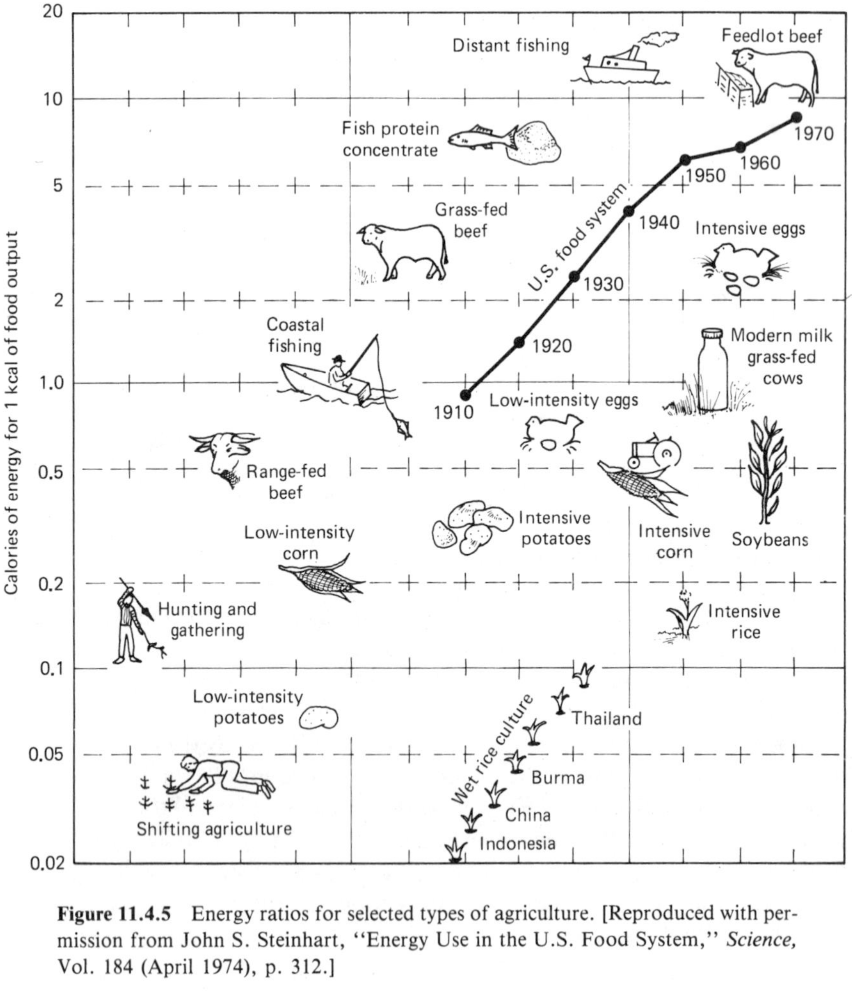

Figure 11.4.5 Energy ratios for selected types of agriculture. [Reproduced with permission from John S. Steinhart, "Energy Use in the U.S. Food System," *Science,* Vol. 184 (April 1974), p. 312.]

11.4.6 Agriculture as a Biological System

Although agriculture is a system of energy conversion, it differs from other systems in one very important respect: it is a *life-based* system. So the question of sustainability in agriculture involves much more than simple energy efficiency. As we have indicated in Chapter 7, ecosystems are complex systems, with many interconnections. When one part of the system is lost, other parts of the system take over some of the functions, and the ecosystem continues to function. But some of the system's resilience is lost, which means that its sustainability has been impaired.

Nearly all agriculture involves a conscious simplification of an ecosystem. The farmer's purpose is to grow as much as possible of a certain crop. To accomplish this, the farmer must eliminate all competitors (weeds) and predators (insects). But in doing so, the farmer also reduces the variety of organisms, and thus the complexity of the ecosystem. When we add unintentional side-effects, such as damage to microorganisms and beneficial insects, we can see that this process of ecological simplification can be quite far-reaching.

The effects of this simplification fall into two general categories: loss of genetic variety and loss of system complexity. Both can have serious consequences.

Genetic diversity. Diseases are caused by biological organisms that attack plants or animals. When a disease attacks a species with a diverse genetic base, disease-resistant strains soon emerge. This is one aspect of natural selection. But when the number of genetic strains has been intentionally reduced through cultivation of standardized genetic types, there is a lower probability of finding a disease-resistant strain. In some cases, the disease can essentially wipe out an entire country's agricultural production. The Irish potato famine is a case in point. Even when the losses are not total, they are likely to be more severe than they would have been in a more genetically diverse system.

System complexity. In nature, soil is a highly complex ecosystem, consisting not only of minerals and organic material, but also billions of organisms, plant fibers, earthworms, and other small creatures. Scientists have estimated that the top 6 in. of fertile soil may contain more than 2 tons of fungi and bacteria per acre. Taken together, these many elements make up a highly stable, interrelated system. The decay organisms feed on waste organic matter, producing important nutrients for plants and earthworms. The earthworms further transform the soil, changing its texture and further enriching its nutrient value. Finally, plant roots feed on the soil nutrients, binding the soil together and producing additional waste materials for the decay organisms. The processes are so intricate and complex that agricultural colleges offer full courses in the study of soil.

This soil ecosystem can be modified in a number of ways. When the same crop is grown repeatedly, the chemical composition of the soil changes. When no organic matter is added to the soil, its moisture-retaining capabilities decrease. When chemical pesticides are applied to the soil, organisms can be killed. The use of heavy machinery can change the physical structure and aeration of the soil. When these changes occur, the ecosystem is no longer stable, and serious consequences may occur. This is what happened during the "dust bowl" of the early twentieth century. The soil changes were so damaging that during periods of drought, tremendous amounts of topsoil were simply blown away by the wind.

11.4.7 Working toward a Sustainable Agriculture

In twentieth-century agriculture, the increased use of nonrenewable energy and the increasing simplification of rural ecosystems have been mutually reinforcing trends. The increasing use of chemical fertilizers and pesticides has changed the character of the soil, and those changes have necessitated the use of ever-greater amounts of fertilizer and pesticides. In recent years, however, some farmers have begun to practice less energy-intensive types of agriculture, relying more heavily on the natural processes of agricultural ecosystems. Unlike energy-intensive agriculture, with its relatively uniform procedures, this type of agriculture uses a highly diverse set of practices, including the following:

- Composting and other "natural" fertilizing methods
- Integrated pest management (use of natural predators, reduced dependence on synthetic pesticides)
- Multicrop plantings (planting small patches of several different species in the same field) rather than monoculture (planting only one type of plant in a field)
- Use of crop rotations to maintain soil fertility
- Use of cover crops (grasses, legumes, etc.) when the field is not planted with a food crop

These agricultural methods require much less nonrenewable energy input than standard, high-energy agriculture, and they are capable of producing comparable yields of many food crops. What is more important, proponents of this ecologically-based type of agriculture claim that their methods preserve the fertility of the soil and protect against erosion more effectively than energy-intensive agriculture. An additional benefit is the reduction of pesticide residues in food and water.

Ecologically-based agriculture requires close attention to the biological process taking place in the fields. On larger farms, there is often not enough time to practice these labor-intensive methods. Proponents of energy-intensive agriculture claim that this makes ecologically-based agriculture unprofitable. But on small farms, the lower costs for machinery and fertilizer sometimes enable farmers to make a reasonable profit on a smaller number of acres. In the long term, low-energy agricultural methods may offer the promise of greater sustainability: an important factor as fossil fuels become increasingly scarce.

11.4.8 Energy Crops

In recent years, there has been a growing interest in fuels produced from plant matter (see Section 4.7 and Table 11.4.1). As energy prices increase, it becomes economically feasible to produce alcohol fuels from plant matter. In fact, gasohol, a mixture of 90% gasoline and 10% grain ethanol, is currently being sold in gas stations all across the United States. As interest in these fuels grows, more farmers may plant energy crops rather than food crops. This competition for cropland could have a serious impact on the world's food supply.

TABLE 11.4.1 Alcohol Yield of Typical Energy Crops

	Alcohol Yield	
Crop	gallons/acre	liters/hectare
Sugarcane	390	3630
Sweet sorghum	380	3550
Corn	235	2200
Cassava	230	2135
Grain sorghum	145	1360
Wheat	83	770

Source: Adapted with permission from Lester R. Brown, *Food or Fuel: New Competition for the World's Cropland* (Washington, DC: Worldwatch Institute, 1980).

The most productive of the energy crops shown in Table 11.4.1 grow most successfully in warm climates. Since most Third World countries are located in tropical or semitropical regions, it seems likely that they will be providing many of the world's energy crops. Brazil, for example, is currently one of the world's leaders in production of fuel alcohol from biomass. But Third World countries also suffer from serious food shortages. As more cropland is diverted to the production of feedstocks for alcohol fuel production, these food shortages can only get worse.

Although the United States currently exports a substantial surplus of grain to other countries, this surplus could evaporate rapidly if energy crops become significantly more widespread. Raising crops for energy consumes much more land than raising traditional food crops, as Table 11.4.2 indicates. As Lester Brown of the Worldwatch Institute points out, energy crops will compete not only for land, but also for all the other resources necessary to agricultural production.

The end reseult may be that world energy prices will directly determine world food prices. The price of a gallon of milk could be determined by the amount of energy that could be produced in place of the milk. (At the moment, that might be a good deal in the United States, where gasoline is cheaper than milk, but this situation could very well change in the future.) In any case, the development of energy farming is likely to have a profound effect on the world's production and distribution of food.

The relationships between energy and agriculture are many and deep-rooted. In most of this chapter, we have looked at agriculture as a process of energy conversion, using both renewable and nonrenewable energy inputs to produce chemical energy in a variety of forms. But to view agriculture only in this way is to overlook its important social and cultural qualities. Farming is not only an economic enterprise; it is a way of life and a source of human nourishment. More than any other energy-related technology, farming keeps humanity in touch with the basic forces of life and death, of the natural energy flows that drive our planet. As we examine the various uses of energy in agriculture, we must bear in mind that the food we eat is the end product of this energy-conversion system. The future of agriculture is our future.

TABLE 11.4.2 Annual Land Requirements for Food and Automotive Fuels

Use of Crops	Crop production (lb)	Acres of Cropland
Subsistence diet (one person)	400	0.2
Affluent diet (one person)	1,600	0.9
Typical European auto (7000 miles per year, 25 mpg)[a]	6,200	3.3
U.S. automobile (10,000 miles per year, 15 mpg)[a]	14,600	7.8

[a]Based on yield of 380 liters per metric ton of grain.

Source: Reproduced with permission from Lester R. Brown, *Food or Fuel: New Competition for the World's Cropland* (Washington, DC: Worldwatch Institute, 1980).

11.5 APPROPRIATE TECHNOLOGY

The United States, western Europe, and Japan are without question the most technologically developed parts of the world. As less developed countries work at developing their own technologies, they naturally look toward the developed countries as a source of technological information and assistance. But the technology of the developed world often cannot be exported directly to less developed areas. The ethnic, educational, cultural, social, environmental, economic, and physical characteristics of Third World communities differ from those of the developed world. It makes little sense, for example, to export nuclear electric generating technology to a country that lacks a central power grid or sufficient electrical demand to use the power generated by the plant.

In response to this problem, many analysts have suggested that technology should not simply be *exported;* rather, it should be developed locally. Mohandas Gandhi in India, E. F. Schumacher in England, and Mao Tse-Tung in China were among the most prominent twentieth-century proponents of this technological approach, which is called *appropriate technology.*

Appropriate technology is not so much a set of technological applications as a method for developing applications. No one technology is "appropriate" everywhere. In many Asian countries, for example, methane digesters are used to produce fuel for cooking as well as fertilizer. This reduces the demand for firewood, which is very scarce, and prevents loss of soil fertility from direct burning of animal refuse. In the forests of the northern United States, where firewood is abundant, this particular application might not be appropriate; however, a facility for producing wood alcohol might be a highly appropriate energy technology.

BOX 11.5.1 AN APPROPRIATE TECHNOLOGY PROJECT IN VERMONT

At Vermont Technical College, Professor Ernest Wright is directing a project which uses the cold night-time temperatures of central Vermont to enhance the food storage capability of the traditional root cellar. The plan, Figure 11.5.1, shows a dirt-floor masonry structure, partially buried, equipped with an automatically controlled ventilation system. The ventilation system admits large quantities of cold night air. This nocturnal air circulation is used during the spring when night temperatures are low but the average 24-hour temperature is above the optimum for produce storage. This maintains a somewhat lower temperature in the food room, compared to a conventional root cellar at the same time of year.

The design includes an ice room where the nocturnal ventilating system directs cold air over containers of water during very cold periods in late winter. This produces a quantity of ice frozen in place. The ice is used to provide additional cooling in late spring when nights are too warm for nocturnal air circulation to be effective.

The cross-section, Figure 11.5.2, shows insulation under the ice room floor slab, while the food room has the traditional dirt floor—necessary in cold weather to provide heat from the stable deep-earth temperature to prevent freezing.

This system takes advantage of the cold nights characteristic of late spring in New England—a productive use of a special local "resource" that might otherwise be wasted.

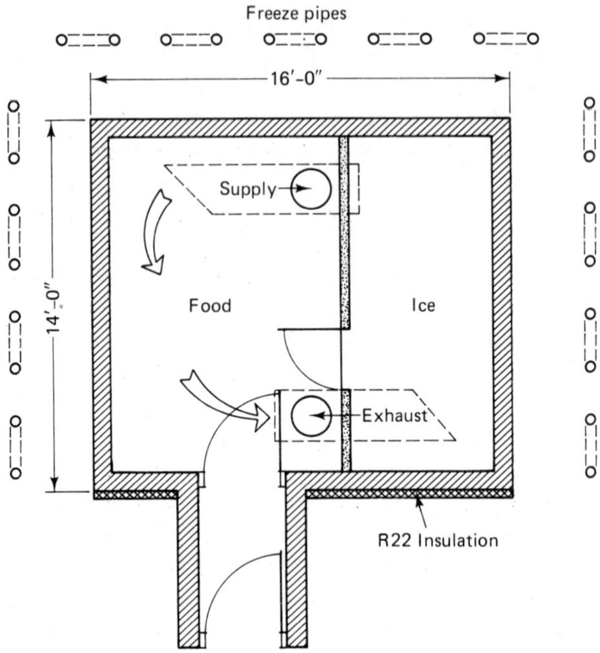

Figure 11.5.1 Floor plan at approximately ¼″ = 1′-0″.

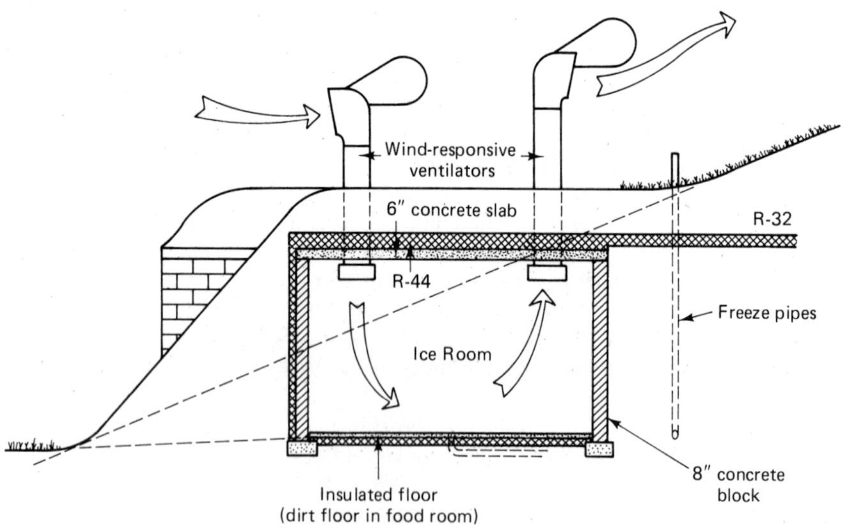

Figure 11.5.2 Cross section through ice room.

11.5.1 Criteria for Appropriate Technology Projects

In evaluating a proposed project, a number of local considerations need to be taken into account, including the following: (1) matching of scale between the project and local needs, (2) local availability of materials, (3) local availability of labor, and (4) cultural appropriateness of the project. Let us look at these factors individually.

Matching of scale. In industrialized countries, large energy-producing facilities are needed because of the scale of industrialization. In the northwestern United States, for example, a number of very large hydroelectric projects have been constructed. The power is needed by the locally based aluminum industry, which in turn provides materials for aerospace manufacturing plants also located in the Northwest. Thus the supply of energy is matched to the need. In a small rural village in a Third World country, however, there may be no need for electrical power—or only for a small amount, which could easily be provided by a small project.

It is also important to consider whether the *form* of energy being produced by a project is matched to the needs of the society. For example, in societies with a need for sophisticated communication and information storage systems, electricity is the only form of energy that can fulfill those needs. A wind-powered electrical generator might be quite appropriate for such a society. But in societies with simpler communications systems, where much information is transmitted through rituals and traditions, electricity may not be such a high priority. A better application of wind energy might be in a mechanical system for pumping and irrigation.

Local availability of materials. When energy technology is exported from one country to another, there is a tendency to export the specific materials used in that technology as well. However, in rural Third World areas, many commonly used industrial materials may not be readily available. Often, locally available materials can be substituted: wood instead of steel for low-load axles and pump shafts; charcoal instead of coke for metallurgical processes. This reduces the need to devote large amounts of foreign exchange to the purchase of imported materials, and it also simplifies the process of finding replacements for worn-out parts. Of course, these substitutions should only be made when safety will not be compromised.

Use of locally available labor. When technology is introduced into nonindustrialized region, it is easy to overlook the skills of the local residents and import foreign workers to complete the project. But proponents of appropriate technology urge the use of local labor for three reasons:

1. By training local workers to carry out the project, it is possible to establish a reliable, long-term local work force. If external funds become unavailable, the project will be able to sustain itself.
2. If local materials are being used for the project, local workers will be more familiar than imported workers with the characteristics of the materials.
3. Cooperation and interest in the project will be higher if local people hold positions of some authority.

Cultural appropriateness of the project. Any technological project must be undertaken with the approval and concurrence of the local residents. If a project is not in basic agreement with people's values, they will not support the project and will even actively oppose it. For example, a hydro project in the southwestern United States can provide local people with water for drinking and irrigation, electrical power, and increased commerce from the recreational business associated with the project. But in some cases, projects have been proposed which would flood traditional burial grounds. Needless to say, these projects have not been accepted by the people of the region.

11.5.2 Conventional and Appropriate Technology: One Example

As early as 1923, it was proposed to use cow dung as the basis of methane and fertilizer production in India. For a number of years, research and engineering studies were carried out, but in the 1950s this concept was abandoned in favor of conventional fertilizer production using coal as a feedstock. However, since the petroleum crisis, the biogas technology has been revived, and several thousand biogas plants are in operation.

From nearly every perspective, the biogas plants are more beneficial than conventional fertilizer plants. For an equivalent fertilizer production capacity, the biogas plants cost about 85% as much as the coal-based plant. All the materials for the biogas plants are available locally, whereas the coal-based plant requires some imported materials. The biogas plants employ over 100 times as many people, and they produce energy as a by-product. By contrast, the coal-based plant is a net consumer of energy.

The advantages of appropriate technology are not always this dramatic. But this example illustrates an important point about integrated projects such as the biogas plants. When we focus too narrowly on a single technological objective, we sometimes fail to see the full range of social and economic changes that are brought about by a project.

11.5.3 Sustainability and Appropriate Technology

An old proverb says, "Give someone a fish and you will keep away hunger for a day. Teach that person how to fish, and you will keep away hunger forever." Perhaps the greatest appeal of the appropriate technology concept is that it focuses on the transfer of knowledge rather than the transfer of hardware. In doing so, it increases the probability that the knowledge will be used intelligently and creatively in situations beyond the initial application.

When people base their energy solutions on an understanding of principles rather than a simple use of established technological designs, there is a much greater probability that the benefits of the technology will be sustained. Many other benefits will also be found: employment for local people, establishment of new industries, greater prosperity, and a general improvement in the quality of life. These are the most important benefits of any technological development, and the true basis of humanity's need for energy.

In this concluding chapter, we have tried to point the way toward integrations of the many analytical principles we have outlined in the text. But we can only hint at the possibilities for further analysis. The examples we have presented are intended to be illustrative rather than comprehensive. We are sure that readers will find many applications of their own. If our opinions may have colored this chapter to a greater extent than earlier ones, that too is intended. In the final analysis, all technology is designed in response to human needs and values. So, in the assessment of technological alternatives, it is essential to make clear what values we represent.

In closing this chapter on integrations, we would like to emphasize the most important integration of all: the integration that you, the reader, can make in your own life using the materials we have presented. If we have tried to emphasize the questions rather than the solutions, it is because we hope that you will be able to supply your own answers in the changing energy climate that will face all of us through the years.

STUDY QUESTIONS

1. What is cogeneration? Explain why cogeneration works best on an intermediate scale (as a project for an institution or neighborhood).
2. Locate an institutional building in your community that might be a good location for a cogeneration facility. Determine the building's electrical and heating needs. Are they well matched? How much of each could a cogenerating project provide? Draw up a cogeneration plan for the building.
3. What fuel would be appropriate for cogeneration in your area? Are there any environmental problems associated with the use of that fuel? How would you propose to deal with those environmental problems?
4. In your neighborhood, find a house that shows some regional characteristics and find another that looks as if it could be built anywhere in the country. Which house is more interesting? Which house is more energy efficient?
5. Locate a building in your community that looks as if it might be energy inefficient. Based on your examination of the building, and an interview with the owner, draw up a plan of energy-saving improvements. How would they affect the appearance and usefulness of the building?

6. Consult with the owner or operator of the building in Question 5. What energy-saving measures has he or she considered? Are any of them the same as your suggestions?
7. Visit a new institutional building or one that is under construction. What measures have been taken to make the building energy efficient? How do they affect the appearance and usefulness of the building? How does the building compare with the building in Question 5?
8. Go to a busy intersection during rush hour and count the number of passengers in each automobile that passes. What is the average number of passengers? What does this suggest about the effectiveness of car-pooling efforts in your region?
9. What percentage of the cars in Question 8 are large, energy-inefficient cars, and how many are small and economical? Which kind of car is more likely to have several passengers?
10. Various possibilities exist for a commuter in a large city. Consider these modes of transportation to work:

    Taking a well-loaded city bus to work
    Forming a van pool at your company
    Commuting in a minicar with three other passengers
    Driving alone in your car
    Using an express commuter train in the rush hour

    Which of these will be more energy efficient? Take an educated guess. Rate the choices in decreasing order of efficiency.
11. Design a transportation plan for your community, using the principles outlined in Section 11.3. Try to consider the needs of all local groups of people, including handicapped and elderly people.
12. Between 1950 and 1970, the number of small farms in the United States declined by approximately 100,000 per year. What are the social and economic effects of such a large decrease in the number of small farms?
13. Interview a farmer who uses traditional methods and another who uses low-energy, "organic" methods. What are the energy costs of the two farms? What are the crop yields? Which type of agriculture seems to work best in your part of the country?
14. Should the United States increase its production of alcohol from crop residues? What would be the consequences of such an action for the environment, for the agricultural system, and for the transportation system in the United States?
15. According to Table 11.4.1, the best energy crops are those that can be grown effectively in tropical climates. Should the United States encourage increased production of fuel alcohol crops in tropical regions? What would be the social consequences of such an action?
16. Explain what is meant by appropriate technology. Find three examples of appropriate technology in your area that seem to meet the criteria outlined in Section 11.5.1.
17. The United States and other industrialized countries depend on the export of technology to help balance their expenditures for imported petroleum and other energy resources. But wider development of appropriate technology in less developed coun-

tries could reduce these exports of high technology. Does this mean that development of appropriate technology is not in the interest of the developed countries? Explain your reasoning.

FURTHER READINGS

BERRY, WENDELL. *The Unsettling of America: Culture and Agriculture.* New York: Avon, 1977.

BROWN, LESTER R. *Food or Fuel: New Competition for the World's Cropland.* Washington, DC: Worldwatch Institute, 1980.

BROWN, LESTER R., ET AL. *Running on Empty: The Future of the Automobile in an Oil-Short World.* New York: W. W. Norton, 1975.

BUEL, RONALD A. *Dead End: The Automobile in Mass Transportation.* Englewood Cliffs, NJ: Prentice-Hall, 1972.

COMMONER, BARRY. *The Politics of Energy.* New York: Alfred A. Knopf, 1979.

COMMONER, BARRY, HOWARD BOKSENBAUM, and MICHAEL CORR, eds. *Energy and Human Welfare: A Critical Analysis,* Vol. 3: *Human Welfare: The End Use for Power.* New York: Macmillan, 1975.

THE CORNUCOPIA PROJECT OF THE RODALE PRESS. *Empty Breadbasket: The Coming Challenge to America's Food Supply.* Emmaus, PA: Rodale Press, 1981.

DUNN, P. D. *Appropriate Technology: Technology with a Human Face.* New York: Schocken Books, 1979.

FLAVIN, CHRISTOPHER. *Energy and Architecture: The Solar and Conservation Potential.* Washington, DC: Worldwatch Institute, 1980.

LONG, FRANKLIN A., and ALEXANDRA OLSON, eds. *Appropriate Technology and Social Values: A Critical Appraisal.* Cambridge, MA: Ballinger, 1980.

LOVINS, AMORY B., *Soft Energy Paths: Toward a Durable Peace.* New York: Harper & Row, 1977.

MERRILL, RICHARD, ed. *Radical Agriculture.* New York: Harper & Row, 1976.

OLGYAY, VICTOR. *Design with Climate: Bioclimatic Approach to Architecture.* Princeton, NJ: Princeton University Press, 1963.

POLIMEROS, GEORGE. *Energy Cogeneration Handbook.* New York: Industrial Press, 1981.

SCHUMACHER, E. F. *Small Is Beautiful: Economics As If People Mattered.* New York: Harper & Row, 1973.

STEIN, RICHARD G. *Architecture and Energy.* Garden City, NY: Doubleday/Anchor, 1977.

STONE, TABOR R. *Beyond the Automobile: Redesigning the Transportation Environment.* Englewood Cliffs, NJ: Prentice-Hall, 1971.

A

Appendix

Tables of Units and Conversions

TABLE A.1 Basic Units of Measurement

Quantity	System	Unit	Symbol	Conversion		
Mass	Metric	kilogram	kg	1 kg	=	2.2 pounds
					=	0.069 slug
	British	slug		1 slug	=	14.5 kg
Length	Metric	meter	m	1 meter	=	3.25 ft
	British	foot	ft	1 foot	=	0.304 m
		inch	in.	1 inch	=	2.54 centimeters
Time	All	second	sec, s			
Temperature	Metric	degree Celsius	°C	°C	=	(°F − 32) × 0.555
		degree Kelvin	K	K	=	°C + 273
	British	degree Fahrenheit	°F	°F	=	°C × 1.8 + 32
		degree Rankine	°R	°R	=	°F + 460
Force	Metric	newton	N	1 newton	=	0.225 pound
	British	pound[a]	lb	1 pound	=	4.45 N

[a]In the British system, there are two units called *pounds*. One is a measure of mass and the other is a unit of force. The two should not be confused. Strictly speaking, we should distinguish between the units *pound-mass* and *pound-force* (also called *pound-weight*).

TABLE A.2 Equivalence Table for Energy Units

Unit	Btu	kcal	ft-lb	hp-hr	kWh	MJ
Btu	1	0.25	780	0.00039	0.00029	0.00105
kcal	4	1	3100	0.0016	0.0012	0.00418
ft-lb	0.0013	0.0032	1	5.1×10^{-7}	3.8×10^{-7}	1.35×10^{-6}
hp-hr	2500	640	2×10^{6}	1	0.746	2.60
kWh	3400	860	2.7×10^{6}	1.34	1	3.49
Megajoule (MJ)	948	239	7.4×10^{5}	0.37	0.278	1

TABLE A.3 Equivalence Table for Power Conversions

1 hp =	746 W	1 kW =	1000 W
	550 ft-lb/s		740 ft-lb/s
	0.746 kW		1.34 hp
	0.18 kcal/s		0.24 kcal/s
	0.71 Btu/s		0.96 Btu/s

TABLE A.4 Energy Content of Common Fuels[a]

Fuel	Btu per unit
Coal	
Anthracite	25,400,000 per ton
Bituminous	26,200,000 per ton
Crude oil	5,800,000 per barrel
	138,000 per gallon
Home heating oil	138,000 per gallon
Gasoline	125,000 per gallon
Kerosene	128,000 per gallon
Natural gas	1,035 per cubic foot
Uranium	500,000,000,000 per ton

[a]Using current technology.

B
Appendix
National Energy and Environmental Organizations:

NATIONAL ENERGY AND ENVIRONMENTAL ORGANIZATIONS: A Partial List

Private Organizations

AMERICAN LUNG ASSOCIATION (Air pollution research and public health education)
1740 Broadway, New York, NY 10019 (212) 245-8000

CONSERVATION FOUNDATION (General environment, energy)
1717 Massachusetts Avenue NW, Washington, DC 20036 (202) 797-4300

COUNCIL ON ECONOMIC PRIORITIES (Environment and economics)
84 Fifth Avenue, New York, NY 10011 (212) 691-8550

DEFENDERS OF WILDLIFE (Conservation, wildlife, legal action)
1244 19th Street NW, Washington, DC 20036 (202) 659-9510

ENVIRONMENTAL ACTION (General environmental policies and education)
1346 Connecticut Avenue NW, Washington, DC 20036 (202) 833-1845

ENVIRONMENTAL DEFENSE FUND (General environment, legal, and scientific work)
444 Park Avenue South, New York, NY 10016 (212) 686-4191

ENVIRONMENTAL POLICY CENTER (Energy, water resources, and public interests)
317 Pennsylvania Avenue SE, Washington, DC 20003 (202) 547-5330

FRIENDS OF THE EARTH (Environment and energy, international issues, education)
1045 Sansome Street, San Francisco, CA 94111 (415) 433-7373

GREENPEACE USA (Wildlife, marine environment, acid rain, nuclear energy)
2007 R Street NW, Washington, DC 20009 (202) 462-1177

INTERNATIONAL UNION FOR THE CONSERVATION OF NATURE AND NATURAL RESOURCES
Avenue du Mont Blanc, CH-1196 Gland, Switzerland

NATIONAL AUDUBON SOCIETY (Energy, nature, wildlife, conservation, education)
950 Third Avenue, New York, NY 10022 (212) 832-3200

NATIONAL CENTER FOR APPROPRIATE TECHNOLOGY (Renewables, appropriate technology)
P.O. Box 3838, Butte, MT 59701 (406) 494-4572

NATIONAL WILDLIFE FEDERATION (Natural resources, wildlife, education, research)
1412 16th Street NW, Washington, DC 20036 (202) 265-2717

NATURAL RESOURCES DEFENSE COUNCIL (General environment, energy, land use)
122 East 42nd Street, New York, NY 10168 (212) 949-0049

THE NATURE CONSERVANCY (Wilderness protection and conservation)
1800 North Kent Street, Arlington, VA 22209 (703) 841-5300

NEW ALCHEMY INSTITUTE (Ecology, appropriate energy and technology, education)
237 Hatchville Road, East Falmouth, MA 02536 (617) 563-2655

OAK RIDGE ASSOCIATED UNIVERSITIES (Energy, environment, research, information)
P.O. Box 117, Oak Ridge, TN 37830 (615) 576-3000

PUBLIC CITIZEN (Health, consumer protection, nuclear energy, general)
1346 Connecticut Avenue NW, Washington, DC 20036 (202) 293-9142

SIERRA CLUB (General environment, education, renewable energy, nature)
530 Bush Street, San Francisco, CA 94108 (415) 981-8634

SOLAR LOBBY (Renewable energy)
1001 Connecticut Avenue, Suite 500, Washington, DC 20036 (202) 466-6350

UNION OF CONCERNED SCIENTISTS (Energy issues, nuclear power, public concerns)
1384 Massachusetts Avenue, Cambridge, MA 02238 (617) 547-5552

THE WILDERNESS SOCIETY (Wild and scenic lands, conservation)
1901 Pennsylvania Avenue NW, Washington, DC 20006 (202) 828-6600

WORLDWATCH INSTITUTE (International issues, education, ecology, energy, population) 1776 Massachusetts Avenue NW, Washington, DC 20036 (202) 452-1999

ZERO POPULATION GROWTH (Population and natural resources)
1346 Connecticut Avenue NW, Washington, DC 20036 (202) 785-0100

Governmental Organizations

DEPARTMENT OF ENERGY
James Forrestal Building, 1000 Independence Avenue SW, Washington, DC 20585

DEPARTMENT OF THE INTERIOR
C Street, Between Eighteenth and Nineteenth Streets NW, Washington, DC 20240

ENVIRONMENTAL PROTECTION AGENCY
401 M Street SW, Washington, DC 20460

NUCLEAR REGULATORY COMMISSION
1717 H Street NW, Washington, DC 20555

TENNESSEE VALLEY AUTHORITY
400 West Summit Hill Drive, Knoxville, TN 37902

Other Pertinent Organizations

ATOMIC INDUSTRIAL FORUM
7101 Wisconsin Avenue, Bethesda, MD 20814

CITIZENS' ENERGY PROJECT
1110 Sixth Street NW, Washington, DC 20001

ELECTRIC POWER RESEARCH INSTITUTE
P.O. Box 10412, Palo Alto, CA 94303

INTERNATIONAL ENERGY AGENCY
2 Rue Andre Pascal, F-75775 Paris Cedex-16, France

RESOURCES FOR THE FUTURE
1755 Massachusetts Avenue NW, Washington, DC 20036

C
Appendix
Glossary of Terms

AC: Alternating current. Electricity that changes voltage and direction periodically. In the United States, ac systems have a frequency of 60 hertz. Other systems operate at 50 hertz.

Acid Air: Air containing excessive amounts of sulfur and nitrogen oxides. These oxides form sulfuric and nitric acid.

Acid Precipitation: Rain, snow, sleet, or fog that is more acidic than normal. Caused in part by coal-burning power plants.

Acidization: A method of enhanced oil recovery. Acids are injected into alkaline rock formations to dissolve the rock and allow the petroleum to migrate.

Active Solar Energy System: A solar energy system that requires external power to make use of solar energy.

AEC: Atomic Energy Commission.

Alpha Particle: A type of radiation consisting of 2 protons plus 2 neutrons.

Amortization: Paying off a debt by installments. Part of each installment is used for interest payment, while the rest is applied toward repayment of the principal.

Anaerobic Digester: A device that converts biomass into more usable forms such as methane. The digestion occurs in an oxygen-poor environment.

Anthracite: A high-quality grade of coal. Burns at a high temperature with a smokeless, blue flame.

Appropriate Technology: Technology that is suited to the particular local needs of the area where it is applied.

Aquifer: An underground reservoir of water.

Atom: The basic unit of all substances; does not change in chemical reactions. Consists of a nucleus and a number of electrons.

Atomic Energy: *See* Nuclear Energy.

Atomic Energy Commission: Agency created to promote and regulate the development of atomic energy. Its duties were assigned to the NRC and ERDA in 1975.

Atomic Number: The number of protons or electrons present in an atom.

Atomic Weight: The mass of an atom, measured in atomic weight units. One proton has an atomic weight of 1.

Auger Mining: A system of mining used on hillsides with exposed coal seams.

Background Radiation: Naturally occurring radiation in the environment due to solar radiation, natural radioactive decay, cosmic radiation, and other natural causes.

Base Load: The level of electrical power demand that must be maintained at all times by a utility.

Bergius Process: A method of coal conversion using hot coal slurry together with char products and hydrogen derived from steam.

Beta Particle: A type of radiation released during radioactive decay or a nuclear reaction. Consists of a high-energy electron.

Biofuels: Fuels derived from plant or animal matter either directly or by conversion.

Biogas: Gas, typically methane, manufactured from biomass.

Biomass: Plant or animal matter suitable for use or conversion to biofuels.

Biosphere: The sum of inhabited or inhabitable ecosystems on the earth.

Bituminous: A medium grade of coal. Bituminous is the most abundant grade of coal in the United States.

Blow-out: A surge of high-pressure crude oil that sometimes occurs when a reservoir is first tapped.

Boiling-Water Reactor: A type of nuclear reactor using water as a moderator and coolant. Has only one coolant loop.

Boundary Layer: A layer of bonded ion pairs found at a *P-N* junction.

Breeder Reactor: A type of nuclear reactor that converts uranium-238 to plutonium-239. Produces its own fuel along with electric power.

BWR: Boiling-water reactor.

CANDU: A type of Canadian nuclear reactor that uses heavy water as a moderator.

Capacity Factor: The ratio of actual energy production to potential energy production. Used to measure the performance of a power plant.

Carbonization: A process used in coal gasification.

Carrying Capacity: The amount of biological activity that can be supported by an environmental system without serious degradation.

Cartel: An international economic organization which attempts to control the price of a commodity.

Catalyst: A material that helps create the conditions for a chemical reaction although it is not involved directly in the reaction.

Chain Reaction: A type of chemical or nuclear reaction in which the first reaction creates the conditions for additional reactions.

Coal Conversion: Coal gasification or liquefaction.

Coal Gasification: A process by which coal is converted to produce combustible gases.

Coal Liquefaction: A process by which coal is converted to produce combustible liquids.

Coal Slurry: A mixture of coal and water that allows coal to be transported by pipeline.

Coefficient of Performance: A measure of the efficiency of a heat pump. Equals the amount of heat transfer divided by the amount of input energy used.

Cogeneration: A system where the waste heat from electric power generation is used for other purposes such as district heating or process steam.

Condenser: The portion of a generator where a gas is cooled and changed to liquid form. Also part of a heat pump or OTEC generator.

Conduction: Method of heat transfer through solid materials.

Construction Work in Progress: A rate increase that is granted to cover the capital costs of a power plant that is not yet in operation.

Control Rods: Metal rods inserted in the core of a reactor to control the speed of the nuclear reaction.

Convection: Method of heat transfer through fluids.

Coolant: The material used to carry heat away from the core of a nuclear reactor.

Coolant Loop: A self-contained cyle of coolant in a thermal power plant.

COP: Coefficient of performance.

Cord: A measure of firewood, equal to 128 ft^3.

Core: The fuel area of a nuclear reactor.

Creosote: A black, tarry by-product of wood burning.

Critical Mass: The amount of a fissionable material required to sustain a chain reaction.

Curie: A measure of radioactivity. Named after Marie Curie, discoverer of radium.

CWIP: Construction work in progress.

Darrieus Rotor: A vertical axis windmill, often compared to an eggbeater.

DC: Direct current. Voltage and current remain constant. Used in many European countries for electrical power.

Declining Block Rate: A rate structure where large users pay less per unit of electricity than small users.

Degree-Day: A measure of need for heating. The number of degree days = 65 − average temperature (in °F) for each day.

Deuterium: A heavy isotope of hydrogen, found in heavy water. Useful in fusion reactions.

Diffusion: A natural process by which atmospheric or water pollutants tend to spread out from their point of origin.

Direct Gain: The simplest type of passive solar space heating system, consisting of south-facing windows, usually with added thermal mass.

District Heating: A system that provides heat to an entire community or group of buildings rather than a single building.

DOE: Department of Energy.

DOI: Department of the Interior.

Doping: The process by which silicon is contaminated with another material (dopant) to change its conductivity. Used in the manufacture of photovoltaic cells.

Doubling Time: The length of time required for a quantity to double in value, assuming an exponential rate of increase.

Dry Hole: A well that fails to produce any oil or natural gas.

Dry Steam: High-temperature steam suitable for use in a turbine.

Duster: *See* Dry Hole.

ECCS: Emergency core cooling system.

Ecosystem: A system of interconnected biological organisms.

EHV: Extra high voltage. Refers to power lines used to transmit electric power over long distances.

EIS: Environmental Impact Statement.

Electron: A subatomic particle with low mass and a negative electrical charge.

Electrostatic Precipitator: A device, sometimes called a "scrubber," that removes sulfur dioxide and other pollutants from smokestack emissions.

Emergency Core Cooling System: A backup cooling system that injects water into the core of a nuclear reactor if the other cooling systems fail.

End-Use Analysis: Looking at energy in terms of supplying a set of desired end uses rather than simply achieving a certain level of energy production.

Energy Auditing: Analyzing the energy consumption of a building or an economic enterprise. Often used to identify potential conservation measures.

Enhanced Recovery: A method of obtaining oil from a well when pumping no longer produces adequate amounts of oil.

Enrichment: A process by which the concentration of uranium-235 in natural uranium is increased to form nuclear fuel or weapons-grade material.

Enthalpy: A measure of the total amount of energy present in a system per unit mass.

Entropy: A measure of energy that is unavailable for future use.

Environmental Impact Statement: An assessment of the environmental impact of a project. Required for all federally funded projects.

Ethanol: A type of alcohol produced by fermentation of organic matter. Used in the manufacture of gasohol and alcoholic beverages.

Evaporator: The portion of a heat pump or OTEC generator where a liquid is converted to gaseous form.

Exponential Growth: The case where a quantity grows by a certain percentage every year (or other unit of time).

Externalities: Costs that are not part of the direct financial cost of an economic undertaking. Environmental effects are often considered externalities.

Extraction: A method of coal conversion in which coal is dissolved in a solvent containing hydrogen.

Feedstock: Raw materials used in manufacturing processes.

Fertile Materials: Materials that can be converted to fissionable materials.

First Law of Thermodynamics: States that energy can neither be created nor destroyed, but can be changed from one form to another.

Fischer–Tropsch Synthesis: A method of coal conversion by burning coal in an atmosphere of steam and oxygen.

Fission: A type of nuclear reaction in which heavy nuclei split into smaller segments.

Fissionable Materials: Materials that are capable of sustaining a nuclear fission chain reaction.

Fossil Fuels: Fuels derived from plant matter laid down during prehistoric times. Coal, oil, and natural gas are the most important fossil fuels.

Fractional Distillation: A process used to separate crude oil into different fuels and by-products.

Fuel Assembly: A group of fuel rods.

Fuel Cell: An electrical conversion device that burns hydrogen fuel. Used in many spacecraft.

Fuel Enrichment: *See* Enrichment.

Fuel Rods: Cylinders containing enriched uranium fuel pellets to be used in a nuclear reactor.

Fusion: A type of nuclear reaction in which two nuclei join to form a larger nucleus.

Gamma Radiation: A type of radiation consisting of very high-energy photons (particles of light).

Gas-Cooled Reactor: A nuclear reactor that uses graphite as a moderator and a gas as the primary coolant.

Gaseous Diffusion: A method of enriching uranium by passing uranium hexafluoride gas repeatedly through a set of tiny openings.

Gasohol: A mixture of 90% gasoline and 10% alcohol derived from plant materials.

Geopressurized Zone: A geothermal area containing hot water with dissolved natural gas under high pressure.

Geothermal Energy: Heat energy found in the earth's interior.

Gigawatt: One billion watts. The power output of a typical new thermal power plant.

Glazing: A transparent or translucent material used in solar heating systems. Glass is the most commonly used glazing material.

Greenhouse Effect: A condition caused when short-wave radiation can penetrate a barrier but long-wave radiation cannot. The basis of most solar heating systems. May also be caused on a global scale by increasing atmospheric concentrations of CO_2.

GW: Gigawatt.

Half Life: The length of time required for the decay of half the atoms of a radioactive material.

Head: The vertical distance between the intake and the discharge of a hydro power plant.

Heat Engine: A device that converts heat energy into kinetic energy.

Heating Coefficient: A measure of a house's heat loss in Btu per degree-day per square foot.

Heavy Water: Oxide of Deuterium, a heavy isotope of hydrogen. Used in CANDU reactors.

High-Temperature Gas-Cooled Reactor: One of the three types of gas-cooled reactors.

Hot Rock Geothermal Field: A geothermal area where no steam is present.

Hot Water Geothermal Field: A geothermal area where hot water is present. Steam may also be present.

HVAC: Heating, ventilation, and air conditioning.

Hybrid System: A passive solar energy system that uses external power to enhance its performance.

Hydraulic Fracturing: A method of breaking up underground rock formations by injecting high-temperature water.

Hydrocarbon: A molecule consisting mainly of hydrogen and carbon atoms. An important constituent of petroleum and natural gas.

Hydroelectric Power: Hydro power used to produce electricity.

Hydrogenation: A process by which the carbon in coal is combined with hydrogen to form hydrocarbons. An essential element of coal conversion.

Hydropower: Power obtained from moving or stored water.

Indirect-Gain System: A type of passive solar heating system in which thermal mass is located between the glazing and the living space.

Inertial Confinement: A method of compressing fuel to produce a fusion reaction. Laser and particle beam confinement are the two most widely studied methods of inertial confinement.

Infiltration: Heat loss due to the escape of heated air through cracks in a building's structure.

***In Situ* Method:** A method of extracting oil shale without removing the shale from the ground.

Insolation: A measure of the intensity of solar radiation.

Ion: A positively- or negatively-charged atom.

IOU: Investor-owned utility company.

Isotope: Any one of several different forms of an element. Isotopes are distinguished by their different atomic weights.

Kerogen: A substance found in oil shale. Can be converted to a petroleum-like material.

Kinetic Energy: Energy in the form of motion.

Laser Confinement: A method of confining fuel for a fusion reaction using lasers.

Lawson Criterion: A condition of temperature, density, and time that must be met in order to produce a sustained fusion reaction.

Light Water: Ordinary water. Distinguished from heavy water.

Light-Water Reactor: A nuclear reactor that uses light water as a neutron moderator and primary coolant.

Lignite: A low grade of coal.

LMFBR: Liquid metal fast breeder reactor. *See also* Breeder Reactor.

LNG: Liquefied natural gas.

LOCA: Loss-of-coolant accident.

Loss-of-Coolant Accident: A nuclear accident in which the flow of coolant to the reactor core is interrupted.

LPG: Liquefied propane gas.

LWR: Light-water reactor.

Magnetic Confinement: A method of concentrating high-temperature plasma, used in fusion research.

Magnetohydrodynamics: A method of electrical generation using an electrically charged gas rather than moving turbines.

Manhattan Project: The secret project that led to the development of nuclear fission during World War II.

Man-rem: *See* Person-rem.

MBD: Million barrels per day. A measure of petroleum use.

Meltdown: A type of nuclear accident in which the fuel assemblies overheat and melt.

Metallurgical Coal: Coal that can be converted to coke for use in the manufacture of steel.

Methane: The most common form of biogas. Also the main constituent of natural gas.

Methanol: A type of fuel alcohol usually formed from cellulose by pyrolysis.

MHD: Magnetohydrodynamics.

Moderator: A material that slows down the neutrons in a nuclear reactor.

MW: Megawatt, or 1 million watts. A measure of power.

NEPA: National Environmental Policy Act of 1970. The nation's basic environmental law.

Neutron: A subatomic particle with approximately the same mass as a proton but no electric charge.

Niche: The particular set of ecological characteristics that allows an organism to survive.

NRC: Nuclear Regulatory Commission.

Nuclear Energy: Energy produced by a nuclear reaction. *See* Fission and Fusion.

Nuclear Fuel Cycle: All the stages of nuclear fuel handling, from mining to waste disposal.

Nuclear Regulatory Commission: Successor to the AEC. Responsible for regulating the nuclear power industry.

Nucleus: The central part of an atom. Composed of protons and neutrons.

Oil Shale: Shale that contains rich deposits of kerogen.

Open-Pit Mining: A type of surface mining practiced mainly in the western United States.

OTEC: Ocean thermal electric conversion. An energy system that uses temperature differences in the ocean.

Overburden: Materials that cover a deposit of coal, oil shale, or other fossil fuel.

Ozone Layer: A layer of ozone gas (a special molecular form of oxygen) in the upper atmosphere which blocks incoming ultraviolet rays.

PAN (peroxyacetylnitrates): The chemicals that make up photochemical smog.

Passive Solar Heating System: A system that uses solar energy without the need for any external power.

Peak Demand or Peak Load: The highest demand for electricity experienced by an electric utility.

Peat: A dense organic fuel made up of partially decomposed organic matter; a precursor of coal.

Penstock: A conduit that carries water from a reservoir to a hydroelectric generator.

Person-rem: A radiation dose, distributed over a population, equivalent to 1 rem of radiation received by one person.

Petrochemicals: Products synthesized from petroleum, especially plastics.

pH: Measure of the acidity or alkalinity of a material: 7 is neutral. A pH greater than 7 is alkaline; below 7 is acidic.

Photosynthesis: The process by which plants use sunlight to manufacture sugars from water and carbon dioxide.

Photovoltaic Cells: Cells that produce electricity from sunlight.

Plasma: A superheated ion stream. Used in fusion research.

Plutonium: An artificially created element; useful in nuclear fission reactions, especially weapons. Plutonium has a half-life of 24,000 years.

***P-N* Junction:** The area of a photovoltaic cell or other semiconductor where the two layers join.

Potential Energy: Energy due to position or structure. Distinguished from kinetic energy.

Power Rating: The maximum output of a power plant.

Power Tower: A central solar electric design where large arrays of mirrors focus sunlight on a central generating unit.

Pressurized-Water Reactor: Type of reactor in which the primary coolant loop contains pressurized liquid water. The heat energy is transferred to a second loop, which operates the turbines.

Price-Anderson Act of 1957: Law that limits the liability of utililty companies to a total of $570 million in case of a nuclear accident.

Primary Energy: Energy in the form of resources. Distinguished from end-use energy.

Process Steam: Steam used for industrial processes other than space or water heating.

Proton: A positively-charged nuclear particle.

Proven Reserves: Supplies of a resource which are known to exist.

Pumped Storage: A hydro plant that pumps water into a reservoir during nonpeak hours to be used for power generation during peak hours.

Pyrolysis: Heating in the absence of air. Used to extract fuel alcohol from biomass.

***R* Value:** A measure of the resistance of a material to heat flow. A higher *R* value means less heat flow through the material.

Radiation: (1) High-energy particles emitted by radioactive material. (2) A method of heat transfer; similar to the transmission of light.

Radiation Daughter: An isotope produced by the decay of a radioactive element. Many radiation daughters are also radioactive.

Radioactivity: The property of a material that emits radiation.

Rankine: A temperature scale in the British system. 0 °F = 459.67 °R.

Ratchet Demand: A charge assessed during low-demand periods. Typically, 75 to 85% of the highest demand for the previous 12 months.

Reprocessing: The process of extracting useful materials from spent nuclear fuel.

Reserve Margin: An amount of generating capacity maintained by a utility beyond the expected peak load.

Retorting: A process of heating shale to extract the shale oil.

Retrofit: Adding insulation, solar heating, or other energy improvements to an existing structure.

Savonius Rotor: A type of vertical-axis windmill, useful primarily for pumping and mechanical work.

Scrubber: *See* Electrostatic Precipitator.

Second Law of Thermodynamics: The principle that in any energy conversion process, there is a decrease in the amount of energy available to do useful work.

Secondary Recovery: A process for extracting oil from a well after simple pumping no longer provides adequate production.

Shale Oil: Oil derived from shale.

Siltation: The process by which reservoirs fill with silt, limiting the useful lifetime of hydro projects.

Solar Collector: A device, usually a glass-covered box, which intercepts solar energy and converts it to heat energy.

Sour Gas: Natural gas that contains sulfur compounds.

Specific Heat: A measure of the amount of energy in kilocalories (or Btu) required to raise the temperature of 1 kg (1 lb) of a given material by 1 °C (1 °F). The specific heat of water is equal to 1 in both the metric and British systems.

Strip Mining: A type of surface mining practiced in the eastern United States.

Surface Mining: The process of removing surface materials to expose a coal seam. More efficient than underground mining.

Sweet Gas: Sulfur-free natural gas.

Synergy, Synergistic Reaction: The process by which two chemicals act together to produce an effect more serious than the sum of their two individual actions.

Synfuels, Synthetic Fuels: Liquid and gaseous fuels derived from nonrenewable sources other than petroleum and natural gas.

Tailings: The waste materials from a mine.

Tar Sands: Sandy deposits impregnated with bitumen, a tarry substance. A synfuels resource.

Tertiary Recovery: A process for producing oil from wells after secondary recovery methods have been exhausted.

Thermal Coal: Coal that is not suitable for use in metallurgical processes.

Thermal Conductivity: The capacity of a material to conduct heat. *See also U* Value.

Thermal Inversion: A weather condition that prevents the dispersal of atmospheric pollutants; caused when the temperature increases with increasing altitude.

Thermodynamics: The study of the behavior of energy.

Thermonuclear Reaction: An uncontrolled fusion reaction; the type of reaction that takes place in a hydrogen bomb.

Third World: The least industrialized countries taken as a group.

Threshold Level: The lowest level of contamination that produces adverse effects. Concentrations below this level are assumed to be safe.

Tight Sands: Highly compressed geological formations containing natural gas.

TOKAMAK: A doughnut-shaped magnetic containment vessel, used in fusion research.

Torus: A doughnut-shaped geometric figure. The shape of a TOKAMAK. Also, a portion of a boiling-water reactor.

Tritium: A radioactive isotope of hydrogen with an atomic weight of 3. Viewed as a likely fuel for fusion reactors.

Trombe Wall: A type of passive solar heating system, consisting of a masonry wall located directly behind the glazing.

Turbine: A device for converting linear motion to rotational motion. Used in most electric power plants.

***U* Value:** A measure of the thermal conductivity of a material. The greater the *U* value, the greater the heat flow through the material.

Uranium: The heaviest naturally occurring element. Used as a fuel for nuclear reactors and weapons.

Vitrification: A process of disposing of high-level nuclear wastes by incorporating them in glassy capsules which are then buried.

Waste Heat: Heat that is produced by an energy-conversion device but not used.

Water Gas: Gas produced by the Fischer–Tropsch synthesis method of coal conversion.

Wet Steam Geothermal Field: A geothermal area that produces low-temperature steam not acceptable for direct use in turbines.

Yellowcake: Uranium oxide.

Index